Solar Energy Applications

K.R. Rao, PhD, PE
Editor

ASME PRESS

Library of Congress Cataloging in Publication Control Number: 2020002288

Contents

Dedication

This *Solar Energy Applications* volume in the ASME Press Renewable Energy Series is dedicated to the individuals, associations and nations in the pursuit of the cause, development and promotion to increase the share of solar power in our global energy supply.

Editor-in-Chief Dr. KR Rao is personally indebted to the late VRP Rao FIE(I) who was the first person who urged me to study and understand the importance of Renewable Energy, especially Solar Energy Applications over a decade back. His urging me to take up my intense research resulted in the 2011 *Energy and Power Generation Handbook*, published by ASME Press of which I was the Editor. My continuing research in Renewable Energy resulted in my authoring the 1800-page 2-volume *Wind Engineering for Power Generation – Meeting the Challenge of Practical Implementation*, published by Springer in September 2019. All of these were possible by the enthusiastic support of the late VRP Rao (FIE) who passed away on October 2, 2019. Thus, it is most appropriate that I dedicate this first book of Renewable Energy Series on Solar Energy to Mr. VRP Rao FIE(I).

Dr. K.R. Rao
Editor-in-Chief
Renewable Energy Series

ACKNOWLEDGEMENTS

My thanks are to all of the original contributors of Chapters 1 through 6 of ASME's Energy and Power Generation Handbook, 2011 and the continuing contributors for the current book whose dedicated efforts made this publication possible.

The singular attention to detail, presentation of graphics, procuring the copyrights for the "artwork" and taking time to research the references to complement the write-up are gratefully acknowledged. The effort in updating is as cumbersome as the initial write-up if not more; as such efforts of Dr. Robert Boehm, Dr. Yong X. Tao, Dr. Hongwei Tan, Dr. Suraj Tallele, Mr. Deepak Yadav, Dr. Rangan Banerjee, Dr. Ram A. Goel and Mr. Jeremiah S. McNatt are immensely appreciated and acknowledged.

It is not uncommon for Past Presidents of ASME to encourage efforts such as these and it was amply demonstrated by Dr. Reginald Vachon who provided an effective Forward for this publication.

This publication was ably supported by Ms Mary Grace Stefanchik and Ms. Tara Smith of ASME Technical Publishing. My appreciation and thanks to them for their cooperation support and patience is greatly appreciated.

Finally, all of this enduring effort, spread over 16 months, would have never been possible had it not been for the constant help and untiring zeal provided by my wife, Dr. Indira Rao, that included all of the sundry editorial chores associated with this project.

Dr. K.R. Rao
Editor-in-Chief
Renewable Energy Series

Contributing Authors

BAILEY, SHEILA GAYLE

Sheila G. Bailey has been a Senior Physicist working in photovoltaics at NASA Glenn research Center for over 25 years. Her most recent projects include nanomaterials and nanostructures for space photovoltaics, quantum wire III v solar cells and quantum dot alpha-voltaics. She has authored or co-authored over 165 journal and conference publications, nine book chapters and two patents. Dr. Bailey is on the Editorial board of *Progress in Photovoltaics*. She is a member of the American Physical Society and a speaker for the American Institute of Physics visiting Scientist Program. She is a member of AIAA Aerospace Power Systems technical committee and the IEE Electron devices Society Photovoltaic devices Committee. She was the chair of the 4th world Conference on Photovoltaic Energy Conversion in 2006. She is executive vice president of the Lewis Engineers and Scientists Association.

Dr. Bailey was an adjunct professor at Baldwin Wallace College for 27 years and is currently an associate faculty member of the International Space University. She has a BS degree in Physics from Duke University, a MS degree in Physics from the University of NC at Chapel Hill, and a PhD in condensed matter physics from the University of Manchester in England. She spent a post-doctoral year at the Royal Military College (part of the University of New South Wales) in Canberra, Australia.

Dr. Bailey is the recipient of the faculty excellence award from Baldwin Wallace College and the Federal Women's Program Award. She is an Ohio Academy of Science "Exemplar". She was awarded the NASA Exceptional Service Medal for her work in space photovoltaics in 1999. She has completed the office of Personnel Management's Executive Potential Program. She was inducted into the Ohio Women's Hall of Fame in 2003 by Governor Taft.

BOEHM, ROBERT F.

Robert F. Boehm is Distinguished Professor of Mechanical Engineering and Director of the Energy Research Center at the University of Nevada, Las Vegas (UNLV). His work has been primarily in the area of renewable and conventional energy conversion. He has been at UNLV since 1990. He holds a PhD in Mechanical Engineering from the University of California at Berkeley. Dr. Boehm is a registered professional engineer, a Life Fellow of the American Society of Mechanical Engineers (ASME), and has received several awards, including the Harry Reid Silver State Research Award, the UNLV Distinguished Teaching Award, the Rudolf W. Gunnerman Silver State Award for Excellence in Science and Technology from DRI, and the John Yellott Award, the highest award of the ASME Solar Energy Division. He has been an invited lecturer at many institutions here and abroad, and he has published over 500 papers in renewable energy and design of thermal systems. He is the author or co-author of ten books. He is the Editor of the *Journal of Solar Energy Engineering*. He also serves as a Technical Editor for *Energy—the International Journal* and a guest editor for *Buildings and Energy*.

BANERJEE, RANGAN

Rangan Banerjee is the Forbes Marshall Chair Professor and Head of the Department of Energy Science and Engineering at IIT Bombay – a Department that he helped start in 2007.

He is on the editorial board of Energy for Sustainable Development, International Journal of Sustainable Energy, International Journal of Sustainable Engineering and International Journal of Thermodynamics and Frontiers in Energy Research. He has been involved in setting up a Megawattscale Solar Thermal Power Testing, Simulation, Research Facility sponsored by the Ministry of New and Renewable Energy. He was the faculty advisor to Team Shunya the first Indian Student team selected for the finals of the Solar Decathlon 2014 held in Europe that involves designing and constructing a fully functional Solar house.

He was a member of the TIFAC Technology Vision 2035 - Apex Committee and Chairman of the Energy Technology Theme. He was a Convening Lead Analyst for Industrial End Use Efficiency and a member of the executive committee for the Global Energy Assessment (2008-2011) coordinated by the International Institute for Applied Systems Analysis.

He has co- authored a book on Planning for Demand Side Management in the Power sector, a book on Energy Cost in the Chemical Industry and a book on Engineering Education in India. He has conducted two international training programmes on solar energy and several National programmes on renewable energy and Energy Management. He is also an Adjunct faculty (Honorary) in the Department of Engineering & Public Policy, Carnegie Mellon University. He has been the Dean (R&D) of IIT Bombay. He has received the Excellence in Teaching Award from IIT Bombay and is a Fellow of the Indian National Academy of Engineering.

GOEL, RAM

Dr. Ram Goel is currently working as a President/Chief Engineer with Soney FM LLC, in Tampa, FL and was previously with Northside Engineering Services, Clearwater, FL; PHA Tampa, FL and FDOT Tampa, FL. He received his Bachelors in Civil Engineering from IIT Roorkee, India with a Masters in Engineering from University of Nebraska, Lincoln. He completed his post graduate courses in structural engineering, finite element, and stress analysis from Columbia University, NY. He earned his Doctorate in Mechanical Engineering from Century University, NM. He is a registered Professional Engineer in 15 states as well as with NCEES.

Dr. Goel is a distinguished Fellow and Life member of ASCE and a member of ASME. He has been serving in codes and standards committees of ASCE for several years along with developing and presenting ASCE tutorial for bridge inspections.

Dr. Goel also has several publications and conference presentations on the Nuclear Power Plant at the Atomic Energy Commission of India and in Technical bulletins. He is a recipient of the prestigious Hind Rattan Award for NRI. This recognition is a honor bestowed by the Ex. President of India. He is also an Outstanding International Aluminous Award recipient from University of Nebraska. One of his proud moments was meeting with President Bill Clinton at an IIT conference in Chicago. .He has more than fifty years of Professional Engineering work experience which covers numerous fields and projects such as nuclear power plants, oil refineries, bridges, schools, Churches, Buildings, Warehouses, Schools, Shade Structures, Awnings, Pool Cages etc.

Dr. Goel has teaching experience at University of South Florida, Tampa; California State University, Los Angeles, and Remington College, Tampa. He is also a notary republic.

Dr. Goel has edited several papers on Early Career Technical Conferences of ASME for several years. He believes that learning while living should be a philosophy of life and growth.

GONZÁLEZ AGUILAR, JOSÉ

Dr. Jose González is Senior Researcher in the R&D Unit of High Temperature Processes at the IMDEA Energia Institute. He received his PhD in Physics from the University of Cantabria (Spain) in 1999 and his habilitation à diriger des recherches from the University Paul Sabatier, Toulouse (France) in 2007. Between 2000 and May 2009, he worked as R&D Engineer-Project Manager at the Center for Energy and Processes — MINES Paristech. In September 2006, he became associate professor at MINES Paristech (or Ecole Nationale Supérieure des Mines de Paris, ENSMP). The main research area of Dr. González is focused on the study and development of high temperature processes for energy and environmental issues, with special emphasis in concentrating solar systems and plasma technologies. His expertise includes process simulation from systems analysis (flow sheeting) to computational fluid dynamics. José González has participated in 15 national and international research projects, published 26 papers in peer review journals, two international patents, and a French patent, and he is author of more than 50 communications in national and international conferences.

MCNATT, JEREMIAH S.
Jeremiah McNatt has been a member of the Photovoltaic and Electrochemical Systems Branch at the NASA John Glenn Research Center in Cleveland, Ohio for the past 17 years and leads multiple R&D programs. He has a bachelor's degree in Physics from The University of Akron and a master's degree in Materials Science and Engineering from Case Western Reserve University. He has worked on a variety of space solar cell related programs including thin film, organic, and high-efficiency cell development and is currently working on solar arrays for unique applications including high temperature/high intensity, low temperature/low intensity, and planetary surfaces. His background includes growth of high efficiency solar cells and packaging of arrays for optimal performance in their environment.

RAO, K.R.
K.R. Rao retired as a Senior Staff Engineer with Entergy Operations Inc. and was previously with Westinghouse Electric Corporation at Pittsburgh, PA, and Pullman Swindell Inc., Pittsburgh, PA. KR got his Bachelors in Engineering degree from Ba-naras University, India, with a Masters Diploma in Planning from School of Planning & Architecture, New Delhi, India. He completed Post Graduate Engineering courses in Seismic Engineering, Finite Element and Stress Analysis, and other engineering subjects at Carnegie Mellon University, Pittsburgh, PA. He earned his PhD, from University of Pittsburgh, PA. He is a Registered Professional Engineer in Pennsylvania and Texas. He is past Member of Operations Research Society of America (ORSA).

KR was Vice President, Southeastern Region of ASME Inter-national. He is a Fellow of ASME, active in National, Regional, Section, and Technical Divisions of ASME. He has been the Chair, Director, and Founder of ASME EXPO(s) at Mississippi Section. He was a member of General Awards Committee of ASME International. He was Chair of Codes & Standards Technical Committee, ASME PV&PD. He developed an ASME Tutorial for PVP Division covering select aspects of Code. KR was a member of the Special Working Group on Editing and Review (ASME B&PV Code Section XI) for September 2007 to June 2012 term.

Dr. Rao was Editor of the *Companion Guides to the ASME Boiler & Pressure Vessel Codes* which he initiated in 2002 that ultimately became "classics" and are enumerated below:

- *Companion Guide to the ASME Boiler & Pressure Vessel Codes, First Edition* (2002)
- *Companion Guide to the ASME Boiler & Pressure Vessel Codes, Second Edition (2006) in 3 volumes*;
- *Companion Guide to the ASME Boiler & Pressure Vessel Codes, Third Edition* (2009) in 3 volumes;
- *Energy & Power Generation Handbook – Established and Emerging Technologies* (2011)
- *Companion Guide to the ASME Boiler & Pressure Vessel Codes, Fourth Edition (2012) in 2 volumes*;
- *Global Applications of the ASME Boiler and Pressure Vessel Code* by K.R. Rao (2014)
- *Continuing and Changing Priorities of the ASME Boiler and Pressure Vessel Codes and Standards* (2016); and
- *Companion Guide to the ASME Boiler & Pressure Vessel Codes, Fifth Edition* (2018) in 2 volumes;

Dr. Rao is currently the Editor-in-Chief of the *Renewable Energy Book Series* to be published by ASME Press in 2019-2020.

Dr. Rao is a recipient of several Cash, Recognition, and Service Awards from Entergy Operations, Inc. and Westinghouse Electric Corporation. He is also the recipient of several awards, certificates, and plaques from ASME PV&P Division including Outstanding Service Award (2001) and Certificate for "Vision and Leadership" in Mississippi and Dick Duncan Award, Southeastern Region, ASME. Dr. Rao is the recipient of the prestigious ASME Society Level Dedicated Service Award. KR is a member of the Board of ASME District F Professional & Educational Trust Fund for 2008 to 2011.

Dr. Rao is a Fellow of American Society of Mechanical Engineers, Fellow of Institution of Engineers, India, and a Chartered Engineer, India. Dr. Rao was recognized as a "Life Time Member" for inclusion in the Cambridge "Who's Who" registry of executives and professionals. Dr. Rao was listed in the Marquis 25th Silver Anniversary Edition of "Who's Who in the World" as "one of the leading achievers from around the globe."

ROMERO, MANUEL ALVAREZ

Manuel Romero received his PhD in Chemical Engineering in 1990 at the University of Valladolid for his research on the solarization of steam reforming of methane. At present, he is Deputy Director and Principal Researcher of the High Temperature Processes R&D Unit at IMDEA Energía. Dr. M. Romero has received the "Farrington Daniels Award" in 2009, the most prestigious award in the field of solar energy research, created in 1975 by the International Solar Energy Society, conferred for his intellectual leadership, international reputation, and R&D contributions to the development of high-temperature solar concentrating systems. In June 1985, Dr. Romero joined CIEMAT, Spain's National Laboratory for Energy Research, working as Project Manager until 2002 with responsibilities on R&D for solar thermal power plants and solar hydrogen. In 2002, he became Director of the Plataforma Solar de Almería, largely recognized R&D facilities for testing and development of solar concentrating technologies, and Director of the Renewable Energy Division of CIEMAT since June 2004 until August 2008 with R&D activities on solar thermal power, photovoltaics, biomass, and wind energy. During his career, Dr. Romero has participated in more than 45 collaborative R&D projects in energy research, 15 of them financed by the European Commission, with special emphasis on high-temperature solar towers. He is coauthor of the European Technology Roadmap on High Temperature Hydrogen Production Processes INNOHYP, contracted by the EC in 2005, and coauthor of the European Technology Roadmap for Solar Thermal Power Plants, ECOSTAR, contracted by the European Commission in 2004. He acted as member of the experts' committee of the Energy R&D Program of the VI and VII Framework Program of the EC until August 2008. Dr. Romero is Associate Editor of the ASME Journal of Solar Energy Engineering since January 2007 and at the International Journal of Energy Research (IJER) published by Wiley & Sons since December 2009. He was Associate Editor of the International Journal of Solar Energy since January 2002 until January 2007.

RAYEGAN, RAMBOD

Rambod Rayegan is a PhD candidate in the department of Mechanical and Materials Engineering at Florida International University. Since January 2007, he has worked as a research assistant at FIU in the sustainable energy area. He is also the president of ASHRAE FIU Chapter since May 2009. He has been a member of prestigious honor Societies like Tau Beta Pi, Phi Kappa Phi, Sigma Xi, and golden key. He has published a number of conference and journal papers in energy and sustainability area. Raised in Tehran, Iran, Rambod now lives with his wife in Miami. He has served as an instructor at Semnan University, Iran, for 5 years. He was selected as the best teacher of the Mechanical Engineering department by students during the 2002 to 2003 academic year and the best senior project supervisor in the 2003 to 2004 academic year. He has served as a consultant in three companies in the field of air conditioning and hydraulic power plants.

TAO, YONG X.

Dr. Yong X. Tao is Betty L. Gordon Endowed Chair and Distinguished Professor of the Department of Mechanical Engineering at the Cleveland State University, OH (CSU). He is an ASME Fellow and Editor-in-Chief of Heat Transfer Research journal with more than 25 years of research and teaching experience. He is also serving as the interim department chair at CSU. Prior to joining CSU, he was Professor of Engineering and founding dean of engineering and computing at Nova Southeastern University in Florida, after serving as PACCAR Endowed Professor of Engineering, director of PACCAR Technology Institute, director of Zero Energy Research Lab and chairperson of the Department of Mechanical and Energy Engineering at the University of North Texas (UNT). Prior joining UNT, he was the Associate Dean of the College of Engineering and Computing at Florida International University in Miami, a professor of Mechanical and Materials Engineering, and director of the Building Energy, Environment, and Conservation Systems Lab (BEECS) and Multi-Phase Thermal Engineering Lab (MPTE) at FIU.

An internationally known researcher in fundamentals of thermal sciences, refrigeration system performance, and renewable energy applications in buildings, Dr. Tao has produced a total of more than 175 journal publications, book chapters, edited journals and proceedings, and peer-reviewed technical conference papers over the course of his career and holds two patents. He has received more than 19.4 million dollars of research funding as a single PI or Co-PI in multidisciplinary teamwork projects from the NSF, NASA, Air Force, DSL, DOE, ASHRAE, and industries. He was the Associate Editor of the ASME Journal of Science and Engineering Applications.

Dr. Tao is founding board director, current Senior Vice-President and incoming President of American Society of Thermal and Fluids Engineers (ASTFE). He is also an active member of the American Society of Heating Refrigeration and Air-Conditioning Engineers (ASHRAE), past chair the Heat Transfer Division of ASME, and past Editor of ASME Early-Career Technical Journal. He has served on many technical committees for ASME, ASHRAE, and AIAA, and was the conference chair or program chair of numerous technical, including international, conferences. In 2005, he was the faculty leader of the award-winning FIU Solar Decathlon entry sponsored by the US Department of Energy. In 2008, as Project Director of the Future House USA project, he led a consortium of academics, builders, industry sponsors, and lobbyists to represent the United States in a ten-country, international demonstration project of renewable energy and environmentally friendly construction that resulted in a 3200-sq ft zero-net energy American House in Beijing, China. On July 16th, 2009, Dr. Tao hosted a visit from the U.S. Secretary of Commerce Gary Locke and Secretary of Energy Steven Chu in the American House and was praised by both Secretaries as playing "vital role in building better collaboration between the United States and China in the area of energy-efficient buildings."

Dr. Tao has a PhD in Mechanical Engineering degree from the University of Michigan, and a BS and MS in Mechanical Engineering from Tongji University in Shanghai, China, and holds a PE license in Texas.

TAN, HONGWEI

Professor Hongwei Tan studied at The University of Tokyo in Japan from 1990 to 1995. He took his doctor's degree in architectural engineering at 1995 and came back to China for working as a professor in Tongji University at 1999.

Prof Tan is currently Executive Director, Research Center of Green Building & New Energy, with the title of Professor, Tongji University

Dr. Tan is in charge of the research center of Green Building & New Energy in Tongji University and teaches in the School of Mechanical and Energy Engineering. He is the Secretary General of China Green University Network (CGUN) and executive member of China Association of Building Energy Efficiency (CABAEE). He has lots of contribution for campus sustainable development in national and international fields.

Prof Tan is mainly engaged in research on building energy-efficiency technology, the application of renewable energy in building technology, energy management system, low carbon energy planning, urban architectural environment, etc. He also is one of key researchers in Chinese national research project on green building and a chief editor of Chinese guidelines regarding sustainable development for campus.

Prof. Hongwei Tan won awards for sustainable development on campus and green buildings, such as:

Technology Promotion Award of 14th SHA SEJ (The Society of Heating, Air-Conditioning and Sanitary Engineers of Japan), 2000

First Scientific Progress Award Chinese high education, 2008

Copper prize by China International Industry Fair 2009

SHASE UICHI INOUYE MENORIAL AWARD (AISA INTERNATIONAL AWARD) 2014.

TALELE, SURAJ HARISH

Suraj Harish Talele currently works as a Thermal Engineer at Marelli Corporation, previously known as Magneti Marelli. His expertise in this role lies in Sunload Analysis using Multi-band Monte Carlo Radiation Model in order to validate the materials used inside Headlamps and Tail lamps for automobiles due to burning glass effect caused because of Solar radiation. His expertise also lies in LED characterization and System-on-Chip (SoC) Printed Circuit Board (PCB) thermal design in order to validate optical output of the LEDs during their operation. He is part of Verification and Validation Team where he validates current designs of Head lamps and Tail lamps for their thermal behavior and verifies if they meet customer and legal requirements in terms of optical output and operational life of materials without outgassing.

Suraj earned his Bachelor of Engineering (BE) in Mechanical Engineering from Mumbai University and went on to earn his graduate degrees of MS in Mechanical and Energy Engineering from University of North Texas and PhD in Mechanical and Energy Engineering from the same institution as well. His research focus was on Solar Chimney and natural ventilation during MS. For his PhD research he investigated different thermal comfort models used to predict human thermal comfort and impact of contextual factors governing human thermal comfort on the comfort prediction. He is co-author on one book chapter on Zero-Energy Buildings and has published conference papers and journal articles focusing on Zero Energy Buildings, natural ventilation and human thermal comfort.

VACHON, REGINALD

Dr. Reginald I. Vachon is an engineer, executive, and lawyer whose career and accomplishments are testimony to his dedication to engineering and to service to our Nation as a member of the US Military and participant in US Government Agencies.

Professional Practice

Dr. Vachon has been a Research & Development Engineer with E.I. DuPont, an Aerospace Engineer with Thiokol, a Design Engineer for Hayes Aircraft and Aerospace Engineer with NASA, among his many engagements. He has consulted on nuclear power for space exploration and power generation on earth. Designed solar energy systems. He is an original patentee for digital image correlation. He co-holds multiple other patents, including optical strain gage measurement technology and for the sizing of coronary arteries at systole and diastole during invasive angiography. The optical strain gage technology operable to 1093.33°C was purchased by General Electric in 2015.

Service to the Profession

Dr. Vachon has offered distinguished service to the profession of engineering in a variety of roles at both the national and international level for more than four decades. These include: Member of the Board of American Society for Engineering Education , President (2003-2004) American Society of Mechanical Engineers; Secretary-Treasurer and a Member of the Board of the International Nuclear Energy Academy ; President-elect of the Pan American Academy of Engineering; former Vice President for North America for the Pan American Union of Engineering Organizations (UPADI) and currently serves as co-Chair of UPADI's Anti-Corruption Work Group; former Chairman of the American Association of Engineering Societies (AAES) and serves chairs the International Activities Committee; Executive VP and former Treasurer and Ex VP of the World Federation of Engineering Organizations (WFEO);chairs WFEO UN Relations Committee; member of the Advisory Council of the WFEO Energy Committee and the WFEO World Engineering Forum Task Force; serves both as the WFEO participant at the International Atomic Energy Agency and as WFEO Representative to the U.N. and serves with the U.N. Scientific and Technological Major Group and the U.N. High Level Political Forum (HLPF) Steering Committee on the 17 Sustainable Development Goals (SDGs).

Academic and Educational Achievements and Contributions

Dr. Vachon has authored or co-authored over 200 unclassified peer reviewed papers and reports: delivered numerous invited lectures; served as advisor and contributor to the ACET anti- corruption film ETHICANA; authored video course on Ethics and Professionalism for engineers may ; hold an endowed chair Professor of Mechanical Engineering at Auburn University ;was Adjunct Professor at Purdue University TPRC; conducted bio-engineering research and developed applications for companies and government organizations He holds the following degrees: B.S. in Mechanical Engineering; M.S. in Nuclear Science, Ph.D. in Aerospace and Mechanical Engineering, and LL.B.

National and International Service

Vachon's career has taken him to countries around the world including Belize, Brazil, Cameroon, Ecuador, Peru, Costa Rica, Egypt, Honduras, Hong Kong, Indonesia, Iran, Mexico, Nicaragua, Russia, Saudi Arabia, Singapore, Sudan, Venezuela and other countries. He has conducted research and development projects for a number of US government agencies; served on the U.S. Department of Homeland Security Science and Technology Advisory Committee and other agencies.; was Resident Engineer and Constructor for the U.S. Strategic Petroleum Reserve; directed multi-disciplinary groups of scientists, economists and engineers for a variety of NASA projects; consultant to the Southern Interstate Nuclear Board for the "Transportation of Nuclear Materials." Dr. Vachon's private sector work has been undertaken for or in collaboration with a broad range of world-class organizations including IBM, Northrop Grumman, GE, NCR, Lockheed Martin, the Polish Air Force Institute of Technology, RUAG, Halliburton, The Port Authority of New York & New Jersey, and others.

Other Notable Achievements

Vachon founded several engineering firms and one communications firm; served on numerous corporate boards; is a licensed attorney and a member of bar of U.S. Supreme Court: co-represented ASME in landmark Hydrolevel case and received a favorable outcome. Dr. Vachon is a registered Professional Engineer in six states as well as a licensed European Engineer; he is also a licensed U.S. Coast Guard Captain. Conducted a NSF Chautauqua series on the design of renewable energy systems in the 70's, co designed an ERDA sponsored solar heated and cooled office building (17000 FT2), and designed a number of solar energy systems.

Membership Honors and Awards

Dr. Vachon is the recipient of numerous honors and awards for his work with the following: Honorary Member, ASME; Fellow, IMechE (UK); Fellow, ASCE; Life Member, ASEE, Associate Fellow, AIAA; Fellow, NSPE; Member, IEEE; Member, ISA; Member, ANS; Fellow, Hong Kong Institution of Engineers; Member, Institution of Engineers Singapore; Member, Sigma Xi; Member, Phi Kappa Phi; Member, Pi Tau Sigma. Dr. Vachon is recognized by the Alabama Society of Professional Engineers as a "Wonder of Engineering in 1975," as the American Society for Engineering Education's "Outstanding Mechanical Engineering Professor in the United States," and as the Western Electric "Outstanding Professor in the Southeast." He is the recipient of the Pan American Academy of Engineering Engineer Statesman Award and a member of the Katerva Awards Panel whose mission is to find, evaluate and accelerate disruptive sustainable innovations from around the world, using these innovations to solve world problems. He is recipient of the ASME 2019 Gold Medal for "eminently distinguished engineering achievement.

YADAV, DEEPAK

Deepak Yadav has a graduate degree in Mechanical Engineering from Mumbai University (2007) and a Master's degree from Indian Institute of Technology Bombay (2009). Between 2009 and 2013, he worked as a project engineer in conceptualisation-to-commissioning of a 1 MWe solar thermal power plant. Currently, he is a PhD candidate in the Department of Energy Science and Engineering, Indian Institute of Technology Bombay (2014-2019). The theme of his doctoral research is 'Assessment of Solar Fuels Production Processes'. He has published papers in the field of energy and sustainability research in leading international journals and conferences. He is also a Bureau of Energy Efficiency (BEE) certified energy auditor and manager. He has participated and presented his research in many workshops, training sessions and conferences. His areas of interest include solar energy, hydrogen energy, energy efficiency and sustainability.

FOREWORD

This compilation of information on Renewable Energy provides a comprehensive story of renewable energy use globally. Solar energy is shown to serve many needs to provide humanity the foundations of human well-being and a sustainable environment. The information serves as input to the UN Sustainable Development Goals and particularly provides input to Goal 7 -Affordable and Clean Energy and Goals: 3 - Good Health and Well-being; 8 - Decent Work and Economic Growth; 11 - Sustainable Cities and Communities; 12 - Responsible Consumption and Production; and 13 - Climate Action. These UN SDGs correlate with the US National Academy of Engineering Grand Challenges. I compliment the contributors and editor-in-chief for an outstanding useful contribution to the literature.

Reginald I. Vachon, Ph.D. PE, Eur. Eng.

PREFACE

Renewable Energy has emerged as an important source for energy and power generation from renewable resources, naturally replenished on a human timescale, such as sunlight, wind, rain, hydro including tides and waves, geothermal heat, and from traditional and modern biomass. Worldwide investments in renewable energy are surpassing expectations, significantly in Europe (Germany and Spain), the US, in Asia (China and India) and in Australia.

Energy, and Power Generation Handbook: Established and Emerging Technologies, edited by K.R. Rao, and published by ASME Press in 2011 was a comprehensive reference work of 32 chapters authored by 53 expert contributors from around the world, with the authors drawn from different specialties, each an expert in the respective field and with several decades of professional expertise and scores of technical publications. Recognizing the need of treating Renewable Energy and Power Generation as a separate field ASME Press initiated "Renewable Energy Series" to address each entity of Renewable Energy in a separate book, revising pertinent chapters of the 2011 Handbook and bringing the coverage up-to-date.

Thus, this book is the first in a series of renewable energy topical books and addresses *SOLAR ENERGY APPLICATIONS* to update chapters 1 through 6 of the 2011 *Handbook* in which Solar Energy was addressed.

This book is meant to cover the technical discussions relating to solar energy source as well as why(s) and wherefore(s) of power generation. A unique aspect of this publication is the scholarly discussions and expert opinions expressed, enabling the reader to make "value judgments" regarding which solar energy technology is applicable for their purpose. This book has the end user in view from the very beginning to the end. The audience targeted by this publication not only includes libraries, universities for use in their curriculum, utilities, consultants, and regulators, but is also meant to include ASME's global community. ASME's strategic plan includes Energy Technology as a priority.

This book could be of immense use to those looking beyond the conventional discussions contained in similar books that provide the "cost benefit" rationale. Instead of picturing a static view, the contributors portray a futuristic perspective in their depictions, even considering the realities beyond the realm of socio-economic parameters to ramifications of the political climate. These discussions will captivate advocacy planners of global warming and energy conservation. University libraries, the "public-at-large," economists looking for technological answers, practicing engineers who are looking for greener pastures in pursuing their professions, young engineers who are scrutinizing job alternatives, and engineers caught in a limited vision of energy and power generation will find this publication informative.

Equally important is that all of the authors have cited from the public domain as well as textbook publications, handbooks, scholastic literature, and professional society publications, including ASME's Technical Publications, in addition to their own professional experience, items that deal with renewable energy and non-renewable energy sources. Thus, ASME members across most of the Technical Divisions will find this book worth having.

Dr. K.R. Rao
Editor-in-Chief
Renewable Energy Series

INTRODUCTION

The discussions in this book cover varied aspects of solar energy in use around the globe. Chapters 1 through 6, deal with Solar Energy in 221 pages addressed by 15 experts from academia, NASA, and practicing professionals from the U.S., Europe, and India. Global interest in solar energy is apparent not only from the current usage but also from the untapped resources and its potential for greater usage.

The chapter revisions are provided by the continuing contributors Robert Boehm for chapter 1; Yong X. Tao for chapter 2; and Rangan Banerjee for chapter 4.

New contributors who joined Yong Tao for revising chapter 2 are Hongwei Tan and Suraj Tallele; Rambod Rayegan who had contributed initially for chapter 2 in *2011 Handbook* did not participate for this update.

Chapter 3 was initially authored by Manuel Romero and José González-Aguilar for the *2011 Handbook* but unavailable for the current publication. The write-up by Romero and Gonzalez being still applicable for 2011 time-frame was retained with the information being updated by KR Rao, editor-in chief.

Deepak Yadav joined Rangan Banerjee in revising Chapter 4.

Chapter 5 appeared in the *2011 Handbook* with the title "Solar Energy Applications: The Future (with Comparisons)" by Luis A. Bon Rocafort and W.J. O'Donnell. This contribution has been completely revised by Ram A. Goel with the subject matter titled as "Future Solar Energy Applications" since Luis A. Bon Rocafort and W.J. O'Donnell were unavailable for this publication.

The last chapter 6 was authored by experts from NASA who elucidated NASA's efforts in both Solar and Wind energy sectors. This is appropriate since both of these energy sources constitute the most popular of renewable energy resources. However for this update Sheila G. Bailey who was the original author with co-author Larry A. Viterna of the Chapter 6 titled "Role of NASA in Photovoltaic and Wind Energy" which appeared in the ASME "Energy and Power Generation Handbook – Established and Emerging Technologies" , 2011, ASME Press were unavailable for the current publication. Jeremiah S. McNatt provided the update for chapter 6 of this publication with the chapter coverage and title "NASA Glenn's Historical Role in Photovoltaic Solar Energy Conversion".

A publication such as this with rich reference material documenting the essence of the contributors' expertise can be a valuable addition to university libraries, as well as for consultants, decision makers, and professionals engaged in the disciplines described in this book.

For the reader's benefit, brief biographical sketches are included for each contributing author. Another unique aspect of this book is an Index that facilitates a ready search of the topics covered in this publication.

Chapter 1, "Some Solar-Related Technologies and Their Applications" is addressed by Robert Boehm. In this chapter, the source of energy that has been available to humankind since we first roamed the earth is discussed. Some of the general concepts are not new, and several particular applications of these technologies are enhancements of previous concepts.

The discussion begins with the special effects that are possible with the use of concentration. For locations that have a high amount of beam radiation, this aspect allows some very positive properties to be employed. This yields a lower cost, more efficient way of generating electricity. Limitations to the use of concentration are also outlined.

Another aspect discussed is the current situation of solar thermal power generation. This approach has been in use for many years. Previously designed systems have been improved upon, which results in more efficient and more cost-effective means of power production. While trough technology has been more exploited than other approaches (and is still a leader in the field), several other systems are gaining interest, including tower technology. Thermal approaches are the most convenient to add storage into solar power generation.

Photovoltaic approaches are described. New developments in cells have both decreased costs and increased performance. Both high and low-concentration systems, as well as flat plate arrangements in tracking or non-tracking designs, offer a variety of application modes, each with certain benefits and shortcomings.

The use of solar-generated hydrogen is discussed. This offers an approach to a totally sustainable mobile or stationary fuel source that can be generated from the sun.

The solar resource can be used for lighting, heating, cooling, and electrical generation in buildings. The concept of zero energy buildings is discussed. These are buildings that are extremely energy efficient and incorporate a means of power production that can result in net zero energy use from the utility over a year's period. Locations with a moderate-to-high solar resource can use this to make up for the energy used. Both solar domestic water heating (a concept that has been applied in the United States for well over a century) and building integrated photovoltaic (PV) are also discussed. South-facing windows that incorporate thin film PV could generate power and allow lighting to penetrate the building. Finally, some exciting direct solar lighting concepts (besides windows) are discussed. The author uses 28 references along with 24 schematics, figures, pictures, and tables to augment his professional and scholastic treatment of the subject.

Chapter 2 by Yong X. Tao, Rambod Rayegan, Hongwei Tan and Suraj Talele deals with "Solar Energy Applications and Comparisons." The energy system applications resulting from the direct solar radiation are discussed. The particular focus is given to the following applications in China, and an example from Mongolia is also presented:

- Utility-scale solar power systems that generate electricity and feed to the electricity grid. There are photovoltaic (PV) systems and solar thermal power systems; the latter can also produce heat for hot water or air, which is often referred to as the combined solar power and heat systems.
- Building-scale solar power systems, also known as distributed power systems, which generate electricity locally for the building, and may be connected to the grid, or may be a stand-alone system which require batteries or other electricity storage units. They are primarily photovoltaic systems.
- Solar heating systems for buildings, which are either used as hot water systems, or hot air heating systems.
- Solar high temperature process heating systems for industrial applications.
- Other special solar systems for transportation such as ships and buses.

During last several decades, China's solar industry has experienced tremendous growth under the government policy support and investment, and academic research efforts. Both solar research and applications have been catching up to the world standards. As of now, China is both the largest producer of solar photovoltaic and hot water systems, and largest exporters. In the meantime, China increases its domestic adoption for solar energy applications by overcoming various challenges.

There are additional solar energy applications in either the appliance category or even much smaller scales such as solar cooking, solar lighting products, and instrument-level solar power sources (watches, backpacks, etc.) the discussion of those applications is beyond the scope of this chapter. Outer space applications of solar energy technology are also excluded. While in this edition Chapter 2 focuses on global applications, discussions are also presented for the investigations undertaken in the United States of America, which cover solar radiation measurements, modeling and solar forecasting. The authors list 29 references along with 33 figures to augment the professional and scholastic treatment of the subject.

In the absence of response from original authors Editor-in-Chief has provided relevant write-up from on-line sources in the public domain for *Recent Solar Applications in Spain, Rest of Europe, Asia, Africa, Middle East and South America* and in addition specifically solar energy pertaining to China, India, Australia, South Africa and Chile covering mostly, from 2009 to date. Original contributors of chapter 3 titled "Solar Thermal Power Plants: From Endangered Species to Bulk Power Production in Sun Belt Regions" authored by Manuel Romero and José González-Aguilar dealt with developments in Spain and Europe up until 2011 which are retained in entirety.

In this update, KR Rao, Editor-in-Chief of ASME's Renewable Energy Series contributed for the "Recent Solar Applications in Spain, Rest of Europe, Asia, Africa, Middle East, South and Central America and Australasia" in section 3.8 of this publication. In this section Rao reflected on considerable changes in solar applications in Spain and Europe since the initial edition of this publication in 2011. The advances in the US have been covered elsewhere in this publication with significant changes in Europe captured in this section.

The discussions by original authors are covered in Sections 3.1 through 3.7 and are summarized in the following paragraphs. Solar thermal power plants, due to their capacity for large-scale generation of electricity and the possible integration of thermal storage devices and hybridization with backup fossil fuels, are meant to supply a significant part of the demand in the countries of the solar belt such as in Spain, the United States of America, India, China, Israel, Australia, Algeria, and Italy. This is the most promising technology to follow the pathway of wind energy in order to reach the goals for renewable energy implementation in 2020 and 2050.

Spain, with 2400 MW connected to the grid in 2013, is taking the lead on current commercial developments, together with the United States of America, where a target of 4500 MW for the same year has been fixed and other relevant programs like the "Solar Mission" in India recently approved for 22-GW solar, with a large fraction of thermal.

Solar Thermal Electricity or STE (also known as CSP or Concentrating Solar Power) is expected to impact enormously on the world's bulk power supply by the middle of the century. Only in Southern Europe, the technical potential of STE is estimated at 2000 TWh (annual electricity production), and in Northern Africa, it is immense.

The energy payback time of concentrating solar power systems will be less than 1 year, and most solar-field materials and structures can be recycled and used again for further plants. In terms of electric grid and quality of bulk power supply, it is the ability to provide dispatch on demand that makes STE stand out from other renewable energy technologies like PV or wind. Thermal energy storage systems store excess thermal heat collected by the solar field. Storage systems, alone or in combination with some fossil fuel backup, keep the plant running under full-load conditions. This capability of storing high-temperature thermal energy leads to economically competitive design options, since only the solar part has to be oversized. This STE plant feature is tremendously relevant, since penetration of solar energy into the bulk electricity market is possible only when substitution of intermediate-load power plants of about 4000 to 5000 hours/year is achieved.

The combination of energy on demand, grid stability, and high share of local content that lead to creation of local jobs provide a clear niche for STE within the renewable portfolio of technologies.

Because of that, the European Commission is including STE within its Strategic Energy Technology Plan for 2020, and the U.S. DOE is launching new R&D projects on STE. A clear indicator of the globalization of such policies is that the International Energy Agency (IEA) is sensitive to STE within low-carbon future scenarios for the year 2050. At the IEA's Energy Technology Perspectives 2010, STE is considered to play a significant role among the necessary mix of energy technologies needed to halving global energy-related CO2 emissions by 2050, and this scenario would require capacity additions of about 14 GW/year (55 new solar thermal power plants of 250 MW each).

The authors discussed with the help of 21 figures, 27 schematics, and tables along with 72 references, the Solar Thermal Power Plants covering Schemes and Technologies, Parabolic-Troughs, Linear-Fresnel Reflectors, Central Receiver Systems (CRS), Dish/Stirling Systems, Technology Development Needs and Market Opportunities for STE.

In addition in section 3.8.1 "Objectives of CSP/STE Research Community in the European Union (EU)" with discussions about concentrated Solar Power (CSP), also called Solar Thermal Electricity (STE) developed in European Union (EU),which plays a relevant role in any future energy scenario of EU because of its capability of base load dispatchability for integrating thermal energy storage are discussed in this update. This is followed in section 3.8.2 about "Recent Solar Applications in Spain" in which the initiative for Global Leadership in Concentrated Solar Power (CSP) with an 'Implementation Plan' developed to meet the energy needs of several parts of the world, for creating an export sector for the European renewable industry in supporting the decarburization agenda of the Paris Agreement have been discussed with supporting graphics - "Figure of SET Plan Key Priority Actions", "Figure of IEA 2050 Roadmap – Generation mix", table showing the "The ranked list of R&I Activities", figure of "The Torresol Energy Gemasolar thermasolar plant in Fuentes de Andalucia near Seville, southern Spain", figure of "The Olmedilla Photovoltaic Park in Olmedilla de Alarcón, Spain", figure of "Solucar PS10", figure of "Three Solar Towers near Seville Spain, PS20, Eureka 5 and PS10", figure of"PS20 and PS10 in Andalusia, Spain", and a figure showing "The first three units of Solnova and two towers of the PS10 and PS20 solar power stations" and a figure of "The 150 MW Andasol solar power station".

Rao thereon enunciated the top 10 Performing Countries for Solar Energy as of March 13th, 2017, with graphics of Solar Energy Panels and A Solar Panel Farm. The discussions enunciate sequentially, first place for solar energy is China, second place for solar energy is Germany, third place for solar energy is Japan, fourth place for solar energy is USA, fifth place for solar energy: Italy with figures of A Residential PV System in Italy, and a figure of Italy photovoltaic capacity in 'watts per capita' by region, 15 June 2014. Italy had set a 2030 solar target of 50 GW, shown in figure Octopus Energy by Emiliano Bellini, January 11, 2019. Rao articulated the sixth place for solar energy is UK, seventh place for solar energy is France, eighth place for solar energy is Spain showing with a figure about the "PV capacity in watts per capita by autonomous communities in Spain in 2013", a figure of Global PV Installations per Habitant, a figure of European PV Installations per Habitant, 2013. There after Rao mentioned about the ninth place for solar energy: Australia followed by tenth place for solar energy: India.

In conclusion Rao mentioned:

- India is committed to tackle power shortages which has woken the PV market in India
- It will be interesting to see China's progress in the coming years.
- Germany, the US and the UK featured near the top and will continue to be there.
- The PV market is looking pretty healthy globally and many governments are committed to sustain growth in PV installations in the future.
- In the latest developments, ISTAT, Italy's National Statistics Institute, said the country's economy grew 0.4 percent in the second quarter and 1.5 percent over the last year, 2017. Those figures would be modest in many countries, but for Italy, it's the most robust economic growth since 2011.

Recent Developments in "Solar Investment Arbitration Disputes" are mentioned in which "Czech Republic and Italy enjoy their first victories, while Spain seeks to annul their first defeat".

These discussions are followed by Solar Power Complex in Morocco, with appropriate graphics, figures of "Solar Radiation Map: Global Horizontal Irradiation Map of Morocco", "Solar GISolar Resources in Morocco", "Map of the DESERTEC EU-MENA Solar Power Plan", a table of "Morocco Renewable Energy projects now to 2030" and a figure of "Noor III Solar Tower of the Ouarzazate Power Station at dusk".

In section 3.8.6 Rao depicted about Solar Power Generation in South Africa with tables of "Growth of Photovoltaics in South Africa since 2011", "Growth of CSP in South Africa". These discussions were followed by Solar Power Generation in Chile, South America along with a figure showing Solar Potential in Chile, and a figure of "Irradiance and Installed Capacity of Chile" and a table of "Chile Photovoltaics Capacity (MWp)".

Chapter 4 titled "Solar Energy Applications in India" has been written by Deepak Yadav and Rangan Banerjee. This chapter details the potential, transition and status of various solar technologies and systems in India. India has a population of 1.1 billion people (1/6th of the world population) but accounts for less than 5% of the global primary energy consumption. The average electricity consumption in India is 1149 kWh/person/year, which is significantly lower than in developed countries.

The scarcity of fossil fuels and climate change problem has resulted in an increased emphasis on renewable energy sources. India has a dedicated ministry focussing on renewables Ministry of New and Renewable Energy (MNRE). MNRE oversees renewable energy installation and generation in the power sector. In 2010, India launched the National Solar Mission (NSM) as a part of its climate change mission to provide an impetus to solar installations in India. After the launch of NSM, the cumulative installed capacity increased from 12.8 MW in 2009 to 21,651 MW in 2018. Globally, India now stands 5th in the total installed capacity of solar power and has targeted 100 GW by 2022. Beyond 2022, there is a potential of achieving 479 GW by 2047. India has also taken a lead in forming the International solar alliance (ISA) on the side-lines of COP21 in Paris in 2015.

As a result of the solar mission, India has achieved 20% reduction in emission intensity against the 33% it has committed to achieve by 2030. India has also committed to have 40% of its total installed power capacity from non-fossil sources by 2030. In 2018, 36% of the installed capacity is from non-fossil sources. Thus, India is on a trajectory to achieve its Intended Nationally Determined Contributors (INDC) targets well before the 2030 deadline.

In this chapter, with the help of 40 schematics, pictures, graphics, figures, and tables, the authors discuss the status of national solar mission, centralized and rooftop photovoltaic (PV) systems, off-grid systems like mini/micro grids, pumps, desalination, street and home lighting. The solar thermal systems have been discussed with reference to technologies like flat plate/evacuated tube and concentrating solar collectors like scheffler and dish. The developments in the use of solar thermal technology for cooking, cooling, industrial process heat, desalination and power generation applications have been discussed.

The chapter provides insights into why centralized PV plants have been a success while rooftop installations have not grown to their potential. The effect of policy interventions like solar parks that provide excellent infrastructure for installing multiple large-scale solar plants is discussed. The success of Bhadla solar park (2255 MW) epitomizes the effectiveness of the solar parks. These large-scale installations and reverse bidding mechanism have reduced the tariff bids from Rs. 11/kWh in 2010 to Rs. 2.44/kWh in 2018. The unintended adverse effect of solar mission on domestic cell and module manufacturing capabilities has also been highlighted. The chapter also explores the factors that prevented concentrated solar power (CSP) technology from achieving its intended targets in India. In addition to India, the authors also briefly discuss the development of solar power in neighbouring countries like Afghanistan, Bangladesh, Myanmar, Nepal, Pakistan and Sri Lanka. The authors provide a detailed list of relevant references and sources that can provide additional details to the reader.

Chapter 5 "Future Solar Energy Applications" covered by Dr. Ram A. Goel emphasizes the primary source of renewable energy in the form of solar energy and harnessing the same. Numerous diagrams and photographs are included, illustrating the technologies and applications. Methods of concentrating solar power are described including parabolic troughs, Fresnel reflectors, solar towers, and sterling engine solar dishes. Methods of storing solar energy to provide continuous power are described, including batteries, fly-wheel energy storage, water energy storage, compressed air, and superconducting magnetic energy storage. Current energy use and production in the United States of America and worldwide are quantified. Efforts of US Department of Energy to promote solar power by introducing various schemes including Tax credits, etc are discussed. Major Applications of solar energy in the form of Solar electric / Thermal power generation, Buildings, Green houses, water heating, pumping, furnaces, cooking etc are benefits of renewable energy resources. Some applications of solar energy we can see in the near future like Photo-biological Cells, Solar Thermal Fuel (STF), Solar Can Power Places That Otherwise May Not Have Access to the Grid, Production of Power through Solar Ponds, Solar Transportation, Solar Desalination, Floating Panels, Floating Solar Farms, Solar Power Harvesting Trees, Solar Charging Stations, Solar Powered Trash, Stored up Solar Energy; and Solar Fashion. Future technologies of solar energy applications have been discussed elaborately which includes Smart Solar Cities, Solar powered vehicles, etc. Solar energy's potential future is illustrated by the fact that it would require less than 1% of the land area of the world to produce all of the energy we need. Of course, solar energy's future lies in its integration into the residential and commercial infrastructure. This challenge is expected to limit the contribution of solar energy to <0.1% of the USA energy consumption over the next 25 years.

The final chapter of this section is Chapter 6 "NASA Glenn's Historical Role in Photovoltaic Solar Energy Conversion" by Jeremiah S. McNatt and Sheila G. Bailey. Since the beginning of NASA over 60 years ago, there has been a strong link between the energy and environmental skills developed by NASA for the space environment and the needs of the terrestrial energy program. The technologies that served dual uses included solar, nuclear, biofuels and biomass, wind, geothermal, large-scale energy storage and distribution, efficiency and heat utilization, carbon mitigation and utilization, aviation and ground transportation systems, hydrogen utilization and infrastructure, space solar power (from space to earth), and nano-structured photovoltaics. NASA in particular, with solar energy, had extensive experience dating back to the 1970s and 1980s and continues today to have skills appropriate for solving our nation's energy and environmental issues that mimic those needed for space flight. This chapter encompasses the historical role that NASA and, in particular, NASA Glenn Research Center (GRC) has played in developing solar technologies. It takes you through the programs chronologically that have had synergistic value with the terrestrial communities and that extend performance of and understanding into photovoltaic technologies. It ends with pointing out the possibilities for future NASA technologies that could impact our Nation's energy portfolio. The technologies that it has developed for aeronautical and space applications has given GRC a comprehensive perspective for applying NASA's skills and experience in energy on the problems of developing a sustainable energy future for our nation. Authors use 26 references along with 15 figures and pictures to augment the professional and scholastic treatment of the subject.

CHAPTER

1

LARGE SCALE SOLAR POWER, HYDROGEN DEVELOPMENTS, AND BUILDING APPLICATIONS OF SOLAR

Robert F. Boehm

1.1 INTRODUCTION

Some introductory comments to this chapter are in order. Primarily these are to explain what will be found in what follows. As the overall title indicates, this is a coverage of three topics from an engineering perspective for the most part, and they cover a brief summary of technical publications. These topics are issues related to large scale solar power (basically topics that are pertinent to large PV or solar thermal plants), hydrogen generation and use (this includes a review of the current work in hydrogen technology from an engineering standpoint) and solar applications in buildings. All of these contain fairly current reviews of the literature. Most cited papers have been published in the time span from 2016 to midway through 2018. A few earlier papers are included. While many sources were consulted in determining the material for this chapter, no claim is made about it being a complete summary of publications. As is typical of reviews of current papers, some of the topics are represented by only one or two titles, while other topics may have a few more.

What was done was to survey a wide variety of types of documents. The bulk of the documents reviewed are from journals, but a few are either monographs or chapters in monographs. As will be seen in this chapter a wide range of topics specific to the three general areas were reviewed. For the most part, one paragraph of information is given for each source. It is thought that this should give the reader enough background to judge whether or not that document should be acquired for further study. Special care was given to the listing of the bibliographic information.

Where figures or tables are given, the references noted on these are where they were found in the current search. The papers where these papers were found for this review may not be the original source of those items. The interested reader should check the source used here for where the particular item may have originally appeared.

Since a large number of recent publications are described in the text, the Literature Cited section consists of four major subparts that cover the four major sections of this chapter. Care has been taken to make these bibliographic citations as complete as possible.

The chapter ends with a paper in the Appendix by the author from 2010 that has been updated to the current time. It has minor revisions to include more recent developments. While it is a standalone document, it is included here to give an overview of the basic technologies that are the subjects of the earlier parts of this chapter.

One overall caution will be noted: There has been some attempt to separate these papers into categories of subsets that are related to one another. However, even though this has been done, in some of the three major sections there is considerable overlap between papers' topics. For example, there is a category for a given topic, and several papers are located there, but many of these involve processes that might allow them to be placed in other categories. There can be quite a bit of overlap between topics. Also, note that the tables and figures that are shown in the chapter are not tied to the text. While they were all reproduced from papers cited, they should not be considered critically important to the reading.

1.2 LARGE SCALE SOLAR POWER GENERATION

This section consists of papers that deal with large scale solar plants that can generate power. Historically this was entirely related to collecting solar heat to drive thermally based engines that could generate power. These kinds of systems are still in the market place. However, they are now competing with solar photovoltaic (PV) systems. In fact, the latter are dominating the market.

A general conclusion is that based on a similar kW rating, PV systems are less expensive than solar thermal systems. They are, in general, more easily installed for the same power rating and their output is less degraded by dust accumulation. There are PV systems in Nevada that have never been washed, while solar thermal systems next door are washed regularly. Not all is lost for solar thermal systems, however. These offer better handling of short term transients (where thermal inertia carries the system for a while), the ability to add thermal storage (much less expensive generally than electrical storage for PV) and generally higher conversion efficiencies. However, the thermal systems usually incorporate fairly accurate tracking, while PV systems do not require tracking or may use less precise arrangements than solar thermal for one axis tracking. Note that there have been concentrating PV systems that track.

This large scale solar chapter is divided into two parts. First will be the historically important solar themal section ending. This is followed by the currently more popular solar PV section.

Both sections more or less review a series of singular papers or chapters from books. The review of each entry is not long, and it is meant to give a "flavor" of the topic. Although the reviews generally handle newer reports, no claim is made about the completeness of the literature reviewed, as it is extemely extensive.

1.2.1 Large Scale Solar Thermal Power

1.2.1.1 General Topics Thermal applications had been the nearly sole methods of generating solar electrical power for many years. In fact, there have been solar power generating projects (or its applications equivalent, like water pumping) for well over a century. Historically, solar generated steam systems were some of the earliest developments as a result of that technology being quite widely used for a variety of "conventional" applications [Butti and Perlin 1980]. Over the years, these types of systems have become more efficient as insights have been gained from operation of more cleverly conceived design approaches. In this section some of the more recent work related to large scale applications of solar thermal technology will be outlined.

Solar thermal power plants can offer a wide range of design possibilities. There are certain benefits to the use of these kinds of plants as was noted in the introduction to this chapter. For one, thermal storage, ranging up to very large sizes, can be incorporated into the design in a relatively straightforward way. In another, the cycle efficiency is governed by the temperatures of the hot and cold ends of the cycles. While most of these types of cycles have been steam-based or a few air-based, newer, higher efficiency cycles working on a supercritical CO_2 base hold promise for improved performance. And over the years many of the many variations of solar thermal plants have been demonstrated. All of the various types that have been considered will not be discussed here. Instead, a relatively recent summary [Chaanaoui et al. 2016] of the numbers of the main types will be summarized in the next paragraph.

Generally, there are four basic types of solar collection systems that have been developed over the last several years and that still continue to be installed. These are dish, tower, trough and linear Fresnel units. According to the paper noted, there were 240 solar thermal plants in operation or near an operational state when the paper was written which were rated at a total capacity of 4.2 GW using thermal generation means. Of these, the trough represents the largest number of plants that have been constructed which numbers 147. Towers, a more recently popular solar thermal approach, comes in second with 60 units with rated power of 460 MW. Dish and Fresnel mirror technologies follow, with respectively 26 and 10 plants and a total global operational capacity not exceeding 50 MW.

Two of the reasons that the trough plant concept is popular is that the system is relatively easily built, and it can be developed more easily than some of the others. The dish system generally has all components (engine and tracking collection system) in each unit. Hence the concept is one of the preferred ways of developing a small system. Of course, if a large system is desired, many of these units can be used to make a system with large power output. But the probability of the need for repairs for the total power system goes up also. This is one of the reasons it has become less favored over the years from an initial high value.

A paper that outlines a possible approach to evaluating the possible installation of a large amount of solar power near a given location has recently appeared [Kassem et al. 2017]. This work considers the installation to be a set of large systems (25 GW) of solar thermal power in Saudi Arabia. Solar thermal was chosen because of the ability to incorporate thermal storage. They applied a SWOT (strengths, weaknesses, opportunities and threats) analysis where the main tool used was SAM (Solar Advisor Model). Troughs were considered for near-term application, while later installations were assumed to be towers. But dish systems and Fresnel reflector systems were also considered. This article illustrates the kinds of analyses that can be performed using the software SAM.

The author of this next paper [Behar 2018] has compared the configurations and performance of *fifteen power generation technologies* ranging from solar driven to fossil fueled, and modeling of each technology is shown. Solar plant aspects are noted in this brief summary. Relative to the solar field, several elements are considered including the optics of the system, heat transfer in the receiver, heat losses in the piping. The Rankine cycle, Brayton cycle and the combined cycles are considered. These three are assumed to be powered by solar parabolic trough technology. The author has used six performance evaluations to rank the cycles. Comparisons of the various plant results are given.

An excellent early (relative to this review) article has appeared and should be considered as an excellent overall coverage of systems [Kalogirou, 2013]. It even includes a review of solar ponds which will otherwise not be covered specifically here.

A more recent article that gives a basic review of solar thermal power generation technology has been given by Santos et al. [2018]. Most power generation technologies are discussed in very understandable terms. However, the literature cited is quite limited and not as up-to-date as it might be.

For diagrams and drawings of most solar thermal systems, see Guney [2016]. These are quite well rendered.

Solar applications to travel in space have been reported by Toro and Lior (2017). Three different engines (Brayton, Rankine, and Stirling) were assumed to be operating on various fluids. The authors indicate that their analysis shows benefits of operating in a low ambient temperature (~3K) even with low temperature heat inputs but where the solar flux may be higher than terrestrial values. Details of the study are too numerous to go into here, but designs of the systems were compared for efficiency (Rankine was the highest at 88.9%,) and power-to-radiator-heat-transfer-area (the Brayton cycle indicated the highest value). The Rankine cycle compared very poorly on this basis.

Work has also been performed on evaluating a range of solar thermal cycles (Dunham and Iverson 2014). In this study, the following cycles were evaluated: regenerated He-Brayton, regenerated

CO_2-Brayton, CO_2 recompression Brayton, steam Rankine, and CO_2-ORC combined cycle. They showed that the Steam Rankine cycle showed the highest efficiency for temperatures up to 600°C, but it requires a change in materials for components above this temperature. CO_2 recompression Brayton cycles show promising performance (over 60%) above 30 MPa and 1000°C. Wet cooling may be required for these conditions. They showed that these advanced cycles may require different considerations above these temperatures compared to the Rankine cycle.

A review of a variety of advanced solar thermal power systems has recently appeared [Stein and Buck 2017]. The systems examined in this paper are, by and large, not being used presently, but they do offer promise for the future. While it is well known from Carnot ideas that raising maximum cycle temperatures can increase efficiency, there are many ways that goal might be met. Several new concepts are evaluated qualitatively, including such cycles as derivatives of the Brayton cycle, like the solar driven supercritical CO_2 engine. It seems that the two more significant development areas lie in the use of Brayton related cycles and the higher temperature solar receivers. Many issues are covered like "power cycles for CSP applications," "supercritical carbon dioxide closed loop Brayton cycle" and "solar gas turbine systems."

A recent paper has addressed prospects and problems of quite common solar thermal plant situations [Xu et al. 2016]. The authors note that this is a quite common application (there are well over 6400 papers related to this topic since 1990). This review addresses many of the details that apply to this solar plant type and location. Not only are they found in the US but in many other locations like the Middle East, North Africa, and Australia. The focus of this paper is the application to desert regions. Some locations have an ideal climate for these types of systems, but may not have grid connection as well as other challenges. A large number of these issues are addressed in this paper.

An important aspect that many solar power researchers are concerned about is the raising of the maximum temperature of the cycle. Simply considering the Carnot engine efficiency relationship indicates that an increase in maximum temperature should lead to higher performance. Of course, this is complicated by the many other effects that also occur.

One of the important issues in raising the temperature of the cycle is the importance of proper design of the receiver. In general, the receiver may be able to be studied separate from the remainder of the plant. Primarily this is because it has a very high thermal radiation input and it is one of the highest temperature components in the cycle. While the temperature of the receiver does influence some of the other material elements of the cycle, proper design of the receiver is clearly a very important aspect of the cycle.

DeAngelis and co-workers [2018] have gone through a preliminary design of a high temperature cavity receiver. They performed a numerical analysis of the receiver to see what variables have the highest impact on its durability and performance. The thermal conductivity of the receiver materials as well as the convection from the receiver cavity and the location of the receiver hotspots had the largest impacts on its performance. Some of the other aspects of the design had little impact on its performance. These included the cavity dimensions, the insulation thickness, the modes of heat transfer at the surface of the insulation, including both the convection and the re-radiation, as well as the emissivity of the inner cavity surface had little or no effect on its performance. The authors cautioned about the thermal stresses that can set up in the design, as these could have a large impact on the performance. Of course, this does not impact the receiver efficiency but does have a possibly large impact on the receiver reliability. They showed, however, that efficiencies of greater than 80% are possible, sometimes reaching values over 90% for utility scale systems. This underlines the possible desirability of using a liquid metal for the working fluid. In the paper, they cite the development of other high temperature components that make these types of systems closer to reality. They indicate that although more work is required, this study points to an optimistic path to high temperature receiver design.

In any kind of solar energy device, there needs to be a surface where the energy is received. While this surface is important in all such devices, it is the solar thermal approach that is most valuable. A paper by Suman et al. [2015] addresses this system aspect. They review the types of solar collectors that are used for a wide range applications, as well as the modifications that have been used over the years to make them perform their energy trapping role in a more effective manner. Some of these developments have included geometrical modifications of the absorber plate, use of selective coatings and nanofluids. Valuable tables are given that list reviews of solar thermal systems, heat transfer enhancement of air heaters, heat transfer enhancement of water heaters, specific examples of experimental studies of solar selective coatings, and reviews of heat transfer augmentation using nanofluids.

One of the more important properties of a solar thermal surface is the absorptivity. If the surface is for a high temperature application, the challenge is even more complicated than if it is for a low temperature application. However, it is definitely important for both. The paper by Suman et al. [2015] has reported a variety of insights about surface treatments to improve radiation absorption. One of the major emphases of this document selective surfaces. The higher the fraction of the solar radiation that can be absorbed, the more desirable is the surface for this application.

Du et al. [2018] consider concentrating solar plants with storage to illustrate the way that these plants can furnish power around the clock. This approach is much better than solar or wind systems without storage, but, of course, the system cost is higher. In this paper, however, the authors show some of the benefits of this kind of system including the operational flexibility. The authors compare systems that they call variable renewable energy (VRE) systems (like a system that operates on solar when it is available and does not operate when sunlight is not available) to systems that use concentrated solar and battery storage systems that they call concentrated solar power (CSP). They describe two separate types of benefits of the CSP system: (1) energy benefit—this is the saving of the VRE cost, and (2) flexibility benefit—the cost reduction effect of substituting CSP for VRE. Also cited is some history of power generation issues in two Chinese provinces and why one appeared more economical in a given province rather than the other. A conclusion is made based on cost criteria for some VRE systems with CSP systems.

When solar radiation is desired to be absorbed by a surface, which is one of the primary ways it is utilized, the thermal absorptivity value of the surface is very important. The higher the value of this, the more of the radiation that is absorbed. Since these coatings can be used on a large area of surfaces, the cost and durability of the high absorptivity coating is important. There have been a variety of approaches to this problem over the years. An article on this topic has recently been published [Karas et al. 2018]. It can be used to examine a variety of articles related to the topic.

To make solar thermal power generation more competitive, it is desired to raise the receiver absorptivity and its lifetime while decreasing the coating cost. In this work, Karas et al. report on an effort to develop black metal oxide nanoparticles comprising of copper-cobalt oxides and copper-manganese oxides. These were selected for investigation because of their low cost and energy efficient manufacturing in large scale manufacturing. As part of the development process, their high temperature stability was also assessed for over 1000 hours. Details of the manufacturing process can be found in the source publication. Results of the materials developed compared quite favorably to Pyromark 2500, a commonly used coating material.

Kumar and co-workers [2017] have reported on a computer analysis they have carried out about a solid particle curtain falling through a high-flux beam of solar irradiation. A key step in the analysis was the use of a Monte-Carlo ray-tracing technique to determine the radiation intensity throughout the curtain. Predictions of the opacity of the curtain were compared to experimental results reported in the literature. Two different sizes of sintered bauxite particles were assumed. Cases representing the two flows were compared on a mass per time basis. In general, the smaller particulate flows showed average absorptance than the larger particale a similar particale flowrates. More efficient absorptance resulted in non-uniform radiation absorption.

A large majority of solar thermal plants use mirrors as the means of concentrating the solar flux. This is not the only way this can be accomplished. It was suggested in a recent paper that Fresnel lenses could be a good way of achieving similar effects [Kumar et al. 2015], and they cite references that document this aspect. The authors of this paper point out that reflectors or mirrors in CSP systems contribute about 50% of the total cost of the system, and that Fresnel systems could be more cost effective. These systems have found limited application in PV/thermal combined systems. The authors list many of the benefits of Fresnel systems to replace mirrors. They include a table that is a "CSP World Map—Fresnel Technology Reflector."

Thermochemical kinds of operations could be of value in a particular application. These will not be reviewed in any detail here. The interested reader could read the paper by Dincer and Bicer on Solar Thermochemical Energy Conversion [2018], as a starting point.

Materials with suitable properties for particular applications are critical to the successful design of solar energy plants. Some of the new materials developed are outlined by Fernández-González et al. [2018]. A review of materials for all fields connected to situations where concentrated solar energy is applied is given. Included in this review are applications from metallurgy, materials processing (including many topics like surface treatment and coatings), as well as nonmetallic materials like ceramics and fullerenes. Because many solar energy applications only operate when the sun is shining, this limits the cost of materials that can be economically applied to these systems. An extensive list of relevant background literature is given.

Operation of receivers in which the aperture size can be varied are described in the paper by Ophoff and Ozalp [2017]. Tests described in the paper were performed with an iris-like aperture covering the opening. It was found that with variable aperture covering used, the optical absorption coefficient improved by 14%, while the re-radiation loss through the aperture was decreased by 2.3%. Some tests using a 7-kW solar simulator at different power levels demonstrated the performance of the device could maintain a steady temperature under variable flux.

It is noted in passing that there are a variety of applications for process heat. This field is less focused on a specific solution compared to power generation because the types of applications can quite varied. They may require high temperature, like those sought to solar thermal power generation, or low temperature heat may be what is needed. Storage may or may not be important. An article that reviews this field briefly has been given by Kedare and Desai [2017].

1.2.1.2 Thermal System Heat Rejection Cooling for thermal power plants is of course required. For Rankine power plants, there are two possibilities for this necessary function: wet or dry cooling. In wet cooling, some water is evaporated to use to bring the cooling water close to the wet bulb temperature. Under certain kinds of conditions, this process can use quite a bit of water. If cooling water scarcity is an issue, the alternative is dry cooling. Basically, this is a little like a radiator on a car

A recent paper [Merino et al. 2018] reports a study of the ground water used in wet cooling in the Mediterranean area of Spain. They considered 50 wet-cooled CSP plants there and analyzed the cooling process in the ones that used groundwater. Because some water is evaporated, the salt concentration in the returned and warmed cooling water increases (Merino and coworkers reported this increases by 2.5 times in condensing the power plant steam). Other influences on groundwater when it is used for wet cooling are also reported. Legal and political implications related to the use of ground water are also described.

Many of the locations where solar energy is quite prevalent also demonstrate a shortage of water. An example of this is the general southwest portion of the US. Of course, wet cooling towers are quite effective in condensing the steam in Rankine cycles, but a fair portion of water in these kinds of systems evaporates into the air and condenses quite effectively. One way of minimizing this water loss is to use *dry cooling*. The heat exchangers transfer heat but not moisture. Liqreina and Qoaider [2014] analyzed a dry cooling situation by using a plant type (Andasol) to base the calculations upon and assumed it was in a Middle East/African location with good solar incidence (Jordan) but small availability of water. The authors simulated an existing plant in Spain that is wet cooled, and they calculated the predicted performance for a dry-cooled plant there. After that comparison of the wet-cooled plant was made to the hypothetical dry cooling plant in Jordan, the similar plant in Jordan with dry cooling indicated that the overall efficiency of the dry cooled plant decreased by 3.1%, the water use was reduced by 92%, energy yield increased by 21.8%, and the LOCE reduced by 18.8%. This demonstrated satisfactory effectiveness of dry cooling.

Over the years, system and component modeling has played very important roles in understanding the operation, as well as the design, of solar power plants. There will be no comprehensive analysis of the literature that has resulted from this work, as it is very extensive. However, a few of the more recent publications will be noted here.

1.2.1.3 Brayton Engines Of course, the Brayton cycle has been a workhorse in a variety of fields, including power generation and aircraft powering, for many years. The basic concept is also the basis of one form of a $SCCO_2$ engine that is covered a little later here. There have been desires to apply the Brayton configuration to a solar form. In this configuration, as well as others, it may be of value to evaluate the engine in a scaled down size for one of many reasons. However, in doing this, the experimentalist needs to pay close attention to many details if there is hope of interpreting results from a small-scale experiment to an ultimately larger scale one. While it is not unheard of to evaluate cycles at a smaller scale during their initial evaluation stages, gas cycles may be more readily operated in this manner. Particularly this may be true in solar-driven units because the setup of large solar driven systems involves considerable effort and space. So, as noted above, it is not unusual to find some newer gas cycles that are designed to operate at very high pressures. Other variations that are found to be applied in the early evaluation stage of solar devices is to use a solar simulator. While these can facilitate the evaluation of operational characteristics in the lab, often on-sun evaluation of devices may be preferred. An excellent example of this kind of work was reported by Aichmayer et al. In this work, a systematic procedure of careful design was followed with particular concern for understanding the scale-up characteristics of the test program. They used a solar dish to furnish solar energy to a micro gas turbine system. Although their main purpose for the tests was to study scaling effects, it does show that if these effects are understood, use of small scale systems to investigate large systems is a reasonable study approach.

An initial phase was to develop a code that represented the performance of an existing solar configuration. The authors understood that several of the physical parameters would be important in the scaling exercise. These were identified and played important roles in the effort. The parameters identified were important in correlating with the boundary conditions applied between the numerical code and the experiment. Because of the system design, the stresses generated were of particular concern. In working through their various steps, a multi objective optimization analysis was performed. Three separate cases were analyzed and compared. Based upon the stress values found, one of the three cases was selected as being the most appropriate, and this was used for further analysis. It was the authors' opinion that there was an excellent representation of the performance that minimized the errors in the calculated results.

Analyses using MATLAB of full size Brayton engines are of great interest to assist in answering various kinds of system performance questions. One such recent analysis considered a solar-driven regenerative Brayton heat engine [Malali et al. 2017]. The authors identified three of the more important performance parameters (Brayton cycle pressure ratio, the concentration ratio of solar dish that powered the engine, and the maximum temperature ratio of the cycle), and they expressed the efficiency of the coupled system in terms of these parameters. The impact on optimal system efficiency of a variety of variables was shown, including the heat exchanger efficiency, total concentrator error, rim angle and non-dimensional radiation flux parameter. The authors developed performance charts that can be used in design efforts.

A great deal of interest has been developing lately to explore possibilities of using higher temperature approaches to solar power generation compared to the traditional Rankine cycle. Much of this interest is focused on variations of the Brayton cycle using CO_2 as the working fluid. An example of this work has been published by Enríquez and co-workers [2017]. Their analysis dealt with three types of these high temperature cycles: the recompression cycle, the partial cooling with recompression cycle, and the recompression with main compression intercooling cycle, with one reheating stage. A large number of aspects of the plant performance-related factors have been investigated. However, they were all for operation at the design point. Additional studies will have to be made to determine off-design performance.

There could be a use for generated steam for a variety of applications. This is an area where solar energy has been shown to work very well as the heat source used for producing this steam. The report of a recent study of the various approaches to this type of end result has recently appeared in the literature [Bittencourt de Sá et al. 2018]. In this work, the authors particularly discuss linear concentrators, which include parabolic troughs and linear Fresnel systems. It does not include dish and tower systems. They also consider phase change systems, where, for example, water is circulated through the solar field, turning into steam as it travels through. If the end application for the steam is to drive an engine or some other higher pressure requirement, then the whole absorber field presumably needs to be at the higher pressure. As the authors show, this approach eliminates a second loop, usually carrying a fluid that remains a liquid through the concentrating system. Elimination of this loop also eliminates one or more heat exchangers where the solar heat is transferred to the cycle working fluid. The authors emphasize that direct steam generation is a relatively new approach. They mention that one of the biggest drawbacks to using this approach is the need for higher pressure and temperature solar components.

1.2.1.4 Trough Plants As was noted early in the paper by Chaanaoui et al. [2016], the solar thermal power concept that has been most popular over the years is the trough type of system. In fact, these are normally configured on a parabolic trough design. Although a major set of system developments was instituted in California beginning in the 1980's with these types of configurations, trough units have been built in small numbers in the preceding 200 or so years. Over the years, the trough elements of large plants have gotten larger, both in terms of width and length. The receiver element is typically made up of a glass tube that encloses a metal tube of somewhat smaller diameter where the collection fluid circulates. The space between the two tubes is treated with a vacuum evacuation. A solar absorbing surface is applied to the metal tube. While the array of troughs can be mounted in a north-south configuration or an east-west configuration, the north-south is generally preferred. Hence, on a clear day the troughs will be moved in a nearly constant manner, following the sun from nearly sunrise to sunset.

Recently a paper has appeared that has documented the evolution of the trough system design [Hafez et al. 2018] from approximately 2004 to 2016. This report covers most of the design elements of the systems and a large amount of details about the configurations used. Given are the objectives of the various studies performed. It also denotes if the study is experimental, theoretical or both. The trough manufacturers, the technologies used, and the system specifications are all given. Details like the design temperatures in many of the studies are given. Several system diagrams are also given. If thermal storage was used, it is indicated. In all, this paper is a good place for either the novice or the experienced worker to find valuable information.

Most solar power codes compute steady-state results. However, an understanding of the transient behavior of system may be quite valuable. This issue for solar thermal applications is examined in the paper by Abutayeh and Alazzam [2017]. Solar thermal plants have fairly long start up and shut down procedures. Of course, the solar flux can vary also, which will impart a transient behavior to a plant. In fact, this is the aspect the authors use to adjust the steady state results to transient operation. Basically, they propose to vary the flow rate such that the exit temperature from the field remains constant. While this a simplification of their approach, it describes the generalities of it.

Details on the design of parabolic troughs have been summarized in a recent paper by Abdulhamed et al. [2018]. It includes topics ranging from geometrical analysis of the receivers to overall system efficiency. There is also a history of the parabolic trough evolution.

A paper from China explores an on-site test method for evaluating key performance characteristics of utility-scale trough loops [Wang et al. 2017]. The procedure consists of both off-focus and on-focus tests performed sequentially. One element of the approach uses some of the collectors to heat the fluid and others are used for determining loss coefficients. Another part of the test involves the on-focus portion to determine the optical performance. Refer to the paper to see the various aspects of the test procedure.

An Egyptian study has explored the possible use of a very short term (15 minute) steady state test to see if parabolic trough collectors meet their defined application requirements with that kind of testing [Elmaihy and El Weteedy, 2017]. To do this they used an existing trough field and made the kinds of measurements that they propose to be used. These measurements were ones that could be readily made on an existing set of units. The meteorological situation needs to include clear sky conditions, and the timing is such that the measurements do not change in any significant way during the test. Most of the measurements are straightforward, and they also define a way to estimate the uncertainties in the measurements. They also discuss how experimental uncertainties can be evaluated.

A recent paper has given some answers to the question how would a conventional parabolic trough collector operate using a variety of working fluids. Okonkwo and coworkers [2018] have used computer simulations with the software Engineering Equation Solver to explore the possibilities. After the various fluids were evaluated, they also considered some modifications to the internal geometry of the receiver. Al_2O_3/oil-based nanofluid showed the best thermal improvement. While high pressure water resulted in the best thermal efficiencies, but required very high pressures, essentially this eliminated it as a possibility. In evaluating the modifications to the receiver, the best option resulted in a 1.13% efficiency increase. Space does not permit commenting on all of the modifications that were evaluated.

Another paper that deals with nanofluids has appeared [Khakrah et al. 2017]. In this one, several of the basic parameters were considered and their effect on the overall performance of a trough was considered. Included in the variation were the wind velocity, nanoparticle volume fraction, inlet temperature and reflector's orientation. To calculate the solar gain, the authors used a Monte Carlo radiation tracing method. One of the main results reported was the increase in efficiency by adding 5% volume of Al_2O_3 of 14.3% and 12.4% for two cases.

1.2.1.5 Dish Systems Like the trough systems described above, solar dishes date to the early days of solar power system development. However, the dish system is generally much smaller, employs two axis tracking, and the generator generally moves with the dish. If large amounts of power are desired using a dish, numerous complete units need to be used. It is also usually the case that engines used with dish systems are air cooled. It is the situation with trough systems that the collector fluid may travel large distances, and in dish systems, the engine working fluid may be a different fluid. In dishes, the working fluid is usually the same as the collector fluid and it travels in a small loop. Several of the dish system designs over the years have driven Stirling engines. This option has great appeal because of the generally high efficiency of these types of engines. However, there have been few large-scale system developments of this type.

A recent paper [Hafez et al. 2017] has given a good summary of the design parameters of dish systems. Included are discussions of the material of the reflector, the shape of the reflector, the receiver design, sizing of the various components, the concentration ratio, and the rim angle. All of these parameters have important implications about how the system will perform. The authors of this paper review various systems that have been built, which were developed for a variety of reasons including generating electricity, cooking, water heating, irrigation, desalination processes and water distillation. They examine what design parameters have the most effect on the performance and find that they include the rim angle, the diameter of the receiver, and the concentration ratio of the dish. Of course, an important performance parameter for these units is the net conversion efficiency. One of the systems (Vanguard) demonstrated a system efficiency of more than 30%. More typically the net conversion efficiencies are in the range of 24-27%.

1.2.1.6 Tower Technologies Solar tower technologies are one of the newer means of thermal power generation. They use the concentration of solar energy by means of sun tracking heliostats (mirrors) located on the ground surrounding a tower. This concentrated sunlight is absorbed at the top of the tower in a fluid which is, in turn, used, either directly or ultimately, to power an engine. To date Rankine cycles have been used for the power conversion. The initial versions of these plants used wet cooling, but because most of the US units have been set up on the arid Western part of the country, more recent work has used dry cooling means to condense the steam. The earliest versions of this type of plant were operated during 1980s and 1990s near Barstow, California. Two versions of this plant operated for a few years each. Solar One operated from 1982-1986 and used steam as the fluid for both capturing the sun's energy and driving a turbine. From 1995 to 1999, named at that point as Solar Two, it operated on molten salt. In 2009, the unit was demolished.

Many details about the makeup of a solar tower system can be found in recent article by Buck and Schwarbözl [2018]. This excellent treatise gives quite a complete summary of the solar tower history and design. Among other aspects it gives a good summary of the existing tower plants.

Turchi and research associates [2017] have examined how towers fit into the various choices of solar thermal power generating devices. While they review a variety of types of systems, their primary goal is to discuss their favored approach: power towers, particular those that operate on supercritical CO_2 (see also a paragraph below). In this paper, they cover a variety of topics including the types of cycles that might be used to the application of dry cooling for this situation.

There has been an interest in pushing maximum temperatures of solar thermal power systems higher. However, there are two results of doing this. First the receiver efficiency goes down. Opposing this effect is that the driven engine efficiency goes up. As a result, the tradeoffs between these two effects will determine an optimum operational temperature. This effect has been investigated by Poživil and Steinfeld [2017] for three different Brayton derivative cycles with pressurized-air solar receivers for CSP tower systems. They found that an optimal preheat temperature of 1300°C and an air cycle pressure ratio of 9 yielded the best solar-to-electric efficiency. All three of the cycles investigated demonstrated nearly the same efficiency. They also investigated a two-stage serial receiver arrangement that incorporated low and high temperature modules. An increase in efficiencies was found.

A combined effort of NREL and DOE personnel has examined possible advances to tower design to see what might come of them [Turchi et al. 2018]. Although a great deal of technical information is given in the paper regarding the analysis of the system, only a small part of it will be reported here. For example, they examine the benefits that accrue if towers using supercritical CO_2 Brayton cycles at 600-650°C compared to the current state of the art towers using direct salt storage at 574°C. They find they are significant. Among other factors, it reduces to the size and thus the cost of the solar field. Conclusions are drawn about many aspects of the plant design.

Another paper has outlined quite a different design than ones that currently exist [Rea et al. 2018]. This design was conceived to address the intermittency problem in the solar resource, so it is dispatchable. The designed unit is quite small compared to conventional solar tower systems, about 3 orders of magnitude smaller. And in contrast with the more general design, their proposed configuration uses a Stirling engine. Their view is that the design and its size should allow rapid deployment rates. Multiple units can be used for larger desired outputs.

One of the newer concepts for a solar receiver is one that generates a flow of particles through the beam. The particles serve as both the absorber material but can also function as a storage medium. This is incorporated with a solar tower system. Sandia Laboratories has been quite active in this concept development. One of their staff members has written an excellent article on this concept [Ho 2018]. These receivers have a window in them through which the concentrated solar flux enters and heats a stream of falling solar particles. The heat picked up by the latter can be used directly, or it can be stored for later use, all typically in the solar particles. Possible temperatures of the particles can reach in excess of 1000°C. The particles can flow directly through the beam or they can be contained in tubes in the beam as they fall. Ho gives a great deal of information about this concept including costs and temperature rises. In the 2016 paper the levelized cost of power from these systems was estimated to be ~$10/MWh$_e$, which was at least an order of magnitude lower than all other utility-scale energy storage technologies when the paper was written.

1.2.1.7 Solar Chimney Power Plants A topic that is relatively new to reviews of this sort is related to the solar chimney power plants. While precursors have been around for quite a while (see Al-Kaylem and Aja 2016), and analyses have been large in number, only one of these types of systems has apparently been built. This is one that uses solar energy to make what amounts to a wind machine. In the usually discussed configuration, typical trapped ambient solar energy is used to heat air around the base of a large tube-type tower. By appropriately channeling the heated air to the inner part of the tower, a substantial air flow can be generated up the tower. If one or more propeller type units are placed in the tower, they can generate power from the heated air flow. While the concept is reasonably simple to envision, to date, less than 2% energy efficiency is anticipated from the fairly large configurations currently imagined.

The paper by Al-Kaylem and Aja outlines many details of the designs that have been proposed over the years. They also outline several aspects of the cost. This paper is quite rich in details about the relative costs of the components and which of these aspects show potential for reductions.

Sangi et al. [2011] have described a relatively early study of solar chimney power plant numerical analysis. This work used FLUENT to consider a 2-d version of a prototype plant in Spain. Reasonably good agreement was shown between the numerical results and actual performance data.

A numerical study has been performed on the effect of a number of physical variables in the performance of solar chimney power plants [Zhou et al. 2017]. Among the parameters varied is the shape of the chimney: They consider a constant area chimney as well as additional analyses of convergent and divergent shapes. A variety of other factors are considered, including assuming constant density throughout as well as variable density, variable mass flow rate, and the variation of the solar flux.

Guo and coworkers [2016] considered this type of plant with soil heat storage. In this analysis, the thermal storage was placed under the collector. They compared a system with storage to one without storage for the first time. They treated the ground properties parametrically and, as would be anticipated, they found that high specific heat and thermal conductivity of the group was beneficial to the plant performance. Of course, with storage, there was a lag in response, so the power production in the afternoon is significantly higher than in the morning.

An interesting two-stage design analysis was performed on chimney power plants by Okoye et al. [2016]. In the first stage, an optimization model was used to determine the layout of the plant. This was done in a way that minimized the cost of the plant while satisfying the plant requirements and of course considering the available solar energy and ambient temperature variations. In the second stage the net present value of the plant is determined using the results of the first stage. This later step takes into account both the cost of the facility and the value of the power produced. The approach was used to evaluate several tradeoffs in the design.

1.2.1.8 Supercritical CO_2 Plants Turning low grade heat into power has always been a challenge. Historically this was usually done with a Rankine cycle using a refrigerant as the working fluid. More recently another cycle is being considered: The transcritical CO_2cycle. This is addressed in a recent paper by Ge and coworkers [2018]. While this cycle is described, it is also the case that researchers dealing with this cycle should be addressing a research aspect in trying to increase its operational efficiency (this has generally been quite low in previous studies). To address this problem, this paper describes a test rig that was set to evaluate the possibility of going to higher efficiency performance. Included in the design is a plate thermal coil CO_2 supercritical heater, a CO_2 transcritical turbine, a plate recuperator, an air-cooled finned-tube CO_2 condenser and a CO_2 liquid pump. The paper contains some preliminary performance results for the system. Using a computational model also developed, more insights into possible improvements to the cycle are being explored.

Various forms of the $SCCO_2$ cycle have been given a great deal of consideration in the nuclear field and, more recently, the solar thermal field. Since this cycle is gas based, it does not suffer the same kind of limitations like, for example, the steam Rankine cycle. Although, as is well known, gas working fluids tend to be less dense than liquids and this lessor heat capacity is a limitation to their use. However, this can be partially overcome in these cycles if they are operated at elevated pressures (which may make the gas denser and hence able to carry more energy per unit volume) and they may utilize somewhat smaller equipment than some competing cycles that do not operate at so high of a pressure.

A paper that deals with Brayton cycles, but using supercritical CO_2 (SCCO2) as the working fluid, has also been published recently [Milani et al. 2017]. Several types of configurations of that engine have been discussed in the literature, and the authors of this paper show several of the more promising configurations. Impacts of the variations of key parameters were discussed. Among the insights they give are 20 plots of how both thermal efficiency and heat exchange area vary with splitting fraction, compression ratio, and expansion ratio. Through this kind of parametric study, they are able to show the optimal operating condition as well as tradeoffs between parameters noted in the previous sentence. Several valuable conclusions are drawn about system design.

1.2.1.9 Thermal Storage Thermal storage is one of the main avenues of saving solar input to be used at a later time. In fact, it may be one of the more cost effective ways of doing this. For this reason, it will be explored in some detail here.

A variety of approaches exist for this mode of storage, and many of these are listed in Figure 20.1 of Bauer et al. [2013]. Of the many that are shown there, the "molten salt" in the left-hand column and the "salts" in the middle column are the main materials currently in use as thermal storage forms. A very important parameter in the choosing of materials for thermal storage is the product of the density and specific heat. This relates the size of the storage to hold a given charge. This chapter contains a great deal of technical information about a variety of salts (phase change or otherwise) used for thermal storage. The authors include a section of the chapter to discussions of system designs for both sensible heat and phase change salt storage systems.

A recent report has given an extensive list of properties for thermal storage mixture material $MgCl_2$-KCl molten salt (Xu, X., et al. 2018). This salt is able to be used for temperatures up to 800°C in concentrating solar power plants. It has a molar ratio of 32%-68%. Not only are values of the properties given, but there are also correlations of many of those properties as a function of temperature. The data given in this paper should be of value to researcher and designer alike.

While there are several studies that include some analyses of incorporating molten salt storage in the analysis, another study considered a different approach where molten salt served double duty as the solar field coolant as well as the storage medium [Trabelsi et al. 2018]. The results of this analysis were compared to those using Therminol, a more typical coolant. It was found that the molten salt configuration performed better.

Thermo-electro-chemical storage has also been addressed in the literature. One example is a paper by Wenger et al. [2017]. The authors propose using thermally regenerated batteries. While the process is somewhat more involved, basically the battery replaces the steam turbine used in CSP plants. The authors discuss the potential cost and performance benefits of this approach.

Storage has long been recognized as a needed addition to many types of solar applications but has been, in general, quite expensive. Domestic water heating has been one obvious exception relative to this statement. With the increasing popularity of solar thermal power installations, with their need to have the power plant portion be able to function for longer periods, a natural application for higher temperature thermal storage has been realized. This is a distinctive characteristic of thermal plants not shared with PV systems. This issue is complicated to a degree because of the range of types of solar thermal plants as well as potential options of storage methods. Luckily, this situation has been recognized by several researchers, and available studies are appearing quite frequently in the technical press. Some of those contributions are reviewed here.

An interesting paper has examined the possible replacement of the two-tank salt systems that are currently in use for solar plant high temperature storage with possible alternatives [Tehrani et al. 2017]. They assumed that the same solar plant (19.9 MW_e Gemasolar tower plant with a solar multiple of 2.5) with 15 hr. of storage for each alternative. They then performed year-long simulations of plant operation and compared the relative performance. The alternatives they considered were: single-medium thermocline (SMT), dual-media thermocline (DMT), and shell-and-tube (ST) systems. To be consistent with recent literature, the DMT and ST systems assumed concrete with a porosity of 0.2 (e.g. where concrete occupies 80% of the system) as their low-cost filler material. It was found that the plant with the two-tank molten salt system yielded the most electricity in a year compared to the alternative. They also considered cost of construction and indicated that great decreases in cost can be made using poured in concrete storage without internal pipes.

A variety of considerations about the incorporation thermal storage was given by González-Roubaud et al. [2017]. While they discuss a large number of issues, their main point of consideration is the use of a 2-tank storage (both direct and indirect) compared to the use of steam accumulators. They point out that the 2-tank indirect salt system is the most commonly used, but that the direct salt system may yield better performance.

The direct molten salt TES system currently used in the solar tower system does offer all the same benefits as the indirect system but it readily allows for a higher collection temperature. Some of the intermediate equipment is eliminated which should require less cost. Both types of salt systems require heat tracing which can sometimes pose operational problems. There are uncertainties in the storage medium prices, and that market is evolving. They indicate that the steam accumulator TES has the lowest storage cost for capacities less than 3 hours. For longer storage times, the direct molten TES is preferred followed by the indirect system.

Two earlier discussions of thermal storage options have been given by Kuravi et al. [2013] as well as Liu et al. [2016]. These give extensive discussions of the various options. Reviews of this literature are highly recommended to people who are familiar with the technology as well as to those who are not.

An extensive review of the situation with phase-change storage for CSP systems was given in the paper by Xu et al. [2015]. The paper is divided into five parts. The first part is an introduction to the types of phase change materials used for high temperature thermal storage, including a survey of encapsulation methods. A second part addresses the application of these materials to CSP systems. The third part discusses mathematical modeling of these systems. Integration of the PC materials into a CSP system is the emphasis of the fourth section. Finally, the fifth section deals with cost issues and as well as comparisons between the sensible storage approach to that of PCM.

A very recent paper gives an excellent review of the many possible applications of phase change materials to solar energy applications. Most of the paper covers thermal examples, but the authors also include some discussion of PV system uses. Both low temperature and high temperature functions are covered [Pandey et al. 2018].

The thermochemical storage option is a "good news/bad news" situation. As outlined by Prieto et al. [2016] the good news is that it offers some almost ideal characteristics, including nearly no losses during storage and a very good volumetric energy density. The bad news is that it is still in its infancy as an application with a great many practical problems yet to be solved. Main development efforts to date focus on three redox reactions. They include sulfur-based cycles, metal oxide reduction-oxidation cycles, and perovskite-type hydrogen production. Metal oxide non-redox cycles are also possibilities due to their similarity to the redox cycles listed. The authors give a good summary of the positives and negatives of each cycle. They conclude that the calcium carbonate cycle has the most evaluation behind it. This cycle should be used for tests on a research scale plant. Probably the manganese oxide cycle is the second most developed.

1.2.1.10 Hybrid Plants Ju et al. [2017] have pointed out that there could be resulting benefits by combining a PV plant and a solar thermal plant. Examples of a couple of the benefits are: compared to PV alone, combined plants may produce power of better quality; compared to CSP alone, combined plants may produce lower priced power. The authors also describe how if the combined plant is hybridized, the operating efficiency may be improved over either plant type alone (over 40%). A review is given of the types of combined plants attempted prior to the writing of the paper. This allows additional insights to be made about how the combining of the plant types might be made. Several other aspects of the combination are also discussed including possible lower cost power and also a broader market.

Described in the paper by Calise et al. [2018] is the development of a dynamic model of high-temperature integrated solar combined cycle power plant. The plant is powered by parabolic solar trough collectors and these drive a three-pressure steam combined cycle system. System performance is analyzed by detailed algorithms. The control system needed to operate the system is described. Both a thermo-economic and environmental analysis is given for both a conventional and a solar-driven combined cycle plants. A 100 MW plant assumed to be located in southern Spain is analyzed in a case study. Impacts of variations of various plant parameters are described. As the result of the analysis given a payback of 15 years is noted.

Another paper relevant to hybrid plants has recently appeared [Zeitouny et al. 2018]. The authors here explore the possibility of operating the PV aspect of the plant at higher than normal temperatures under concentrated solar energy. This could result in a significant fraction of power being generated from thermal means. An issue is trying to keep the loss small from operating the PV cells at higher temperature when trying to boost the relative thermal power. The authors note several positive outcomes of doing this. Work such as this may make high temperature hybrid system operation a reality.

Hybrid systems are also the primary focus of a paper by Pramanik and Ravikrishna [2017]. They have considered CSP being hybridized separately with a variety of renewable and nonrenewable energy sources. In doing this they categorized the systems they considered to be high, medium, and low-renewable hybrids. Extensive tables are given that delineate the existing systems. They show that the low-renewable hybrids offer higher efficiency and capacity factor at lower cost.

1.2.1.11 Environmental Solar plants are being developed in many places on the globe. Some of these plants are quite large scale, filling millions of acres of land globally. A few studies have addressed the potential environmental impact of these plants.

Solar plants can have significant environmental impacts if the siting of utility scale solar energy plants, as well as their construction details, are not carefully considered. Several pertinent aspects have been examined in a recent paper [Hernandez et al. 2014]. In carrying out this review of utility-scale solar energy (USSE) development, they have considered things like the impacts on biodiversity, land-use and land cover change, soils, water resources, and human health. On one hand, this presents some good news because USSE systems have some low environmental impacts compared to other forms of energy, but there are areas that need the careful attention of developers. Several of these are examined in the paper.

A paper that covers ways that more solar energy development can have a positive impact on the world has more recently appeared [Shahsavari and Akbari 2018]. After discussing the major countries that are both underdeveloped in energy as well as polluting, the authors then discuss methods by which solar energy can be of environmental benefit. However, since most of the pertinent solar applications that can be of benefit also require up-front financing, they discuss the approaches that can be used to accomplish this.

In general, it can be done by assistance from the richer nations. Financial tools that can be used are also covered.

As a thought follow-on to the paper just noted, a Japanese contribution [Shoai-Tehrani et al. 2018] outlines a specific way that decisions on low-carbon investments can be made. Although this report does not address only solar energy, it does offer a method for following more cost-effective, low-carbon approaches specifically for utilities. The focus of the paper is the European energy sector. Results show that the utilities are, by and large, not easy to move from their "business as usual" point of view. Among other findings is that a negative constraint on coal is much more effective that positive incentives on low-carbon alternatives.

The environmental impact of solar energy development has also been explored by Breeze [2016]. He discusses several of the concerns about the development of this resource. Of course, land use is one of the major concerns. Since solar energy is diffuse, large amounts of land may be used for the development of this resource. Another important issue is water use. However, this aspect has some alternatives. Wet cooling in solar thermal power plants does use a great deal of water. However, dry cooling can cut this amount by about 90%. It could also impact various animal species, particularly bird populations relative to concentrated solar power plants. Breeze also discusses the chemical emissions issue related to solar power plants. This is important in the manufacture of the solar components (e.g. solar cells or selective surfaces). He suggests that if the generation plant life time can be extended, this would lower this impact per unit of power produced. He also touches on the intermittent nature of solar energy and impacts this has. Other topics are also discussed beyond what is mentioned here.

Another report is noted here, but not reviewed. It is the paper by Kylili et al. [2018], and this deals with environmental assessment of solar thermal systems for industry. Generally, this addresses lower temperature kinds of applications than are necessary for solar power generation.

1.2.1.12 Economics, Reliability When comparing the economics of solar systems, there are many ways that systems can be separated for evaluation from an economic standpoint. Of course, there are distinguishing features between several options: PV vs. CSP, air vs. liquid systems, high concentration vs. little or no concentration. However, one meaningful method is to distinguish between systems with storage compared to systems without storage. Storage should allow for more system output but at an increased cost. The size of the system can also impact the variation of the output. Small PV systems can be quite variable, but large systems may do some filling-in of low sun situations such that the output is more uniform than small systems. Also, for example, one-axis or two-axis plants would have different amounts of output, all other factors being the same. Dispatchability, where the plant output is time controlled, can be a favorable feature to the solar customer. A common parameter for evaluating solar plants, which can be applied to any plant, is *capacity factor*. Another factor is *net cost*, which includes all the various costs and benefits that accrue. Capacity benefits ($/MWh) of marginal solar resources decrease with increasing % of solar energy penetration. If energy storage is added to power system, but not integrated into the plant desired operation, the value of the power decreases with large outputs.

Plant operators will need to be quite mindful of their plant operation as the percentage of solar power grows. The economic situation is going to be more complicated in the years to come. Many of these aspects are touched on in the article by Helman [2017].

1.2.1.13 Niche-Regime Conflicts As distributed solar has become more prevalent in recent years, there has been a negative attitude develop toward it by some electrical utilities. In general, there is the view that solar energy can be a good thing, particularly in light of its decreasing cost. However, the issue is one about systems owned separate from the utility that are taking business away from it. Utilities do not like to see business taken from them in what amounts to approved alternatives to the utility function. Part of this stands to reason, because the utility still must be able to supply power to almost all customers, whether they have their own solar system or not, and this carries with it a cost to the utility. While various governing boards have tried to address this issue, it remains a sticking point. When this is combined with a decrease in net revenue for the utility, it definitely becomes an issue. It has been recognized and goes by this name (niche-regime….) in many quarters. The utilities usually cite some grid stability issue associated with too much solar energy on the grid. This situation may be difficult for niche organizations to refute.

A recent paper by Hess [2016] has examined this situation. As pointed out by him, niche organizations that are facing conflict with the utility have some means of at least partially overcoming the problem. This can be done in one of three ways, including: a) form coalitions with political parties that support the niche technologies; b) gain support from potentially helpful industrial organizations; and c) form coalitions with social movements that may mobilize political protests.

1.2.1.14 Analytical Tools A paper that recently appeared has examined a variety of solar related performance evaluations using energy and exergy analyses [Sansaniwal et al. 2018]. As is well known, an energy analysis considers, as the name implies, all forms of energy on essentially an equivalent basis. On the other hand, exergy analysis treats all components on a work basis. So, for example, a heat transfer process is converted to its work equivalent using Carnot ideas. In this paper, the authors consider the following solar processes: drying, refrigeration and air conditioning, water heating, cooking, PV, and CSP, and they show how an analysis is carried out. The information in this paper could be handy to someone who is performing a theoretical analysis and wishes to do it on the uniform exergy basis.

1.2.1.15 General Solar Systems (not specifically CSP or PV) The question of how easily is a coal plant replaced by solar and/or wind plant is addressed by Leonard et al. [2018]. They picked a geographical location that had a reasonably large amount of wind and sun to see how large a plant that used those resources would be required for replacing a coal plant. As part of the evaluation they determined they needed to include storage as part of the evaluation. For this they found that generating hydrogen would be a good storage path to follow. They discovered that substituting for the coal generated power with wind required significantly higher storage capacity than that using solar, but that with the combination of the two requires even less.

A paper has appeared that predicts how PV and CSP will fare in the years that come [Pietzcker et al. 2014]. The authors used the code REMIND 1.5 which is a general equilibrium growth model that is described in the paper. They predict that PV's lead over CSP will continue for a period, but then CSP will move to a dominant role for reasons that they describe. Their end prediction is that solar will supply somewhere between 19% and 48% of the total world energy in the period 2010-2100. Global warming will be limited to 2°C in that period because of the solar applications.

Brief reviews of PV and solar thermal have appeared. The first of these was authored by Khan et al. [2016], while the second one was written by Kabir et al. [2018].

In series of papers (only a few noted here) Trainer has made the case the renewables (and solar particularly) cannot be used for supplying all of the energy in even the most favorable climates [Trainer 2014, 2017a, 2017b, 2018]. Although his main work has been in Australia, his conclusions are farther reaching than that, he argues that at best it would be very much too expensive to try to reach this goal. Those conclusions are posed for times in the future and may indeed prove to be true. However, time will tell. Further comments will not be made about these papers here.

A study of a special goal to install 100 GW of grid interactive electricity to be generated by solar energy by 2022 in India has been reported in a recent paper [Hairat and Ghosh 2017]. The authors have evaluated the techno-economic feasibility of adding this amount of solar power to their grid. For a number of reasons based on empirical evidence and outlined in the paper, the authors anticipate the effort would jeopardize the country's energy security. Several steps are outlined that would allow the country to better capitalize on the effort. For example, they conclude that India should put greater emphasis on CSP technology and de-centralize applications for solar power.

1.2.1.16 Tracking Tracking can be very important in solar systems where maximum collection is sought. This could be either PV or CSP systems, but it is particularly important with either of these if concentration is sought. A recent paper addressing this aspect of solar energy systems has appeared [Nsengiyumva et al. 2018]. Most tracking systems are either single axis or double axis. Another distinction is that some can be active trackers and others can be passive trackers. Many other distinctions in systems are also possible. A caution is that a number of tracking errors are possible, and this paper outlines some of these briefly. It is noted that the double axis solar tracker in the form of a polar-axis and azimuth/elevation that incorporate the solar movement models and the dynamic closed loop feedback control are the most effective. When applied to PV applications, a tracker using this type of control will give more than a 40% incidence improvement compared to fixed panel results. It is also shown that if no optimization is used on the tracker, it can show significant energy use compared to the optimized situation which shows 2-5% of the collected energy to operate the tracker. A large portion of the paper compares various tracker configurations.

1.2.1.17 Land and Water Limitations Several states have put forth renewable portfolio standards for power generation. These vary in detail from state to state where they exist, but basically, they specify how much of each state's electrical power needs to come from renewable energy. In the case of some states, there may be an additional requirement for a certain amount of a specific energy source (e.g. solar energy), but in other states there is no particular source specified beyond being renewable.

A recent study has considered the ability of solar technologies to furnish the energy required by the renewable portfolio standards in six southwestern US states from the 19 solar energy zones in those states [Burkary et al., 2018]. This was evaluated by considering the land and water required to accomplish this, and considering if these two aspects were available. Of course, the ability to accomplish this was dependent upon the solar power generation technology being used in addition to the land and water available. An additional dimension of the study was that of estimating the reductions of the carbon emissions in each state as the result of using the solar resource.

The results of this comprehensive study revealed several important conclusions. An important one was that there was no land problem associated with meeting these requirements. This was true for every state considered in this paper. In comparing land use, photovoltaics used less land than most solar thermal concepts. While there are many detailed conclusions drawn in the paper, a few overarching conclusions can be drawn from information given there. Generally, PV was favored over solar thermal when land and water considerations were taken into account. Wet cooling of any type thermal power plant should never be used if water is a scarce resource.

1.2.1.18 Soiling An uncertainty that often exists when a new solar plant is going to be put in is how much soiling will take place. Of course, a key element of any solar facility is there will be a large amount of collection equipment interacting with the local environment. Will there be environmental effects such that various foreign matter will gather on the solar collection elements and decrease their effectiveness? This could include a great deal of possibilities, but two of the major culprits could be environmental dust and bird droppings. Some papers have addressed a few of these issues.

One such recent paper has been published by Costa and co-workers [2018]. This is an extension of a similar paper written by the same authors for the period 2012-2015. This particular paper reviews the coverage of 184 papers published in the time period approximately 2016 through a portion of 2017. Among other information, it includes 4 major tables. These tables are:

1. Noteworthy review papers and reports addressing soiling and related issues published in 2016.
2. Photovoltaic Technology Applications: Summary of dust and soiling papers published in 2016 indicating primary focus, device/materials investigate, conditions and findings.
3. Concentrating Solar (Thermal) Power (CSP) Technology Applications: Summary of dust and soiling papers published in 2016 indicating primary focus, device/materials investigate, conditions and findings.
4. Photovoltaic Technology Applications: PRELIMINARY summary of dust and soiling papers published in 2017 indicating primary focus, device/materials investigated, conditions and findings. This is a partial listing for year 2017 (up to the time of their paper submission).

This and its earlier related paper are extremely handy sources of information for solar workers who are concerned about soiling.

1.2.1.19 Solar Thermal Literature Cited This is given in the section near the end of the chapter in 1.5.1.

1.2.2 Large Scale PV Power

1.2.2.1 General Items Over the years, system and component modeling has played very important roles in understanding the operation of, as well as the design of, solar power plants. There will be no comprehensive analysis of the literature that has resulted from this work, as it is very extensive. However, a few of the more recent publications will be noted here. Also, note that the PV field has been changing quite rapidly, so the dates that some of the details that may be listed below should be placed in perspective when considering earlier papers, as there could be changes with time. Although the papers noted below are predominantly from 2017 and 2018, there are some earlier papers noted. Hence quantitative details in any of those papers might have changed.

The book chapter "Solar Generation and Photovoltaic Technology" written by Ramalingam and Indulkar [2017] covers a great deal of basic PV information. About 2/3 of the way through the chapter, the authors write about PV grid connected systems and what makes up a system. This section includes some worked problems about this latter topic. Near the end of the chapter, coverage is given to new trends in materials and operational modes. This is recommended reading for someone becoming familiar with PV cells and systems.

In the brief chapter by Ashby (2016), an overview of issues related to photovoltaics is given. After an introduction to the topic, the author moves on to a chapter entitled "Stakeholders and Their Concerns", where he outlines who the stakeholders are, and lists some significant concerns about the topic. The latter include cost, visually objectionable settings, and the intermittency of the generation. He then presents several facts. A significant one is materials: most PV units are made from silicon, and the concern expressed is that 70% of the production comes from a single nation (China). Another material involved is silver. Cell manufacture takes energy and produces CO_2, which have been going down substantially per Watt of cell rating. However, he hypothesizes a situation where the growth period stops and a later "break-even" situation exists. He predicts the pay-back period is 29 years if the power is sold by the producer, but could be about 8 years if the producer uses it. So, the value of PV is mixed in his analysis.

A chapter in a recent book deals primarily with PV systems [Dincer and Bicer, 2018]. In this chapter are several items, some of which are less pertinent to PV systems. The topics that are relevant deal with crystalline silicon, thin films, optimum band gaps of solar cells, current-voltage characteristics, and losses in solar cells. There are sections on concentrating PV (CPV) systems and combined PV and thermal systems. Topics also include optics issues, temperature effect on CPV performance, and tracking and control. They also deal with spectral distribution modeling, photonic calculations, and the System Advisor Model. The latter was developed by the National Renewable Energy Laboratory and can be used to calculate performance of various solar systems, including a variety of PV types. Other topics are included.

Modern grid design codes must have specified that utility grids with a high penetration of PV plants be outfitted with reliable fault ride-through design [Al-Shetwi and Sujod, 2018]. This brings up two issues for the designer: i) ac over-current and ii) dc-link over-voltage. Either of these aspects could cause grid disconnection or damage to the grid inverter. A solution is proposed in this paper that consists of a control scheme that includes a dc brake chopper circuit and a current limiter to protect the inverter. A hypothetical situation has been modeled with SIMULINK to assist in explaining the scheme. Via the simulation aspects, the overall success of the proposed system is demonstrated.

Another general chapter on solar energy has been written by Helman [2017]. It deals with a variety of topics. The ones that pertain to PV are: Photovoltaics, dispatchable solar without storage, overview of valuation methods, resource costs, calculation of economic value, methods for scenario development and baselines for measurement of benefits, capacity benefits, energy benefits, ancillary service benefits, integration and curtailment costs, and total economic benefits.

An excellent brief treatise on utility scale solar has been published by Wolf [2018]. Among many questions that may bother people is, how is "utility scale solar" defined? Wolf indicates that there could be a variety of answers to that question, but he indicates the smallest system bearing this name is probably 4 MW_{AC} which is equivalent to approximately a 5 MW rating. He deals with many issues like "Site Selection," "Local Electricity Network and Usage," the types of arrays that could be used, project development, "Grid Connection," "Financing" and "Regulatory Issues for Solar Parks," to name only a few of the sections. Clearly this is a source of information that should be read by a first-time developer.

Another paper [Cabrera-Tobar et al. 2016] is one that deals with the configuration of the grid and all the equipment connected to it. They have separate sections where they discuss the electrical components, the internal PV plant configuration and the collection grid topologies. Numerous papers are referenced from the literature in each of these cases. All of their discussion is pertinent to large scale systems as indicated in the title of paper. This appears to be a very good discussion of the design and operation of the many issues with large scale systems.

As installed PV systems become more numerous, the reliability of the various components becomes more important. Issues related to reliability are dealt with in the paper by Hacke et al. [2018]. They review test protocols for various pieces of equipment for PV systems. The authors point out that the largest reliability problem is with inverters. For this reason, the inverters and their evaluation are discussed. It is suggested that many of the test protocols could benefit from more inclusion of stress factors in the natural environment including wind driven rain, dust and grid disturbances. Other factors should be incorporated like how ambient temperature, humidity egress, and voltage bias affect the inverters. Data showing how the inconsistency quality of inverters and the durability of other components increase the operational cost for the owner.

According to Ludin et al. [2018], life cycle assessment (LCA) evaluations need to be conducted of the environmental and energy issues associated with PV technologies to see if these issues are being carried out in a sustainable manner. A variety of these types of studies of the key types of PV cells are reported in this paper. Three different LCA methods were used: cumulative energy demand, energy payback time, and GHG emission rates using data from the literature. Their studies showed that silicon PV technology has the highest energy consumption, longest energy payback time, and highest GHG emission rates.

A chapter in a book that covers many solar energy topics but is especially related to PV systems will be noted. Some focus will be given here to a couple sections in the first chapter. In a section that deals with the difficulties of harnessing solar energy, the first topic noted is that the solar resource is dilute. To underline this fact, it is noted that to produce 1 MW_E with PV in an extremely favorable high flux area, it would take 20-25 km^2 of area. In a temperate climate, perhaps twice that area might be required. A PV plant there might produce power 20% of the time, while in extremely favorable climates, a plant might produce power 35% of the time. In another section that compares PV production with or without batteries to other renewable forms of energy, pumped hydro and wind produce more power but solar is rapidly closing the gap.

Several thin film CdTe cell systems have been installed, and a recent paper summarizes the reasons that may have prompted that to happen [Munshi et al. 2018]. Some comparisons between this technology and thin film crystalline silicon, also a popular option in the PV market, are given. The latter is the older and most installed technology, but the former has lowest levelized cost of electricity and the most installation growth. The two technologies were installed with approximately the same nameplate capacities side by side in three different configurations: roof-top, floating on water and ground mount. After a three-month monitored performance period in Thailand, their outputs were compared. CdTe performed better in a large variety of comparisons that are outlined in the paper.

Experiences with the Design-Bid-Build process that Dougherty and Boyd [2017] experienced have been summarized. In the initial phases of the project outlined, the contractor used some numerical tools to estimate the size of a PV facility. The authors then performed their own analysis of this system. It was found in the authors' study that the design could be improved and made recommendations that yielded a 2.5% gain in yearly power output. This paper outlines the steps that took place in the development of the project. Actual performance of the system was compared to both the contractor's and the author's estimated output.

Many of these papers have outlined system modeling approaches, while others have used a variety of techniques to study the actual performance, dynamic or otherwise, of the solar plants. One of the modeling approach papers has been authored by Chiacchio and coworkers [2018]. They have used an approach that integrates both the deterministic and the stochastic nature of a renewable power plant. The basis for this approach is reliability engineering, specifically the Stochastic Hybrid Fault Tree Automaton. While space limitations do not allow for an examination of the details of this technique, a few comments will be made about the approach and the findings.

One of the key elements of the paper is that the authors present the current state of the art in system modeling relative to PV systems. This is focused on highlighting the limitations of existing approaches. They indicate that a shortcoming is that existing approaches are unable to combine both the deterministic process of energy conversion with the stochastic behavior of the plant availability (which includes the intermittency of the solar resource) into a single model. Their approach overcomes this limitation using a dynamic reliability based approach. However, because it uses a Monte Carlo approach, it often takes long computation times to achieve reasonable accuracy.

In a paper that deals with large commercial rooftop PV units, Sueyoshi and Wang [2017] address the issue of scale effects on the units' efficiency and returns to scale (RTS). The study considered 855 large units in California. Data Development Analysis was used to measure the scale effects. Considered were all of the uncontrollable factors' (like weather) impacts on the generation. Using this analysis technique, they find that the systems below to increasing (48), constant (807), and decreasing (0) RTS, which indicates the systems are designed at their appropriate sizes. They use this data to conclude that operational inefficiency is not due to the systems' sizes, but rather managerial inappropriateness. Results of the study indicate ways to improve the operational efficiency.

Another paper begins by pointing out that PV cells can only use part of the solar spectrum (Guiqiang Li et al. 2018). The authors of this review paper suggest that utilizing the additional energy in the sunlight wavelength spectrum for thermal energy generation beyond those used by PV might be a good approach (the result of this is called a PV-TE application). Systems of this combined type are reviewed along with typical performance and some optimization results.

In one of his latest evaluations of the possibility of energy being supplied by 100% renewable sources for Australia, Trainer [2018] considers energy return on energy invested (EROI) to aid his insights. This would use very large systems to produce the energy. He indicates that EROI estimates for individual sources of energy do not provide a reliable guide to the overall system value because of the large amounts of redundant plants that must be available. He concludes that 100% renewable sources are not realistic approaches.

Durusu and Erduman [2018] have provided what they term as an improved methodology to design large scale solar PV plants. In this approach, they use commercially available panels and inverters. All of the physical parameters like the weather and the location of the plant are input parameters. The various parameters are evaluated using a generic algorithm to estimate the number of panels and inverters required as well as the optimum panel tilt and other physical factors. Case studies using the approach are given in the paper.

A computer study of enhancing PV power production using a solar chimney for cooling has been reported by Jamali et al. [2018]. This concept requires the use of cells that are semi-transparent and a design of the system that help realize the desired goal. The cell array covers a thermal collector that heats the air, and thus cools the cells. A convective flow is enhanced by the use of a solar chimney. Estimated performance shows that for a solar flux of 500 W/m^2, an enhancement of 29% in power generated compared to a conventional cell array results.

1.2.2.2 Electrical Energy Storage Ding and colleagues' [2015] contribution could have been included with the solar thermal discussions as generally thermal storage is considered the prime way to heat while electrical storage can be considered with PV. A variety of storage forms can be considered in this category including electrical and magnetic forms, chemical approaches, mechanical means, and thermal methods. Strictly electrical storage usually is assumed to include the first two of these but a broader definition could apply. The authors distinguish between "coupled energy storage" (which might include rechargeable batteries and flywheels, generally "power" applications) and "decoupled energy storage" (which might include pumped hydro and hydrogen fuel storage, generally "energy storage"). The authors go on to describe several of these applications. A significant portion of the chapter is devoted to a discussion of a variety of battery types. The authors give some very good information in graphical form in the paper, and the interested reader should examine those graphics. It is clear why Li-ion batteries are selected to be used in many applications although their capital cost is quite high. Of the various approaches shown in the paper, hydrogen fuel shows both good and less so characteristics. Hydrogen fuel storage is addressed in more detail in another section of this chapter.

A good review of batteries and their application is contained in the paper by Zhang and co-workers [2018]. Several types of batteries are profiled in this paper. Also touched upon are some aspects of the power conditioning system. The paper contains insights about grid applications. In addition, new generation wide-bandgap semiconductors for the power electronic applications are discussed. There is also some emphasis on the design, construction and operation of energy storage systems for this type of generation.

In a recent paper, the authors [Leonard et al. 2018] explore replacing coal fired plants with renewable resources, both solar and wind. They observe that there is not a natural close match to what is needed to do this. Hence diurnal or seasonal storage becomes necessary to help fill in the gaps. Hydrogen storage could be an alternative approach. The authors choose to analyze a region of North Texas, which has high levels of wind and solar, to see what more is needed to supply the demand. They find that a significant amount of storage is required. Details of their thought process and calculations are discussed in the paper.

In the future, it is anticipated by many people who view renewable energy favorably that we will see many very large PV plants. Another aspect that the future could hold is large scale electrical storage. It is clear that large sizes of PV plants can be purchased for less expense per kW than small scale systems. It is also probable that large sizes of electrical storage will benefit from improved technology and reduced cost per kWh over the coming years. Hence, it is of value to investigate key aspects of large sized plants with electrical storage.

One paper that has done this is that of Lai and coworkers [2017]. In this paper, the authors evaluate a number of issues. They first delve into some brief discussions of the PV developments in several countries around the world. Then they discuss PV technology. This includes both currently available cells as well as ones that are being developed. Issues related to PV stability and integration are explored after that. Following this, solar forecasting is discussed, and this includes listing the four main ways that forecasting is accomplished: statistical basis, use of artificial intelligence, physical approach, and hybrid techniques. They then discuss electrical energy storage (EES) schemes for PV power systems. Key sections of the latter include insights into EES costs, discussion grid storage growth, and the types of EES available (this includes an outline of various types of batteries available).

Another paper covers lead acid batteries (LAB) in particular [May et al. 2018]. While these are quite common in smaller systems they could be used in larger systems also. For utility scale systems, they enter into a realm where there are numerous competitors, including Li-ion, sodium sulfur, and flow batteries, but these latter types will not be discussed here. The technology of LAB and how they can be better adapted to energy storage application is described. They are capable of long cycle and calendar lives. The latter have shown great improvements over the last 20 years. Li-ion batteries have better energy density and specific energy, but these are less important for static applications. LABs are efficiently recyclable in a total sense. In the authors' paper, a selection of larger LAB installations is examined and lessons learned identified.

A recent paper from authors in Cyprus has addressed cost of storage for a variety of renewable energy forms [Nikolaidis and Poullikkas 2018]. Since the paper addresses a variety of renewable energies, some of these forms may not be pertinent to large-scale PV systems. For medium-to-large scale systems, advanced lead-acid and molten salt batteries can be considered both capable of being of value and are the most cost effective. In the paper a variety of criteria is evaluated ranging from energy capacity to round-trip efficiency for most types of batteries as well as other forms of electrical storage.

It is well known that PV systems can show rapid ramp up and ramp down in power output such that it can be almost instantaneous. Depending upon the circumstances, this can cause problems for the grid. As a result, some utilities are limiting the maximum ramp rates that a system can encounter. The paper by Jiang and coworkers [2018] gives insights to this problem with the use of batteries to help limit these rates. The authors use lithium-ion batteries for most of their studies for two primary reasons: a) they have high energy and power densities, and b) they are being manufactured in larger and larger capacities. A great deal of experimental data is shown in the paper, but results will not be repeated here. The interested reader should refer to the paper for the various test results.

1.2.2.3 Demand Response A new algorithm for managing solar PV intermittency has been put forth by Sivaneasan et al. [2018]. To do this it utilizes the building HVAC system to compensate dynamically for any deficit due to the power generated by the PV system. If the PV power drops, the system slows the HVAC fan to bring the grid load to the same value as before. This kind of system would not affect grid stability. Additional issues need to be taken into account including the occupant health and safety and factors such as controlling CO_2 level, and these would be linked to the load shedding aspect of the HVAC system. The viability of the system has been tested using a thermal model of a structure. Some of the benefits of the system include it maintains occupant thermal comfort and uses a much smaller battery than might otherwise be required.

1.2.2.4 Environmental Impact A number of environmentally-related publications related to solar products and applications have appeared. It is not the intent to review a large portion of them here. Rather we select just a few more-recent ones to give some flavor of this particular aspect of solar environmental impacts here. In the present section, a few papers related to photovoltaic production, use, and retirement are given. Although there are several types of PV cells with varying chemistry, the PV industry has been treated as a whole in a limited number of papers. Some of these are considered in this section.

The way powering by solar energy takes place in villages in India, Senegal, and Kenya is reported in the paper by Ulsrud et al. [2018]. At this scale, the electricity provision is part of the village infrastructure, policies, support mechanisms, how the systems are designed and replicated. These developments allow the villagers to find relevance in a number of areas. The findings of the study emphasize a need for institutionalization of decentralized electricity and other issues that are defined in the paper.

An article has appeared that outlines several of the issues related to the possible environmental impacts of large-scale solar plants [Hernandez et al. 2014]. It is important to understand them because they are a mix of positives and negatives. Some of the issues include biodiversity, land-use and land-cover change, soils, water resources, and human health. They indicate that increasing the environmental compatibility can mitigate climatic and global environmental change.

One recent paper has addressed many aspects of the life of PV cells and their manufacture. This is the paper by Srinivasan and Kottam [2018]. In this paper the authors have traced the life cycle of PV cells from a large number of manufacturers. For example, the paper contains a table that lists nearly 50 worldwide manufacturers and their six-year history (some manufacturers do not have data scores throughout this full period) through 2015 of their solar scores awarded by the Silicon Valley Toxics Coalition. The authors analyze 9 companies in this list in some detail. A variety of data is presented in the paper that the reader may find of interest, including the energy payback time for the manufacture of the cells. This number has been decreasing and may go as low as 1-2 years by 2020 for some manufacturers.

A recent paper presents a relevant issue. In a state that is considered prime for the solar natural resources, how much land is available for development for solar power generation (Majumdar and Pasqualetti 2018)? They find that less than 2% of the land is considered excellent for this purpose, and this is either privately held or it belongs to the state trust. However, the authors estimate that if this land is developed in the next several years with solar generation, it could make the state a regional energy hub. Since the state population is growing, there may competition for the land.

Another environmentally-related paper has been put forth by Breeze [2016]. Among the various pieces of information he gives, he cites values for carbon emissions. For Europe, estimates suggest that the lifetime carbon emissions of a solar PV power plant are between 20 gCO_2/kWh and 80 gCO_2/kWh. He notes that this is around 10 times less than the emissions for a fossil fueled power plant. He then notes that these emissions are equivalent to an energy payback time for a commercial plant in Southern Europe of 0.7-2.5 years. The equivalent values for a coal-fired power plant of around 900 gCO_2/kWh and for a gas fired power plant they are around 400 gCO_2/kWh. He notes that solar thermal power plants show a better lifetime performance of 30 gCO_2/kWh. Power output for solar power plants can vary considerably with time of year. Breeze cites numbers for Europe where winter time output is about 20% of the summer time output. Of course, weather can have a very large effect on output.

A recently published paper has dealt with aesthetics of solar development, a topic not usually discussed, but that could become more important: How the plant looks [Sánchez-Pantoja et al. 2018]. This topic has been addressed in the literature and is reviewed in this paper. The authors have considered the objective factors, aesthetic perception, and methods for evaluating impact of the assessment factors. Three factors impact these evaluations: land use, solar system energy and glare. There has been a lack of consensus on how to evaluate these aspects. These are considered a basis of how to proceed further with the evaluation research.

There are few papers in this section that detail issues over what might be the application of land: Solar generation or another use. This is noted in this environmental section of PV technology, but it is probably not precisely an environmental issue. In a paper related to land use, Späth [2018] has applied the Q methodology to inquire how affected stakeholders viewed the solar development. This gave him the insight to come up with possible ways to minimize conflicts between supporters of the agricultural use and the supporters of the solar use. There was a difference of opinion between the strict defenders of protecting agricultural land and those who considered it ok to develop agricultural land for solar uses. The author also recommends the use of both small and large scale solar developments to cater to the various opinions of land owners.

A paper that addresses the impact of wind direction on the power generated has been put forth the Vasel and Iakovidis [2017]. This study utilizes some carefully selected solar data to study this aspect. It involved a single solar farm in the UK which is made up of fixed orientation panels facing south at a 20 angle. In a six-month period from midwinter to midsummer, 42 pairs of data were identified. Data points in each pair had the same solar irradiance, equal ambient temperature, and equal wind speed, but they differed in wind direction. All had one datum point in the pair that had wind blowing from the south and the other datum point having wind direction not from the south. It was found that wind blowing from the south always yielded the highest efficiency.

The environmental performance of a small standalone PV power plant designed for an off-grid community has been described by Akinyele [2017]. Life cycle energy use and the emission-related impacts, as well as a number of other things, are analyzed. Both the PV array and the BOS elements are considered in the analysis. A series of calculated results of variables ranging from global warming potential to energy payback times are given. The author is hopeful that this paper may be useful for planners of grid independent energy systems in developing countries.

Changes are occurring to the agricultural sector of the world. While this statement may have multifaceted drivers, this is often due to the shifting from crop raising to other functions such as population growth. Pollination seems to be one of the bio-functions that may suffer in this situation. In some locations, this may be due to the installation of PV plants that may have more potential for profit than agricultural functions. This paper proposes a way of dealing with this problem. The authors [Semeraro et al. 2018] suggest that it is a way to harmonize energy production, agriculture and the enhancement of ecosystem services, looking for a synergy between different economic activities and stakeholders. Part of the solution is that the authors propose dual use of PV areas, hopefully enhancing the capability of pollination over what it might be if not used in this way.

A recent paper describes a planned installation of a large-scale PV plant in China [Chang et al. 2018]. While the specific site results will be of only limited interest to most readers, the approach the team used could be more valuable. The physical characteristics of the location could be of interest to readers with plants planned for similar types of locations. However, they address possible changes to the local thermal environment due to the installation of the plant. Thus, the paper focuses on the local thermal conditions, their possible change, and how the authors estimated it.

Shahsavari and Akbari [2018] have written a solar paper related to emissions, and this focuses particularly on developing countries. They make the case that solar is the way to go through PV and, to some extent, solar thermal. The claim is made that PV is the appropriate technology for many poor countries that have people living off the grid. Use of PV can greatly reduce emissions of CO_2, NO_x, and SO_2. The authors note the barriers to developing the solar resource for widespread use in these needy countries.

A quite different paper from those listed elsewhere in this brief review is one that has the title: Potential lives saved by replacing coal with solar photovoltaic electricity production in the US [Prehoda and Pearce 2017]. To accomplish this, the authors first determined the mortality rate in states burning coal. Then the energy generation rates and current life cycle mortality rates due to coal combustion in those states are calculated. Following that, the deaths/kWh/year of coal and PV separately are found. They needed 755GW of US PV installations to accomplish this, and they estimated that 52,999 US lives would be saved. From using this approach, they found that "...for most estimations of the value, saving a life by offsetting coal with PV actually saved money as well, in some cases several million dollars per life."

Most of the other papers in this category refer to the impact on the environment due to the presence of PV construction or installation. However, in a different aspect, a paper that discusses the effect of the environment on the performance of a PV system is reported by Waqas and Jie [2018]. In this paper, they were using a phase change material to cool a PV panel in a hot environment. They used a computer simulation to estimate that and compared the results to those measured on an actual system. It was shown that this was an effective treatment, increasing the cell efficiency up to 10% in the hottest summer month. The authors suggest that the PCM melting point be chosen to be 10-12° C different than the average ambient minimum temperature. They have posed a solution for this situation.

1.2.2.5 Tracking The paper by AL-Rousan et al. [2018] covers many of the issues related to use of tracking for PV systems. They start with the statements like "the current solar photovoltaic systems have significant drawbacks," and "using photovoltaic systems is challenging." However, they go on to talk about utilizing tracking PV panels. Among the topics discussed are best panel orientation and comparing simple tracking systems to state-of-the-art approaches. They sort the tracking systems into five categories, including: Sensor driver systems, microprocessor systems, open- and closed-loop driver systems, intelligent driver systems and combinations of these four. They indicate that intelligent driver systems that use learning algorithms are some of the most promising systems.

1.2.2.6 Grid Related Issues Nghitevelekwa and Bansal [2018] address the issue of large-scale grid integration and the role of generation dispatch in solving issues that may develop from it. Although there has been adoption of stability strategies for the intermittent voltage, frequency, and angle variations for PV applications, generation dispatch techniques can equally be adapted to large scale systems for this purpose. This has been shown in other types of systems. It is felt that this will be adapted by utilities in insuring their continuity of supply and assist in cutting costs. Batteries can be used with inverters to limit fluctuations from PV systems. Through ramp-rate techniques, batteries can be used in parallel with PV systems to control drops in output power.

A recent paper has explored the issue of grid parity for solar energy, considering the situation in both the US and Australia [Webb et al. 2017]. Historically, solar energy had required a premium cost compared to conventionally generated power. This situation is changing, and it is the trend explored in this paper. They indicate that the cost reduction of solar power puts it below coal and in many cases below natural gas plants. Part of the reason for this more favorable cost for solar compared to coal is due to transmission which is estimated to be about 40% of the cost for the latter. They also point out that solar is in a favorable situation compared to wind, primarily because the lead time for wind is on the order of twice as long as for solar. They also examine the effect of PPAs in their impact on the market. While much of the paper's discussion is related to PV, the authors also indicate that battery storage and CSP will also have major effects. Impact of community-based microgrids is also explored.

Outlined in this paper is a novel strategy that can be applied to improve the management of a grid system of active power, generation reactive power compensation and power quality. This is applied to a large grid-connected system through a cascade of DC-DC converter and a PWM inverter. Management of the system issues is in such a way as to result in stable DC voltage and sinusoidal grid power. However, priority is given to production over power quality improvement. Many other details of the design of the system are given but it results in exploiting the generation aspects to full capacity of the system. They present simulation results that indicate the effectiveness of the developed control arrangement.

1.2.2.7 Forecasting Ramp rates in generation with wind or solar are of concern as the percentage of renewables on the grid grow. In fact, at high penetration, these fluctuations need to be managed. Accurate forecasts of ramp rates are very important if compensating kinds of functions are to have some success. This report by Abuella and Chowdhury [2018] advocates using a post processing adjusting approach that is developed in this paper for improving hour-ahead forecasts. Results of a case study indicate how well this works.

This paper [Chaudhary and Rizwan 2018] includes a model for solar forecasting using neural networks and a wavelet transform. The accuracy of the model is evaluated using both root mean square and mean absolute error approaches. Comparisons are made to other developed approaches and are found to meet the desired accuracy limits. Pumped storage is included in the model to satisfy the demand requirements. Results of the approach are validated using a MATLAB/Simulink platform. With these approaches, it is shown that the model can be applied to large-scale systems.

In the paper by Rodríguez et al. [2018], a future prediction of solar power production technique is discussed. The tool they developed uses an ANN which they developed using MATLAB software. To validate the model, they trained the ANN by using the root mean square error of the prediction for days outside the database. They showed the difference between the actual and predicted power generations ranged from 0.5 – 9%.

Being able to estimate what the upcoming weather might be is something that could be of great value to a solar facility owner/operator. In general, this benefit increases with the size of the plant, it is ultimately of worth to electrical grid overseers, and it could apply to solar thermal plant operators and PV plant managers as well. There have been numerous attempts to develop valid prediction methods over the years, and this type of effort will continue. In general, as is expected, the more advanced in time the predictions are, the less precision results. Another distinction in predictions is whether they are developed for PV plants or CSP facilities. This is because PV facilities are able to use the total solar input (both beam and diffuse) while CSP typically uses only the beam component. It will be noted that most of models reported here are for PV systems. Note that no claim is made here as to the relative accuracy of any given prediction method.

The most common types of models are those used to predict next-day weather. Predictions of this type are quite valuable because, among other reasons, they are usually the more accurate compared to longer term forecasts. One such version of this approach has been reported by Creayla and co-workers [2017]. This analysis is developed specifically for PV system forecasts. Machine learning played an important role in this study. It is interesting to note that the estimation approaches that are cited in the paper are quite good for predicting the solar flux but not very good at forecasting the power generated.

As is pointed out by Antonanzas and co-workers [2016], due to issues like the desire for grid stability and variations in the solar resource, the ability to forecast the solar flux at given location at some future time is quite desirable. Thus, time and location are both variables that impact the solar flux and need to be taken into account when formulating future estimates. The paper presents motivations for accurate projections, it then follows this with a summary of the main techniques that are used for making predictions. Currently the most used estimating techniques are for day ahead. The authors then go into point vs. regional forecasts and deterministic vs. probabilistic predictions. This is followed by a classification of authors based upon the forecast horizons and the input origins they represent. Day-ahead predictions are the primary tools used currently, but more emphasis on hourly distinctions may develop.

Behera and coworkers [2018] have proposed a PV power forecast method that uses an optimized modified extreme learning machine technique. Of course, the more accurate the predicting technique, the more valuable will be the power if the predictions are used in a smart manner. While the details of the technique will not be repeated here, the authors indicate it is more accurate in short term modeling than are the more traditional models. Comparisons are made to a physical system near the authors, where their model is compared to actual performance as well as predictions from other approaches.

In another recent paper, Bright and coworkers [2018] have used a satellite-derived short term forecasting approach (the authors refer to this as "nowcasting") to improve prediction of future power production. They combine satellite data and up-scaling only fleet estimate approaches, and compare that to two newly developed correction and hybrid cases. They then used results of a single reference PV system for predictions to many PV systems near their location in Australia to evaluate accuracy of their technique. They conclude that a combination of satellite data with estimates from reference PV systems are always more accurate than either of the approaches by themselves.

A model of solar power generation as a function cloud cover (in units of oktas), ambient temperature, latitude and longitude has been described [Gandoman et al. 2018]. The technique is claimed to be easy to use. Although the authors say that both oktas and ambient temperature are required, the paper noted does not include corrections for ambient temperature. The concept seems to be quite easy to apply.

Another approach to predicting weather relative to PV solar power options was given by Raza et al. [2017]. Here a multivariate neural network ensemble forecast framework is proposed. It is shown that some significant prediction improvements result from the use of this technique.

A way that has been developed for predicting weather for PV systems on so-called model output statistics (MOS) has been reported by Verzijlbergh et al. [2015]. This routine is based upon a large set of meteorological variables that are available from standard numerical weather prediction models and a clear sky model. The technique focuses on the forecast error which results from a linear regression algorithm that uses a set of variables that best explains that error. The full explanation of the approach is beyond this short description of the method.

1.2.2.8 Economic Studies A study of the impact of policy incentives for commercial solar power growth in the Northeast US has been given by Crago and Koegler [2018]. They examined existing county-level data for 13 states in the time period 2005 to 2013 to do this. In particular, they considered the relationship between incentives and PV capacity while accounting for insolation, electricity price, market factors, demographic characteristics, and other factors. The first two aspects named were highly significant. Environmental aspects and familiarity with solar technology were not nearly as important, in contrast to the residential market. This leads to the conclusion that the commercial market is driven primarily by possible financial returns. The authors predict that the continued growth of this market segment will likely depend on incentives and decreasing installation costs.

Considering the facts that PV has experienced significant decrease in costs and greatly increase in deployment, in this article the authors (Comello et al. 2018) assess where this energy source is positioned in the marketplace and how this may change in the future. To do this they determine the current cost competitiveness of this power source in select US locations using a levelized cost of electricity approach. In doing this they are able to quantify the influence of supportive public policies, time-of-use pricing, and anticipated future developments that will affect to the price of PV. They also consider the possible effect that batteries may impart to the overall situation. After all of their analysis, they conclude that PV and battery storage will have an increasing impact on the power field in the future.

A fairly detailed paper regarding cost analysis of PV plants in various European countries has been given by Honrubia-Escribano and co-authors [2018]. In this paper they give a listing of the top 10 PV manufacturers in the world in 2015, with Trina Solar and JA Solar (both companies in China) leading the list, and Hanwha Q CELLs of South Korea in third place. In this paper, they have examined specific PV technologies in seven countries of Europe in this comparative study. Quite a range of costs for these countries are shown, with the highest costs in the UK and the most favorable values in Germany. One point the authors make is that the most economic plant in one country could have considerably different costs in some others. More recent lists of have appeared in various trade magazines that indicate that Jinko Solar and Trina Solar (both of China), and Canadian Solar have led these types of lists at other points in time.

1.2.2.9 Electronics All electrical generating equipment uses some kind of electronics in it for control and operation. Some applications are for power generation. Hence, power electronics is important in systems like those used in photovoltaics. A part of a recent book chapter has addressed this topic in a chapter called "Power Electronics for Renewable Energy Sources" [Islam et al. 2018]. Applications to both PV and wind systems are covered here. In the PV portion of this chapter, the following topics are covered: basics of photovoltaics, types of PV power systems, stand-alone PV systems and grid-connect PV systems. There is also a section that covers hybrid energy systems that includes the following topics: series configurations, switched configurations, parallel configurations, control of hybrid energy systems and grid-connected hybrid energy systems. Both fundamental information (for example, descriptions of solar cells) as well as system insights are covered. This is a good basic description of control systems for the included topics.

1.2.2.10 Soiling As part of the requirements for considering the construction of a new PV plant, some locations require the performance of a study where soiling potential can be assessed. This report [Capdevila et al. 2016] describes a study of the dust accumulation over a period of 4 months for a site location in a region of the Nubian Desert (Egypt). The researchers used a time series of Aerosol Optical Depth (AOD) data, which is based upon satellite images. A Soiling Loss Factor, which compares the ratio of the effective energy to the cell (module) for a soiled cell compared to a clean cell is key to the method. Not all the details given in the paper will be repeated here, but suffice it to say that soiling loss factors correlate with changes of three variables: the AOD, wind speed and rainfall.

1.2.3 References

See section 1.5.2

1.3 HYDROGEN DEVELOPMENTS

1.3.1 General Topics (Several Per Document)

A paper deals with production and also addresses topics in storage and transportation of hydrogen [Abdalla et al. 2018]. The authors point out that hydrogen is relatively easily produced from renewable methods, but this is not the way that most is produced at the current time. It is more expensive currently on an energy basis compared to a fuel like gasoline. It can be produced from water, and the latter covers 70% of the earth's surface. So, in spite of having many positive qualities, it suffers from a variety of hindrances to the near-term application of it for a variety of purposes that it could serve. Many of these issues are explored in this paper. Note that the paper includes nearly 300 references to pertinent sources in the literature that can be of value in gaining further insights.

When all the characteristics of hydrogen are considered as a source of energy, it has applications to renewable energy utilization when environmental factors and climate issues are important [Baykara 2018]. It is shown to have the least impact on both of these latter factors but it is not favored on economic grounds. However, its carbon-neutral characteristics make it possibly a desirable solution in a period of increasing global temperature. This is a fairly concise review that includes nearly 200 references to recent literature.

Chinese authors have prepared a significant paper related to the production of hydrogen [Xu et al. 2018]. Referring to 8058 hydrogen-related papers, they have characterized the development of this product to the present. The authors hope that it gives guidance for the transition from a fossil fuel-based society to a hydrogen-oriented economy. Several development stages are defined by the authors. In addition, five indices were defined to assist in assessment and to set value. They identified disruptive technologies, and they found that solar nuclear hybrid hydrogen systems were possible future configurations as well as concentrated photovoltaic – thermoelectric systems.

Another basic has been given by Abad and Dodds [2017]. This is a chapter from a set of texts on sustainability. Both the commonly used methods for doing this today as well as newer, less used, techniques are summarized. As might be expected in a chapter like this, the authors start with "Steam Reforming of Gas and Light Oils." The basic technologies are outlined before they move on other topics, including surveys of sources like coal gasification, biomass gasification, various forms of electrolysis, solar thermochemical water splitting, and biological hydrogen spitting. An interesting section relates to geographical limitations to hydrogen plants, where they point out that some concepts may not be appropriate for some locations of the world.

A summary of large-scale hydrogen production methods has been reviewed in another recent paper [Aasadnia and Mehrpooya 2018]. In addition to reviewing many of the existing plants, the authors propose a novel way of classifying plants in terms of their heat exchange configuration and their expansion process. They also introduce a section on the renewable generation of hydrogen, which is gaining interest. In addition, they give some insights to the efficiency and cost of systems. They examine the situation of incorporating constant entropy expansion processes compared to the typical constant enthalpy processes. However, they do not anticipate that new methods of hydrogen liquefaction will be introduced in the near term. They indicate that the trend in specific energy consumption for the large-scale processes will remain in the range of 5-8 kWh/kg_{LH2} for the near future, while industrial systems could range up to double that amount.

Nikolaidis and Poullikkas [2017] have reviewed 14 processes for producing hydrogen using a variety of sources and methods. Steam methane reforming is the most used technique, followed by coal gasification. Since both of these methods require fossil fuels, additional techniques of a renewable nature, of which there are several, need to be better developed. In the near term, process fuel reduction through integration of a membrane reactor may be possible as could be combined cycles utilizing alternative energy sources. Water splitting techniques will not become competitive unless carbon taxes are imposed. Several issues related to storage, transportation and utilization need to be addressed.

The paper by Dincer and Acar [2017] sets out an evaluation scheme to weigh different approaches to hydrogen production. It is hoped that the use of this scheme will help define how to go to a new approach based upon the evaluation of its many characteristics. An idealized approach, perfect in every element, would show as a circle with unit radius. A number of considerations are included in the evaluation and are a means of evaluating concepts. Some conclusions reached via this evaluation concept include: geothermal, biomass, hydro and solar concepts in that order have the highest sustainability performance. The lowest ranking source of H_2 is coal, followed by nuclear and natural gas. When H_2 production systems are compared, thermochemical rank the highest. Other comparisons are shown in the paper.

The book chapter by Acar and Dincer [2017] gives a great deal of background about hydrogen, and its use in a wide variety of ways, including: as a combustion fuel, in electrochemical conversion, for chemical conversion (other than combustion) as well as just physical conversion. Some comments will be made here about topics discussed in the chapter. Use of hydrogen as an internal combustion fuel is generally well developed. Fuel cells are a potentially sizable application for hydrogen, and these are in various stages of development depending upon the type of fuel cell and its use. Most other general applications still require general development to be widely accepted. Related to the combustion aspects of hydrogen, the following are general characteristics: wide flammability range, low ignition energy, small quenching distances, high autoignition temperatures, high flame speed, high diffusivity, and low density. In terms of chemical conversion systems (other than combustion), it can be used to make water, fuels, plastics, and pharmaceuticals. For example, it can be used to make NH_3 which is a common base for fertilizers. On the physical front, hydrogen can be used as a gas or as a liquid. Note that the latter has a temperature of -253°C, and liquefaction of hydrogen is a very energy intensive process. Hydrogen will probably be used in a large number of transportation applications. Refer to the original article for a large variety of other information about hydrogen.

It is well known that hydrogen is an energy carrier that is quite flexible in its applications. Because of this flexibility, it can support a variety of applications including energy storage and electricity production. Wang, D., et al. [2018] explore the interactions between fuel cell electric vehicles, hydrogen production facilities, and the electric power grid. To quantify the flexibility aspect, the authors project system wide balancing challenges for California out to 2025. More renewables will be deployed and electricity demand will increase. The authors model the various interactions named earlier. They develop control algorithms that can achieve a variety of objectives. Topics including peak shaving, valley filling, and ramping mitigation are proposed. They indicate that oversizing electrolyzers can provide considerable benefit in addressing these topics. Meanwhile this helps the transportation sector decarbonize.

1.3.2 Production--Electrolysis

An overview of the hydrogen production has been given by Coutanceau [2017], which is a chapter from a book by the same author. This gives a great deal of basic information about electrolysis. Basic data pertinent to this process for a number of compounds is given. One very interesting aspect pointed out in this chapter is that theoretical efficiency of the fuel cell which is taken as the ratio of the electrical energy produced (which is the negative of the Gibbs energy of water formation) to the chemical combustion energy (which is the negative of the enthalpy of formation for water) gives an overall energy efficiency of the fuel cell of 83%. This is twice as high as the efficiency that is normally calculated for an internal combustion engine (about 40%).

Another paper about water electrolysis that can be applicable to renewable energy has been written by Ci and Yu [2018]. In this review paper, the authors summarize the main ways of doing this. The techniques reviewed include the following water electrolysis approaches: alkaline, proton exchange membrane, solid oxide, and alkaline anion exchange membrane. A fairly complete description of these various techniques is given in the paper.

Recently a number of papers have appeared that analyze systems that accomplish a multitude of tasks. One example of these is an article published by Siddiqui and Dincer [2018]. The configuration they analyze involves a number of processes in a single unit: Solar energy supplies the system via a tower that powers the system that provides desalinization, hydrogen production and power generation. A reverse osmosis unit and a PEM electrolyzer are used in this configuration. Analysis is carried out on both an energy and exergy basis. Also, included in the analysis were the impacts of charging system and state properties on system performance.

Water splitting by monolithic solar devices pose problems if up-scaling is desired. This is caused because the ion transport distances are long which results in high ionic Ohmic losses. If up-scaling is desired, this can be a significant bottleneck in trying to reach high efficiency while maintaining optimum performance. In the paper by Trompoukis et al. [2018], a new device design is proposed. This involves solar water splitting based on porous multi-junction silicon solar cells. Simulations showed that porous monoliths can benefit from lower Ohmic losses compared to some other designs. They show that pore sizes on the micrometer scale can greatly reduce the Ohmic losses. The authors describe the fabrication of an array of small holes that resulted in performance improvement. This led them to propose a novel device that they describe.

In another paper related to electrolysis of water, Chakik et al. [2017] have considered the effects of the amount of various operating parameters on the hydrogen production. A particular focus was the zinc materials from which the cathode is manufactured. Several different alloys were used for this and the electrolyte was various concentrations of NaOH.

Certain alloys are much more productive than others. It was also determined that increasing the electrode current at constant voltage increased the production rate, as did increasing the voltage at constant current.

Stern [2018] describes a fuel cell system that he has developed with several unique aspects. It includes a fuel cell system that utilizes hydrogen of sufficient purity that is generated on demand from an apparatus where a controlled chemical reaction is taking place between water and sodium metal. Sodium hydroxide is a byproduct of the reaction which is collected within the apparatus for later processing by electrolysis, with the Na reactant recovered. The author has performed detailed analyses of the system showing that it is scalable and can operate under a range of climatic conditions.

Arsalis et al. [2018] have described and analyzed a total energy system that is powered by PV for a residence. The model developed uses an electrolyzer and hydrogen storage with a fuel cell subsystem to furnish energy on a yearly basis. Quantification of the energy and power flows are part of the description of the system as well as the details of the costs. They report an estimated electricity cost as 0.216 EUR/kWh (currently higher than the cost of electricity in Cyprus, home of the authors) for a system with capacity for 100 households. The authors estimate that the cost can be made competitive if the electrolyzer/fuel cell lifetime is increased and the costs of electrolyzer and PV are reduced.

One of the best methods for quick hydrogen production is the methanol electrolysis process, and this topic was the focus of a paper by Tebibel et al. [2017]. This process generates high purity gas and can be used with renewable energies. Because of its low theoretical potential, it can save a great amount of energy over other means of hydrogen generation. For this and other reasons, this concept is positioned to become one of the preferred options for hydrogen generation in the authors' opinion. In this paper, hydrogen produced using this concept powered by PV is described. The mathematical models used to analyze this system are described as are case studies using it. The approach has many positive aspects about it.

1.3.3 Biomass Related

The ability for municipal solid waste to be a source for production of hydrogen has been evaluated [Sharma and Melkania 2018]. This was done with bioaugmentation in the organic fraction of the waste. This augmentation was accomplished with three bacterial species: *E. coli, Bacillus subtilis and Enterobacter aerogenes.* A range of bacteria/sludge ratios were used ranging from 0.05 to 0.40. A Gompertz model was used to evaluate the cumulative hydrogen production, lag phases, and maximum hydrogen production rates. The several values for these data are listed.

Generating hydrogen from biomass is being given a great deal of attention. A paper recently has reported on some examples of this [Salkuyeh et al. 2018]. The paper gives the evaluation of two major categories of gasification approaches: fluidized bed gasification and entrained flow gasification. It was shown that the entrained flow technique yielded the higher efficiency as well as the lower minimum price of hydrogen. When a carbon capture and liquefaction system is incorporated, of course the efficiencies of both systems decrease. The authors make some estimates for what the conditions have to be to make this approach cost competitive with steam methane reforming.

Certain kinds of biomass can be used with a fermentation process to produce hydrogen. This is a lower cost method than many of the other approaches. Using these materials results in clean energy production, as well as a means of disposing of agricultural and municipal wastes. Two drawbacks result from these approaches: low production rate and substrate degradation. The authors of this paper [Wang, J., and Yin 2018] review the related work that has been performed. Since hydrolysis of smaller molecules is rate limiting for biomass degradation, a pretreatment step can enhance both the production efficiency as well as the biomass degradation rate. Pretreatment can aid in a number of ways. A variety of the pretreatments (including physical, chemical, biological and a combination) are discussed in this paper. Still much work needs to be accomplished to realize the potential of these approaches.

Show et al. [2018] have published a review paper assessing various developmental aspects of biomass energy. There are many types of fuels from this source, but the one focused on in this paper is algal biomass which is characterized as being abundant, clean and renewable. The problem is that this form is in its early stages of development but is extremely promising. This paper gives a good review of the status of this bio resource.

"Dark fermentation" is a fermentive conversion process of an organic substrate into biohydrogen. It has been applied in a number of processes. This plays a major role in the concept described in the paper by Sołowski et al. [2018]. In this paper, many of the ways that dark fermentation has been used are described, particularly with carbohydrate rich wastes, in the production of hydrogen. Many of the kinds of bacteria that can be used in this process are considered, characterized, and compared. As part of this discussion the pretreatment processes as well as the various reactor designs to accomplish this are analyzed and discussed. An additional topic discussed is the use of a membrane to purify the resultant hydrogen from the dark fermentation process.

A paper related to the one just noted was recently published by Jarunglumlert et al. [2018]. This is based upon a technique called "dark fermentation." This concept is presently in the laboratory scale, so there is still quite a bit of development work to do. But since this approach uses less energy and chemical materials compared to other chemical routes, it is appropriate to examine it here. This article covers a myriad of details about the current situation with this approach. Concepts examined include the carbohydrate content (more is better), fermentation condition (ideal initial pH 6-8, temperature range 50-55°C), continuous operation preferred, and two stage production results in savings. Of course, there are many complications since food waste can be quite variable in composition.

The most dominant and certainly the most primitive form of photosynthetic activity on the face of the earth is furnished by microalgae. Recently some scientists have reported on its use for hydrogen production [Khetkorn et al. 2017]. This is certainly a path to a clean energy power generation with simply water being the sole product. Techniques are available that can increase the hydrogen generation for this source. Analysis indicates that this should be an economical approach to hydrogen production. Recently some large-scale bioreactors have been developed and are operating. This paper highlights recent technical developments that add additional promise to these approaches.

1.3.4 Production—Other Means

With sorption, enhanced reforming gasification for hydrogen generation, a process that is nitrogen free, high caloric product gas can be produced [Schweitzer et al. 2018]. Because gasification temperatures can be low and a limestone bed material can be used, in-situ CO_2 capture is possible. Comparisons are made between results from a bubbling fluidized bed gasification model to those found in a 200 kW_{th} pilot plant. This allows a study to be made for efficiency and feasibility. Capital costs, operational expenditures, and hydrogen production outlays are calculated for an economic assessment. The authors give some estimates of what the market situation needs to be for this technology to move ahead.

A paper that deals with the production of hydrogen from methane has been given by Parkinson and coworkers [2018]. In it, steam methane reforming is technically compared to other pathways with low CO_2: (i) Electrolysis coupled to sustainable renewable electricity sources; (ii) Reforming of hydrocarbons coupled with carbon capture and sequestration and; (iii) Thermal dissociation of hydrocarbons into hydrogen and carbon (pyrolysis). If no value is set on the carbon, steam methane reforming is the least expensive approach. However, if low CO_2 processes are desired, methane pyrolysis might be considered. For methane pyrolysis, a process based on a catalytic molten Ni-Bi alloy is described and used. If carbon gains enough value (about \$21 t^{-1} CO_2 equivalent) pyrolysis may be the best choice. Several aspects are discussed in the paper.

Although thermal reforming is one of the major ways that hydrogen is produced, it does not typically use solar energy. An article that considers some of the alternative options to this type of approach has been written by Coutanceau [2018]. First a few words need to be given about "what is thermal reforming?" This is a term that is familiar to many but not to all. To illustrate this, consider steam-methane reforming, one of the primary ways that hydrogen is generated. While the details can be quite involved, it will be described in simple terms here. The first step is an auto-thermal reforming at about 1000°C over a ceramic in the presence of oxygen. This is followed by a steam reforming at about 900°C over a NiMgAl oxide catalyst in the presence of steam. The mixture that results contains a great deal of carbon monoxide that has to be removed. Removal of the CO is quite a complicated step. It can be seen from this brief incomplete description that reforming is a chemical intensive set of processes. What is important is that about 96% of the hydrogen in use today is generated from this kind of approach, and that is primarily because this approach is much less expensive than is electrolysis. In the future, there may be an emphasis on eliminating carbon from the approach, and this could give electrolysis a boost.

A new approach for producing hydrogen was designed and fabricated to evaluate the use of aluminum and solar energy as part of the process [Sheikhbahaei et al. 2018]. The process uses the fact that when aluminum is added to a solution with sodium hydroxide, the metal is quickly dissolved. From this basis, numerous experiments were performed to evaluate the hydrogen production rate. The authors were able to develop an optimal and stable hydrogen production rate based upon system scale and consumed materials. They found that at low temperatures, the hydrogen production volume, efficiency, and COP of the system increased at higher sodium hydroxide molarity, but the opposite effect occurred at high temperatures.

Photocatalysis is a technique that can aid in the production of hydrogen. Emphasis of much of this work has focused on the improvement of photocatalytic efficiency in the visible light region by solar light utilization. In a recent paper, Okemoto and coworkers [2018] have reported on some work they have done in this particular topic. While most of the details of this work are beyond the scope of this summary, a few comments will be made. Semiconductor based photocatalytic technology is an important tool in this area. A key to this approach is the development of new semiconductors that are more pertinent to this effort. The work by Okemoto and coworkers, using urea as a reducing agent, has developed a new semiconductor that extends the light absorption region and results in a high activity for hydrogen production.

When alternatives are considered to replace fossil fuels, then photocatalytic production of hydrogen should be considered as is done in this paper [Colón 2016]. This approach, derived from oxygenates, is physically very important and can be highly efficient. The information given here will help the reader understand the principles of photoreforming reactions. In this paper, the author discusses the various photocatalytic materials proposed in the literature and the role they play as different co-catalysts.

The authors of a paper on producing hydrogen with green chemistry principles [Çelik and Yildiz 2017] make the point that while hydrogen gas is usually considered to be "green", some methods used to produce it are not so. They make the case that because of this, there is a need to redesign the production processes so that the life cycle of hydrogen gas does not contain the undesirable processes. Twelve principles were applied toward this goal and the analysis considered four groups of processes according to energy sources: electrical, thermal, hybrid and biological. The sources that fair the best in this evaluation are described.

One of the newer approaches to hydrogen generation to be noted here is chemical-looping technology. A very thorough description of many of its aspects has been given by Luo and co-workers [2018]. It can be used in several ways. For example, one of the primary ways that hydrogen is generated today is via steam-methane reforming. However, this process is quite involved, and it is difficult to get thorough carbon cleaning. On the other hand, hydrogen generation using chemical looping technology accomplishes its end goal with much higher process efficiency, and it can capture carbon dioxide. The authors of this paper give extensive background information on this concept. They describe both chemical-looping reforming as well as chemical-looping hydrogen production.

A paper has been published that deals with hydrogen generation from "clusters" [Attia and Samer 2017]. This is term that may not be universally understood, but it refers to matter that exists in size between isolated atoms and bulk metals. They are the second most elemental building blocks after atoms, and they establish a bridge between atomic and nanoparticle performance. Because of the size of these particles, use of them can accomplish certain functions. They can act as a photocatalyst for hydrogen production from water subject to certain limitations.

In simple terms, reforming is a way to generate a set of gases from appropriate other ones by applying heat. A paper from 2014 by Brown et al. reports on reforming a steam-methane mixture to a hydrogen-carbon monoxide mixture using a solar parabolic dish system. A significant part of this study was an experimental solar-to-chemical energy process that yielded a conversion efficiency of 69%. The authors list some projected economics for this kind of process. Of course, this depends upon the cost of the reactants and the value of the products.

Dagdougui [2018] has published a paper that considers many of the options for fuel that the world may use in both the short term and the long term. The one common thread is that the fuel needs to be hydrogen. This is denoted for three reasons: i) there are many ways that it can used, ii) there are many sources of it, and iii) it appears to be the only energy source of sufficient magnitude to be used ultimately by the world. And the author addresses many of the apparent shortcomings. For example, although there are not a lot of hydrogen fueling stations, that is almost entirely due to a lack of current demand. This is part of the "chicken and egg" enigma. Because there is not a lot of demand, there is not much product to service the current demand. It is pointed out that the costs of hydrogen from renewable sources is currently quite high, but they still deserve consideration because they are the least environmentally damaging and their costs are coming down. The author points out the problems that have to be overcome but it is felt that end goal is worth the effort (and the price).

Many survey articles have been written on a variety of issues. Some of these deal with hydrogen subjects only, while others cover a wider range of topics pertinent to this review. One of the latter types is a paper that deals with a broad range of topics, in this case electrochemical energy conversion [Dincer and Bicer 2018]. In this piece, a large section deals with fuel cells. After giving some general aspects of fuel cells and briefly dealing with comparative characteristics, the authors give some valuable descriptions of the various types. The types covered in this second aspect are the following fuel cells: alkaline, proton exchange membrane, phosphoric acid, solid oxide and molten carbonate. They then give an analysis and assessment section which has several subsections. A short section on general environmental impacts follows. Winding up of the chapter includes a Case Study (deals some analysis of a molten carbonate fuel cell stack) and an Illustrative Example (this focuses on a PEM fuel cell) that shows how the various parameters discussed in the earlier section affect performance.

This paper reports on a study of an integrated, coal-based gasification system developed for both hydrogen production and power generation [Seyitoglu et al. 2017]. The components of this system include an air separation unit, gasification unit, gas cooling and cleaning unit, pressure swing absorption for hydrogen production, high temperature electrolyzer for hydrogen production, Brayton cycle, steam Rankine cycle and organic Rankine cycle (ORC) system for power generation. Several varieties of coal were included in the analysis. The authors performed both energy analyses as well as exergy analyses. Both Aspen Plus and Engineering Equation Solver were used. Overall energy and exergy efficiency values were tabulated.

1.3.5 Storage, Transporting

A good source of technical information about hydrogen storage is the book chapter written by Baetcke and Kaltschmitt [2018]. The major sections of the chapter after the Introduction include: Hydrogen Storage in Pure Form, High Pressure Storage, Liquid Hydrogen Storage, Cryo-Compressed Hydrogen, Material-Based Storage, Metal Hydride Storage, Liquid Organic Hydrogen Carrier (LOHC), Metal Organic Framework (MOF), and Activated Carbon. After these sections, some more comparative sections are given, including: Comparison, Technical Values, Economic Figures, Market Aspects, R&D Status, and Final Considerations. Many of initial sections like High-Pressure Storage are broken down into sections like State of Technology, Characterization, Markets and Perspectives. It appears to be very good source to find concise descriptions of many of the technical aspects of hydrogen and its uses for mobile applications.

Hydrogen storage in its various forms is considered in a paper authored by Sreedhar et al. [2018]. Physical, chemical and electrochemical forms of storage are considered in various forms like gas, liquid and solid. In this paper, the various types and forms are compared in many aspects ranging from efficacy to safety. The authors summarize many of the relevant equations that have been used in the literature to analyze storage over the years. Both kinetic and thermodynamic aspects of hydrogen storage are summarized, and some special emphasis is given to techniques using metal hydrides. Various aspects including the mechanisms, rate controlling aspect and energy elements are also discussed. A look to the future completes the paper.

In the paper by Wulf et al. [2018], they point out that although hydrogen in fuel cell vehicles results in vehicles without emissions, the hydrogen supply chain, from production to fueling station, can vary considerably. This involves both economic and environmental aspects. The authors claim that the first of these aspects has been investigated extensively, but not so the second one. So, this paper addresses this shortcoming. In doing this, life cycle assessments of three scenarios are examined: (a) liquid organic hydrogen carriers and storage, (b) compressed hydrogen storage in salt caverns with pipelines, and (c) pressurized gas truck transport. For most cases (except for short distances), pipelines are the best solution. Seasonal storage has almost no impact. They indicate that strong scaling effects underlie the good performance of pipe networks.

Three delivery systems for hydrogen are considered, and performance and cost assessment evaluations are given in the paper by Demir and Dincer [2018]. Variations are considered in the conditions that the hydrogen is stored, as well as the way the gas is transported, the latter being done in three different ways: in pressurized tanks, in a cryogenic liquid tanker or via pipelines. Several factors are given as a result of analysis: hydrogen production capacity and the levelized cost of energy distribution. Also, the environmental impact and items like the number for hydrogen fuel stations, the distance between stations and length of the distribution lines given. The costs are listed for each situation analyzed.

As every hydrogen researcher knows, one of the ways hydrogen can be stored is in an appropriate porous material. Minuto et al. [2018] have considered how the possible choices of porous materials can be tested for this application, and they have developed a new outlook for evaluating this property in materials near ambient temperature. They have shown why the absolute adsorption capacity near ambient temperature is not a good indicator to evaluate various materials to choose which might be more promising. One of the reasons is this property might change with temperature. They have also shown that by using a quick cyclic adsorption isotherm or thermal desorption spectroscopy it is easy to identify those materials that are more suitable for this purpose near ambient temperature. This helped identify which of three commercial activated carbon materials they evaluated, it is the one with higher content of ultramicroporosity that is the most promising.

1.3.6 Applications

With hydrogen now being considered for a wide range of applications, several new kinds of plants are being proposed. One example of this is a hydrogen production plant consisting of coal gasification, thermochemical water decomposition and hydrogen compression systems [Al-Zareer et al. 2018]. The technical details of this are beyond this brief description, so refer to the original source if interested.

An approach for hydrogen generation for mobile vehicles and ships may be to use a discharge plasma in liquid fuels. This approach does not require storage vessels, for example. The technique is of fast response, large capacity, small scale, and high degree of portability. A microwave discharge reactor with standing wave coupling is one such approach [Sun et al. 2017]. Experiments using this approach by the authors on ethanol showed that in addition to the hydrogen product, there was carbon dioxide as well as small amounts of methane and acetylene present. The technique used appears to be adaptable to vehicles.

An experimental study related to a solar hot water system that uses fuel cell heat is reported by Assaf and Shabani [2018a]. It is part of a solar-hydrogen system to furnish 100% of the power demand for a remote household in Australia. It was built to validate a model developed earlier. The analyzed solar-hydrogen system furnished 100% of the required power in summer and a little less in the other seasons. Measurements on the actual system showed similar performance. It was found that the fuel cell furnished heat that was used about 97% in the winter and 60% in the summer. With its available different modes, it is shown that the system can be effectively used for remote households.

In a paper related to a previous one, hydrogen may be a good fuel to adapt for sea transportation, and a recent paper makes this case [Bicer and Dincer 2018]. In the paper both hydrogen and ammonia are examined as possible replacements for heavy fuel oils in engines of sea transport vehicles either in total, or as a component of a mix involving both fuels. The authors consider the life cycle cost of switching fuels. Of course, this could involve impacts on many factors from the vehicles that transport goods to the cost of generating the replacement fuels. It is important to note that they also consider mixtures of these new fuels with the heavy fuel oils. Of course, the greenhouse gases can be reduced quite substantially depending upon the end formulation, both on the seas as well as in port.

A recent article has discussed some of the increased understanding of the possible use of formic acid to store hydrogen [Wang, X., et al. 2018]. The basic acid has quite benign characteristics, and it can be used both to generate as well as store hydrogen. Its chemical nature results in less CO generation. Many different catalysts can be used with formic acid to generate hydrogen, unfortunately as well as some CO_2, around ambient conditions. This particular paper reviews current developments of the investigation of catalysts relevant to this application.

Report of a high temperature study of generating hydrogen from hydrogen chloride has been given by Khalid and two co-workers [2018]. This electrochemical, analytical study assumed a temperature 1073 K and pressures around standard atmospheric values in a solid oxide electrolyzer based on a proton conducting electrolyte. The model accounts for all major losses, namely activation, concentration and ohmic. Using the model, the energy and exergy efficiencies are found. The prediction for them is better than that found for a steam electrolyzer at similar temperatures, giving hope that this may be a new way of electrolyzing at high temperatures. With these types of results, it is hoped that it could be coupled with a high temperature nuclear gas cooled reactor power plants.

Suleman et al. [2015] consider the various approaches for the generation of hydrogen relative to a comparison of the techniques used as well as their environmental impacts. Studies of both fossil fuel as well as renewable approaches are summarized. Of course, when approaches using fossil fuels are considered, steam methane reforming is the major approach to consider, since it is the source for most of the hydrogen used in the world today. For renewables, the authors consider several approaches including the sodium chlorine cycle. Another approach considered is electrolytic production using various types of cells. When environmental impact is considered, steam methane reforming is clearly one of the largest impacting approaches. The second largest abiotic depletion is electrolysis using a mercury cell.

Some of the more valuable products that can be used in a fuel cell are hydrocarbons that have been reformed or partially oxidized [Sengodan et al. 2018]. This is because it is along the route to production of hydrogen or syngas. Hydrocarbons are also of value in high temperature fuel cells and these are available through one of the reforming processes. Reforming and partial oxidation catalysis to convert hydrocarbons to hydrogen rich syngas plays an important role in fuel processing technology generally. Currently research is being carried out in the area of reforming and partial oxidation of methane, methanol and ethanol and it includes catalysts for reforming and oxidation, methods of synthesis of catalysts as well as effective utilization of fuel for both external and internal reforming processes. Several modern issues in this general area are reviewed in this paper.

A review of work with hydrogen enriched compressed natural gas (HCNG) being used as a fuel in SI IC engines has been given recently by Yan et al. [2018]. Engines using this gas have demonstrated superior performance compared to ordinary gasoline engines. The vast amount of natural gas available is another reason to consider it seriously. This paper gives a comprehensive survey of studies of power, efficiency and combustion, and emission characteristic of HCNG fueled engines. The authors also correlate fuel properties with engine performance. In this study, the properties of HCNG are first reported. Then summaries of notable effects on engine performance are given. Some discussion of numerical models of engine operation is also given. The article ends with brief reporting of the effects of other hydrogen enriched fuels such as hydrogen-gasoline, hydrogen-diesel, and biogas.

Use of integrated utility and biogas supply networks to develop a resulting network model to use to analyze for hydrogen production is described in the paper by Hwangbo et al. [2018]. The utility network might be what is found in a large petrochemical plant, while the biogas supply consists of streams that may be found in a wastewater treatment plant and anaerobic digestion. Steam reforming, a well-known process, integrates the two networks and produces hydrogen. An algorithm that is used for solving mixed integer linear programming models is applied here. This minimizes the total costs and satisfies the hydrogen demand. A case study for a Korean application is included.

1.3.7 Hydrogen Combined with a Named Renewable or Nuclear Energy Source

A brief article has recently appeared that considers the current state of affairs as well as the future for solar fuels [He and Wang 2018]. Even though work in this area has gone on for some time, the article concludes that there is still a great deal of work required until "solar fuels" become a realistic marketplace product. The authors (two people from the chemistry department at Boston College) point out that there is still considerable scientific work required to develop the kinds of materials needed for this, and after that the engineering development must also cover much ground.

Solar spectrum splitting is a technical concept where the solar spectrum is split into bands that can drive photo-quantum and photo-thermal processes to produce hydrogen and distilled water. This has been applied to brine water in a paper by Baniasadi [2017]. This approach decreases the energy needed to accomplish the given task. The author has explored the sensitivity of the cycle performance to the solar insolation, electrolyzer temperature, and the operating pressure, and the results indicate that the use of solar cells with concentration with a multi-band gap mirror can greatly increase the productivity of the cycle. The author also reports both energy and exergy efficiencies for the cycle.

A paper [Sorgulu and Dincer 2018a] has appeared that addresses a possible application of concentrated solar power and wind energies for a residential system that provides heating, cooling, power, and hydrogen. The system consists of an integrated electrolyzer, fuel cell, and absorption cooling unit that are put together in an environmentally positive and efficient manner. Excess power is used to produce hydrogen that can be stored when there is more than is instantly needed. Thermodynamic analysis is used to assess the unit's performance. Ambient conditions are taken into account in the system. Sustainability of the system is assumed to be achieved because it is efficient and environmentally friendly.

Output results of a 20 kW PV power station were used to make a comparison of those computed from a PV code and with actual data [Fereidooni et al. 2018]. The authors then used this data to evaluate the amount of hydrogen that could be generated from the output. An assumed power requirement and efficiency of an electrolyzer were made. Some shading of the actual system occurred resulting in a 5% decrease in what could be generated, and this was ignored in some of the remaining calculations that were performed. It was estimated that about 373 t/yr of hydrogen could be produced. The authors then followed this by considering what the situation would be if the panels were adjusted monthly where an increase in hydrogen output of nearly 7% was shown.

Illustrating the analytical capabilities of the researchers, a paper has been published that treats a solar-hydrogen combined heat and power system integrated with solar-thermal collectors to supply both power and heat, including hot water demand, to an application [Assaf and Shabani 2018b]. The authors use a generic algorithm to optimize the system. Components of the system include a hydrogen-based storage consisting of an electrolyzer, a tank to store hydrogen and a fuel cell. The solar thermal system includes an evacuated tube collector and a hot water storage tank. MATLAB was used to simulate the system. Local values were used that define the energy demand and meteorological data. The overall goal was to determine the sizes of the main components of the system subject to concern about the reliability of the system, minimizing the levelized cost of energy used, and minimizing the excess energy from the PV system that is not used. A few percent of the thermal demand were not met in the optimized solution.

Recently Omar and Altinişik [2016] have published a paper that describes a simulation of a remote combined hydrogen generation and water heating system. A key element of this is a hybrid collector. This collector simultaneously generates electricity and hot water. Borland Delphi 7 programming language is used for the simulation. Weather data for three locations in Turkey are used to determine the best location for the system. Two separate sets of cases are considered: in one the system temperature is fixed and pressure is separately evaluated for cases, while in the other, the pressure is fixed and temperature is stepped through various values. Good comparison to some previous work is shown.

A paper by Liu et al. [2018] has reported on a new fuel-fired distributed energy solar-hybrid system. This system incorporates a thermochemical reaction that uses mid to low temperature solar and exhaust heat to drive it. The solar energy produces syngas via a thermochemical process and it, with supplemental methanol decomposition reaction syngas, is used to drive an internal combustion engine. Solar thermochemistry and chemical recuperation are used to improve the overall energy efficiency as well as efficiency of the solar to electricity conversion. The result is improved performance even in the off-design conditions. Calculated results indicate that the proposed system is quite efficient.

The driving of a photo-electrochemical hydrogen production system and a PV cell system in tandem with a varying concentrated solar beam to over 10x light intensity is reported in a paper by Bicer and Dincer [2017]. After the light is concentrated, the beam is split into two portions. Losses caused by reflection and transmittance of the light in the system are estimated. Efficiencies for each process are approximated as is the overall efficiency and compared between non-concentrated light compared to concentrated light. It was found that the results for the PEC with concentrated light showed an efficiency of 49.3% and this is attributed primarily to splitting of the spectrum.

Monnerie and coworkers [2017] have reported a study they have performed that considered two cases (higher and lower amounts of hydrogen production) of a high temperature and pressure electrolyzer driven by a solar tower power plant. The tower used a molten salt coolant because this would best control the receiver temperature to a constant value which was needed for the electrolyzer. Hydrogen storage was also included. Software used were Aspen Plus 8.4 and HFLCAL. Two possibilities to consider are that more power is generated than needed, or less power generated than is needed. In both cases the plant would interact with the grid. It is shown that the consumption of grid electricity is reduced for increasing solar multiple but the efficiency is also reduced. The paper has good documentation of the plant design and operation.

Solar photocatalytic hydrogen production is a topical area that is stimulating a significant amount of research. One paper [Cao et al. 2018] is an example of this. The thrust of the work focuses on a novel CPC reactor. Two models were developed: one a natural convection situation, and the other a gas disturbance model, however, the unit is operated in the first mode most of the time. All of the pertinent local meteorological conditions are estimated from the site. The paper gives the details of physical system considered. A range of expected productivity is given.

Yüksel [2018] has published a paper where an analysis is described of an integrated system that consists of a parabolic trough collector, a modified organic Rankine cycle, a single effect absorption cooling system and a PEM electrolyzer. Heated water from the parabolic collector allows the organic Rankine cycle to produce electricity. The electricity is then used for hydrogen production. Hence, the outputs of the system are electricity, cooling, and hydrogen.

A paper that reports the effects on solar energy production of pure hydrogen and pure oxygen of a variety of variables using alkaline water electrolysis has been given by Sakr and coworkers [2017]. In separate tests, an acrylic separator and a polymeric membrane were each used to separate the electrodes. Several variables were examined for their impact on the amount of hydrogen produced: electrolyte concentration, solar insolation, and electrode separation. From this the overall electrolysis efficiency could be calculated. Higher hydrogen production resulted with the use of the polymeric membrane and also with the electrode separation being 5 mm compared to 10 mm.

Recently a very practical study has been reported of solar produced hydrogen [Badea et al. 2017]. The authors assumed electrolysis in their analysis and considered four different technical solutions for production and storage as well as a few assumptions related to process efficiency. They showed the maximum production was a result of hybridizing the solar hydrogen system that used both electrical and thermal energy.

Some papers have addressed the application of clean energies like hydrogen into a conventional hydrocarbon industry. One such paper is that written by Likkasit el al. [2018]. It describes the application of solar-aided hydrogen production to the oil and gas industry. In this application, the hydrogen can be used in oil upgrader processes. In the study described in this paper, three solar aided processes are considered and analyzed: (i) solar steam methane reforming using a volumetric receiver reactor, (ii) solar steam methane reforming using molten salt, and (iii) solar thermal power generation coupled with water electrolysis. Option (i) had the lowest levelized cost, option (ii) had the highest efficiency, and option (iii) had the lowest greenhouse emissions. None was selected, but the authors hypothesized about some trends for the future.

A study has been performed to understand better how commercial nuclear reactors can be used to drive hydrogen generation systems [Sorgulu and Dincer 2017]. Two nuclear plants are planned for construction in Turkey, and they are considered to be the motivation of the adding of hydrogen generation to them. In doing this, they used the software package Hydrogen Economic Evaluation Program (HEEP) developed by International Atomic Energy Agency (IAEA). Many of the costs were evaluated with the software.

Hydrogen producing nuclear power is an option that can result in large yields, and it is examined in a paper by Sorgulu and Dincer [2018b]. This might be accomplished either by electrolysis or thermochemical cycles, and it could be a key element of having nuclear energy play a large role in the potential hydrogen era. In this paper, the authors consider two nuclear pressurized water plants that are planned for Turkey, and these are evaluated for possible hydrogen production. They used the software "Hydrogen Economic Evaluation Program (HEEP)" developed by the International Atomic Energy Agency. The program contains numerous options for hydrogen production, and it includes the ability to include most of the costs associated with building a reactor. For the two planned Turkish reactors, the authors found that the predicted cost of hydrogen varies from 3.18 to 6.17 \$/kg H_2.

If hydrogen is increasingly used in the future, many aspects of its generation and use must be developed. A recent paper by Gökçeka and Kale [2018] has reported on the analysis of two versions of a renewably-based hydrogen filling station, that could be put on a small island in Turkey to refuel 25 vehicles per day (125 kg each day) all year long. The two versions are wind-photovoltaic-battery and wind-battery systems which were assessed with HOMER software. It was determined that the cost of product for each of these systems was, respectively, \$ 8.92/kg and \$ 11.08/kg. Several other studies of this situation were part of the study.

Solar and wind are two most commonly recognized sources of energy that are considered renewable. While papers report on tapping either one of these sources to perform a variety of tasks, the paper considered here deals with wind generated hydrogen [Mohsin et al. 2018]. It discusses how they adapted a Wiebull probability function to quantify the quality of the wind. They combined this with a log law that accounts for ground height of the wind.

The value of a highly efficient concentrated photovoltaic (CPV) system when used for electrolytic hydrogen production is explored in the paper by Burhan et al. [2018]. This is a prime possibility for long term energy storage. But, as pointed out in this paper, the large size of the PV system limits where this can be used. In this paper, a compact CPV-hydrogen system for residential applications in urban areas is considered. The design details are given for such an arrangement. In addition, several aspects of operation are described, and some operational data are given.

While there are a number of positive characteristics of geothermal energy conversion, one of the negatives is that large amount of hydrogen sulfide might be emitted. This emission is one of the main concerns about geothermal power plants. Because this is an issue in many situations, there has been an increasing interest in developing suitable methods and technologies to produce hydrogen from the hydrogen sulfide. One possible route is AMIS technology (AMIS is the acronym from the Italian language for "Abatement of Mercury and Hydrogen Sulfide.") A unit using this technology could produce hydrogen gas from the hydrogen sulfide emissions from a plant. Several details of this concept are evaluated in this paper, showing that hydrogen sulfide emissions can be decreased to a very small value [Karapekmez and Dincer 2018].

Recently a paper has been published that reported the analysis of a geothermal power plant that is combined with a hydrogen generation and liquefaction system [Yuksel et al. 2018]. The system is assumed to consist of a combined geothermal power system, a PEM electrolyzer, and a hydrogen liquefaction and storage system. Energy and exergy balances were performed on the overall plant. As part of this, some of the plant parameters are studied parametrically. It was found that the parameter that had the largest impact on the results was the geothermal source temperature.

1.3.8 Other Topics

A paper related to fuel cells and hydrogen technologies with the stated goals of demonstrating the importance and usefulness of multiscale computational approaches and tools, reliable input data, and verification procedures has appeared [Stamatakis et al. 2018]. To accomplish this, the authors use three typical case studies of simulation models and tools at various scales. The first case involves the materials design for fuel cells via a process-based methodology that the authors outline. A second case is related to consideration of hydrogen safety related issues using Computational Fluid Dynamics studies. Finally, the third case shows the optimization of renewable energy – hydrogen systems for remote or isolated communities where cost effectiveness of the produced energy is of primary concern. The authors emphasize the synergies between numerical and experimental research.

Best practice guidelines for numerical simulations and CFD benchmarking for hydrogen safety applications have been set forth by Tolias et al. [2018]. Because correct use of CFD tools is critical in order to have confidence in the results, the authors have given a comprehensive set of Best Practice Guidelines (BFG) in numerical simulations for Fuel Cells and Hydrogen applications. These BFG are designed to satisfy the practical needs of researchers performing simulations. A larger document was apparently generated from this work, and this paper gives the essence of the original document. It covers all of the hydrogen safety-related phenomena. The authors furnish several CFD benchmarking exercises to show appropriate modelling strategies.

A paper authored by Brennan and Molkov [2018] deals with a phenomenon that some people may not have heard of: pressure peaking. They are the words that describe what occurs when a gas lighter than air (hydrogen and helium are typical examples, but the former is of importance here) is released into an enclosure where the vent(s), volume, and leak rate occur but no air can flow in because the source flow out will not allow that. The peak in pressure that occurs at some point in the situation is called the peaking pressure. Relatively high pressures can occur. Obviously, this situation may be relevant to indoor applications of hydrogen where, for example, fuel cells may be operating. This paper gives analytical equations and a nomogram for the flow and pressure build up situation.

Unconfined release of hydrogen is explored computationally in a recent paper by Edelia et al. [2018] with the goal to predict if it might explode. As part of doing that, parameters that can promote cloud explosions (e.g. initial pressure, time to ignition, and leak height position) are explored. A key finding is that unconfined hydrogen has the potential to build up a large vapor cloud and explode if the leak is near the ground. Based on this finding, the authors warn that, in the case of leaks near the ground, it is not always safe to assume free flow jet release. They give insights through included graphs about how far from an occupied building high pressure storage should be located.

In a paper by Leonardi et al. [2018], the development of a hydrogen dual sensor is described. One major application of this is for use with fuel cells. Configuration of the sensor includes a ceramic platform head with semiconductor metal layer on one side and a Pt serpentine resistance on the back side. Design of this device allows a rapid reading response and is capable of sensing very dilute concentrations of hydrogen with the semiconductor layer. Because of its characteristics, it is ideally suited for application in leak detection. The back side of the sensor preforms as a thermal conductivity sensor. The authors indicate that the unit appears to serve as an accurate and inexpensive device to be used for safety and efficiency control applications for fuel cell systems.

Hydrogen can degrade the condition of steel (this is important because it includes many pipelines) under certain conditions, and usually this is something of considerable concern. A recent review paper [Ohaeri et al. 2018] has explored this situation in some detail. The pipeline does not need to be transporting hydrogen to be a problem, as it can be released by the presence of moisture as well as molecular water reduction resulting from cathodic protection. Hydrogen can be the cause of cracking due to embrittlement, which can be generated both by the hydrogen present in the pipeline as well as the stress level there. Outlined in the paper are sources of hydrogen attack, as well as their induced failure mechanisms. This document also cites several past and recent studies of these phenomena and the mitigation trials and performance tests in the media at various conditions.

Authors of this paper [Dincer and Acar 2018] make the case that "smart" energy solutions are needed in the modern world. This might involve cutting edge technologies and artificial intelligence applications. In doing this, five categories might be considered involving energy, namely: fundamentals, materials, production, conversion, and management. This particular study introduces eight smart energy solutions: exergization, greenization, renewabilization, hydrogenization, integration, multigeneration, storagization and intelligization. The authors go on to discuss newly developed smart energy systems that combine hydrogen systems and renewable sources to provide more sustainable solutions. The authors present a wide range integrated of systems to show how this can be accomplished.

1.3.9 Hydrogen Literature Cited

See Section 1.5.3.

1.4 BUILDING APPLICATIONS OF SOLAR

1.4.1 Hardware, General Equipment, Material, Building Design Elements

An excellent chapter from a book on building systems covers "Advanced Building Energy Efficiency Systems" by Krarti [2018]. This goes over a number of pertinent topics including: implementation of energy efficiency (which examines different types of energy analysis principles), indoor thermal comfort (which discusses various thermal comfort models), whole building energy models (self-explanatory), dynamic insulation materials and systems (discusses modeling of such systems), thermoactivated foundations, and finishes with an introduction to LED automated controls (discusses optimized controls for lighting systems). This is a good source for insights to many of the newer energy concepts related to buildings. Not only is the chapter filled with concise information, it includes a number of graphics, including graphs and diagrams. It is highly recommended for insights to the modern energy efficiency building world.

While almost all buildings include some kind of blinds, the formal evaluation of their use has not occurred often. However, a recent paper by Avarice and Ghazi [2018] about their effect on the energy balance of the building explores one version of that kind of effort. Their project started out with a questionnaire of occupants from which three behavioral patterns were identified: one is keeping the blinds always open, another is keeping the blinds always closed, other types of behavior are also noted like opening the blinds on arrival and closing them on leaving (the latter is denoted as an active user). Some software (DIVA for Rhino plug in) was used to analyze the energy flows for the different situations. The authors underlined the need to inform the building occupants the effect of blind position.

Few studies have been performed on assessing quantitative dynamic shading systems. But interest has been growing in this area. A recent paper [Elzeyadi 2017] reports on the important yet understudied topic related to comparing the performance of different dynamic shading technologies. This involves assessing shading impacts on building energy savings, daylighting distribution, glare control, and solar insolation management for a typical office space in ASHRAE Climate Zone 4C. There are numerous variations in these comparisons. Some performed better in reducing solar load and building energy consumption, while others were better at reducing glare and giving better daylighting performance. The paper concludes with insights about the application of shading devices.

The effect of adjacent trees relative to cooling the local area has been studied and reported in the paper by Hseih et al. [2018]. While the concept of cooling from trees has been common knowledge, quantifying the effect has not been extensively studied. Addressing this shortcoming is the point of this paper. To do this, the role of trees was quantified using microclimate field measurements and sap flow rate of trees. These were then entered into EnergyPlus to study the impact on buildings in the area. Four scenarios were investigated. Scenario 1 was for a treeless condition. Scenario 2 with existing trees was for the situation like is found in Nanjing City. Scenario 3 includes more trees than the second one includes, while scenario 4 included transpiring trees. Scenario 2 resulted in a reduction of cooling of around 10%.

A paper that could be quite valuable to solar designers has been put forth by Jakica [2018] related to daylighting and the solar potential for building integrated PV. This is important at all times but as buildings become more clustered, this challenge becomes even greater. This paper was written within the framework of IEA-PVPS Task 15 – BIPV, and it is hoped it illuminates all elements of the design process. Almost 200 design tools are comparatively described to some degree, and the description covers functions like accuracy, complexity, scale, computation speed, representation as well as the building design process integration into a variety of software environments. Related tools from other fields are also considered. The author hopes this will open new perspectives to application tools.

Net metering is where consumers who have solar energy systems can "sell" excess generation back to the utility generally at the same price as they purchase it. The policy has attracted criticism because some say it benefits system owners and is paid for by a general "tax" on all electrical users. Because of this, many public utility commissions have begun to consider this potential problem. In the paper by Comello and Reichelstein [2017], the authors have examined the impacts of the amount of this payment on the installation of PV systems in three states: Hawaii, California, and Nevada. They find that if the payment is set at or above the levelized cost of electricity, this would have a positive effect of more customer owned systems. However, a lower price would have the opposite effect.

A recent paper by Chemisana and Zacharopoulos [2015] has explored the use of high concentration PV in buildings. The concept of building integration is explored as part of the paper. They indicate that concentration was developed as one means of economy, but building integration can be an additional way to save cost. Case studies regarding these systems are given in the paper. One issue is that tracking systems are required for installations of this type, and this can be a complication. Two examples of truly building integrated high concentration PV are described.

"Distributed energy system" is a term used in a paper that many will recognize as being an alternative to a "centralized energy system", in which the latter makes up by far the majority of the systems in use today for heating and cooling of buildings. The authors of this paper [Kang et al. 2017] claim that quantitative performance assessment, the influential factors, and the impacts of these kinds of factors for cooling dominated climates are not well understood for distributed energy systems. So, these authors have performed studies to quantify how these systems perform in Hong Kong. Reports of the tests of these systems are given. Both the energy savings and the payback period are determined. They report on the impacts of the major design parameters and energy policies. They conclude that these systems show energy savings only when performing certain functions in large scale buildings. However, performance can be improved by optimizing the equipment capacities. It is shown that better economies are realized when the system owner is allowed to generate and sell electricity. It is noted though that gas prices have a significant effect on the economics of the systems.

In a paper by Chwieduk [2018] (note that there are two papers by this author in this review), a series of simulation results are presented relative to the performance of attic rooms. Where it used to be assumed that attic rooms were quite lossy, this is, in general, not the case now, even at high latitudes. This is particularly true if the rooms contain large windows. Cooling may be required for a large portion of the year. In fact, rooms facing any of the four major directions may require twice as much cooling as they require heating, even at high latitudes.

Solar reflectance of a building's skin has a large effect on heat transfer into the building. This is a key observation for the paper by Wu et al. [2018]. Ordinarily, deposition of airborne black carbon takes place on the building surfaces causing the absorptivity of the building to increase. However, what is demonstrated in this paper is that a photocatalytic coating that contains TiO_2 will maintain the solar reflectance of the building and decrease the energy absorbed over time compared to an untreated skin. The authors performed some experiments where they coated Portland cement mortar specimens with different combinations of material, including the photocatalytic composition that is featured in the article. While the full story of the results is too involved to recount here, the new treatment seemed to work quite well.

This paper by Vanaga et al. [2018] describes an experimental and numerical study of the dynamic behavior of a façade module that can accumulate solar energy. This module is made up of a point focus Fresnel lens that concentrated a solar beam on a copper plate with fins which is used as a heat transfer enhancer to a phase change material. Studies of this concept indicate the processes are quite complex to describe because of the usual highly transient nature of the system. A study of the temperature difference between the opaque reference wall and the solar module vary between 0.5°C and 9°C depending upon the time of day and the season.

In a review paper, Shirazi et al. [2018] have examined in detail thermally driven cooling devices. There are several concepts that exploit this function, including absorption and adsorption chillers, solid and liquid desiccant systems, and ejector cycles. However, the absorption approach is the most accepted because of its relative maturity, higher efficiency, and reliability, and they have been applied in large systems. All of this yields good levelized costs compared to the other solar options. In this paper, a historical and literature review of previous experience with these types of machines is given. A great deal of the previous work has taken place with single effect machines driven by low temperature collectors. To address the poorer performance of the single effect machines, the authors point out that these have suffered not only from the low temperature operation, but also because of storage tank and piping losses. However, they may be the best choice in lower-flux regions. Fewer applications of the multi-effect machines available have taken place, and they are inefficient when the solar flux is low. For high flux locations, they are the best choice. Economic studies have shown that solar cooling cannot compete with conventional cooling systems. So, performance of the solar driven absorption machines should be improved, and these authors believe this can be done with a focus on the balance of plant.

An investigation of a developing cooling technique has been given by Jani et al. [2018]. This is the solid desiccant – vapor compression hybrid cooling system, and in this case solar energy is the regenerative heat source used as the driver for its operation. In this paper, the energy savings relative to a conventional vapor compression system is noted where the energy saving of the newer system can be 40-50% compared to traditional VCR systems. The authors note that desiccant based hybrid space cooling techniques are more sustainable, as well as economically and environmentally more feasible for climates with high outdoor humidity or hot-humid summer characteristics. Researchers who wish to pursue work on this system should find this paper a good starting point for that.

Solar thermal applications in buildings have experienced rapid market growth and the heating and cooling aspects of this are explored in a current paper [Ge et al. 2018]. Not only is this progress reviewed, but some possible future developments and markets are discussed. The evacuated tube collector has really bolstered the solar hot water market. For various types of buildings generally, how to integrate collectors has been the topic of many studies. The district heating literature is increasing as are solar PVT systems. Relative to building cooling, a great deal of attention has been directed to work on adsorption and absorption systems. In addition, economic analysis of all of these systems generally is discussed. Return on investment period has been used extensively in evaluating systems, and low initial cost and advantageous allowance are more efficient ways to make these technologies economically attractive.

A paper that reports performance and other comparisons of various residential cooling units for hot, arid climates has appeared [Ali 2017]. The cooling units include a residential scale solar thermal system driving a silica gel-water adsorption chiller, and this is compared to a conventional cooling system. Both of these are compared to an off-grid system. The experimental systems were of a rating of 8 kW. Experiments were carried out in Egypt in a 90 sq m building with people inside. Comparisons are made on a number of bases, including the Total Equivalent Warming Impact. Results show that all systems were able to meet the cooling load. As might be expected, the off-grid system used very little grid electricity. Many other detailed results are listed in the paper.

Relatively a large amount of energy is used to control the temperature of a typical residence. It has been proposed in a recent paper that this amount of energy could be decreased if individual heating/cooling units could be used, each one just to heat/cool a single person. Each person in the household could use different settings for achieving comfort. The description of personal temperature management techniques is the basis of the paper by Zhao et al. [2018]. In this paper, a personal heating/cooling system is devised. It consists of a thermoelectric device that supplies conditioned air through a network of a tree-like set of rubber tubes knitted into a thermoregulatory undergarment. Heating and cooling energy is estimated first. Then weight vs. performance of heat sinks is studied separately to minimize the weight of the system. Thermal manikin tests were used to evaluate the total power draw and it appeared that a little less than 25 W of cooling energy and a little less than 19 W of heating energy could be used to satisfy the design requirements.

Dincer and Bicer [2018] have written an extensive chapter that not only covers PV as the title indicates, but also gives a great deal of coverage to solar thermal applications. The reader is directed to this chapter for a significant amount of information on both general categories of topics. Some of the topics covered in thermal area include solar passive heating, flat plate collectors, evacuated tube collectors, and discussion of a number of technical analyses.

Granqvist and Niklasson [2018] have examined the development of materials that work well with a variety of solar thermal applications. The materials need to deal with the characteristics of the radiation and the desired performance in their applications. The first of these aspects include the solar radiation's spectral distribution, angle of incidence, intensity, thermal emission, atmospheric absorption, photosynthetic efficiency, while the materials can be metallic, semiconducting, dielectric, glassy, polymeric, gaseous, etc. Of course, surface coatings have been developed to show desirable properties, possibly including appropriate types of substrates. The authors consider a number of applications in this discussion, including solar thermal converters, transparent thermal insulators, clear sky radiative cooling, as well as windows with a variety of performance aspects. A number of materials are discussed in this article.

1.4.2 Storage Issues

Energy storage is a concept that is becoming more and more important. While there are many forms of this, a limited view of a paper by Del Pero and colleagues [2018] will be listed here. In general, there is a need for storage whenever the energy source variation does not match the time variation of the need for the energy. There are many forms of this storage, including thermal, mechanical, electrochemical, etc. The applications considered here are for buildings, which is limiting its scope. Carrying out a technical assessment of a type for a specific application can be quite difficult because of the broad range of sources. This paper summarizes the types of storage that may be appropriate for building applications, and it gives some Key Performance Indicators that could help in identifying solutions. In the paper these Indicators are applied to case studies from a division of the International Energy Agency.

Since many national energy strategies encourage the use of solar, the storage of this intermittent phenomena could be quite valuable. This addition offers a variety of opportunities not possible without it. As outlined by Krese et al. [2018], thermochemical storage offers performance benefits not available with sensible heat or phase change forms. When this is combined with solar energy, a very good renewable, widely applicable, energy source is seen to exist. In this paper, background is given on thermochemical storage with particular emphasis to its application in buildings.

Solar and wind energy utilization need some type of energy storage for the source variation to match the load variation. This is the topic of a paper by Jones and Powell [2015]. They consider building thermal energy storage (TES) to meet this need as a load shifting technology rather than traditional energy storage. This issue is investigated as wind and solar power generation penetration increases. As part of this analysis they also investigate building side and smart grid enabled utility side storage. A key finding of the work is that cooling by wind related TES value remains approximately constant but generally decreases with increasing solar energy penetration. The authors also make the points that a storage management strategy is vitally important as well as the fact that utility side operating methods perform at least 75% better than building side systems. In a surprising finding to the authors, increases in solar and wind energy input does not increase the value of storage and sometimes a decrease in value occurs. Results of the study show that the connection between load shifting energy storage and renewable energy may not be nearly as strong as originally thought.

In another paper that deals with thermal storage, the phase-change temperature of an insulation is taken as the major variable in an optimization process to see how this parameter influences the results [Saffari et al. 2017]. This paper draws upon the fact that some previous papers have shown that various phase change temperatures give the "best" solution for different specific situations. In the current paper, a simulation-based optimization coupling.

EnergyPlus and GenOpt with an *h-T* function to determine the PCM optimum peak melting temperature for a variety of building climate control processes, including: cooling, heating, and the annual total heating and cooling energy performance of a residential building in various climatic conditions. For the several cases considered, the results indicated the most cooling energy benefits occurred at a melting range of 24-28°C, and for the heating dominant climates the optimum temperatures ranged around 18-22°C. Notable savings were indicated for various energy processes at the optimum temperature.

Another paper [Saxena et al. 2018] reports work on phase change material for building insulation, and this is for cooling applications. The guiding principle of the reported study is that if an appropriate phase change temperature is chosen, it will minimize the amount of exterior energy that has to be furnished for cooling. A significant limitation on the work is that it is all developed for a location in India. However, the paper is reviewed here so that readers who wish might apply the approach to other locales. Three distinct insulations with unique transition temperatures were evaluated and the results are reported. Heat addition to the insulation is due to the exterior temperature effects, and heat rejection for the insulation takes place to the nighttime conditions. A reduction in amplitude of the temperature swings for the building is reported. Only one of the three materials noted accomplishes the desired duty, resulting in a unique temperature for the phase change transition.

Thermal energy storage materials are the focus of a recent paper [Lizana et al. 2017]. The authors consider three options: Sensible heat storage, latent heat storage, and chemical storage. Ultimately, they consider chemical storage not to be an option at the present time because of its state of development, although it has some positive qualities. In the paper, physical, thermodynamic, kinetic and chemical properties are addressed, as well as costs. Of the other, more conventional storage materials, their volumetric heat transfer coefficients range up to 430 MJ/m^3 for phase change materials and to 250 MJ/m^3 for sensible heat materials. The conventional thermal storage systems that find applications in "free-cooling" ventilation systems, solar energy storage solutions for short and long-term storage periods, and demand-side management strategies all hold promise. Their application can lead to a reduction of energy consumption of more than 30% in some instances.

An article that deals with the placement of phase change materials in the walls of buildings describes how far this kind technology has advanced in the last ten years. The authors [Stritih et al. 2018] indicate that this is due to three aspects, including the technological advancement of the materials themselves, the way they are encapsulated, as well as their method of being manufactured. Two basic strategies have been used in this context. One is to trap the heat in its movement through the walls, and the other is using plasters that can collect the heat effects from the living space. The authors' conclusions, based upon their location, is that summer heat should be intercepted in the wall with phase change materials, but that the winter operation can be dealt with conventional insulations. They used TRNSYS to analyze the situation. Data related to the various insulations considered were used. They restricted the latter to products with low flammability values so that the PCMs could be applied to commercial buildings. One of the main limitations they indicate is that they not be applied where meteorological conditions in summer would do not allow the PCM to convert back to a solid during the night.

Another paper has dealt with issues about placing PCM in the walls of a building (Gounni et al. 2018). Here the authors placed a layer of PCM at different distances into the wall thickness. In every case, the PCM was the same material and the same thickness, but it was placed at various locations from the outer edge to the inner edge of the wall. There were differences noted among various locations. The location on the inner surface of the wall showed that the peak heat flux was reduced most there.

The common way that most residential electrical PV systems are currently installed convert sunlight to electricity when the sun is shining, and draw power from the grid when there is not enough generation to serve the building's needs (primarily at night). This can be rectified somewhat by using battery storage. The batteries would be charged during the day from an over production of electricity and discharged at night to furnish the building's needs [Vieira et al. 2017]. The use of lithium-ion batteries for this purpose is the major point of this paper. Typical power use and PV power availability for a location in Portugal is considered. As would be expected, this can result in a significant saving of grid power.

One of the promising elements of solar electric generation is the incorporation of battery storage with the PV panels. A particular version addressed in the paper by Üçtug and Azapagic [2018] uses lithium-ion batteries, one of the prime storage media. The authors indicate that the general concept offers many positive benefits, but it is "underexplored." It is claimed that this is the first time life cycle environmental impacts have been evaluated for a variety of locations within Turkey. 0Impacts on both local and the national level were considered. The system studied integrated multi-crystalline PV and a lithium-ion battery. Simulated modelling assumed a smaller, rather than larger, system and performance was calculated hourly for a normal sized house. The system considered included a 1 kWp solar PV with a 1 kW inverter and a 2.1 kWh lithium-ion battery. Details of the simulation as well as calculated results are given in the paper and will not be repeated here. However, the lifetime aspects of the system are described in the makeup of the materials that result at the end of product life. A variety of resulting factors are shown for each of the areas Iran, global warming potential ozone layer depletion.

A review that focuses only on lead batteries has been given by May et al. [2018]. This paper discusses the improvements that have been made in these more-common type batteries. As stated in the paper, these batteries are capable of long cycle and calendar lives, and there have been great improvements in these variables in the last 20 years. As is well known, the lead batteries are totally recyclable with facilities for doing this across the

US, which is not matched by any other batteries. Some contrasts are given to Li-ion batteries which are the lead batteries biggest marketing rival. The paper also describes some the larger lead battery installations.

1.4.3 Single Building Energy Analysis etc.

Building energy performance can be influenced to a large degree by geometric configurations of the building, but minimizing the energy use of those configurations by may be quite a difficult task [Kheiri 2018]. Hence, a variety of methods have been used to accomplish this goal. This paper deals with how this procedure might be simplified by having the building design parameters be treated as optimization independent variables, and the building energy consumption can be treated as the independent variable. Optimization procedures are reviewed. Also, decision making techniques are discussed. In doing all of this, current optimization tools are outlined.

As is the case with other papers considering the energy retrofitting of existing buildings, Martinopoulos [2018] has reported on the conversion of a building with solar energy utilization. A design was made such that solar supplied 100% of the energy needs. A life cycle, "cradle to grave," assessment (LCA) was made over the 25 years of assumed useful life on the solar equipment. The author used a dynamic energy simulation for predicting the energy required and the SimaPro software and Ecoinvent database were used to determine the LCA. The LCA showed that the building could cover 100% of its energy needs while also breaking even, in terms of emissions, in less than two years for the worst case.

A design for a demonstration building that met the 2012 International Energy Conservation Code is described in a recent paper [Fanney et al. 2017]. In addition to improvements to the building envelope, several other aspects of the house were designed with a vast array of energy conserving and renewable energy efficient technology. These include a special heat pump system, special heat recovery unit, a whole house dehumidifier, a PV system and a solar domestic hot water system. Net total energy use (energy furnished compared to energy required) was determined. The first year the energy surplus was 102.3 kWh. In the second year after changes to the house were made, the energy surplus was 2241 kWh. Several details of the house and its operation are discussed in the paper.

A case is made for underground thermal energy storage coupled with solar district heating by Reed et al. [2018]. While it has been used in a few situations previously the authors tout the desirable characteristics of this, particularly for this application. It offers all of the benefits sought in a solar energy system including time-shifting of grid-based power such that the supply is readily available to meet the demand. Europe is clearly a leader in this application. Although its benefits are generally recognized in North America, it has not experienced the same acceptance. One of the problems is that the financial institutions use different criteria than are put forth in engineering studies. This study considers what capital costs systems of this type would have to have to be able to appeal to the North American financial world.

This paper by Mohajeri et al. [2018] discusses that the shape of a building's roof has an impact on PV located there. The reason for this is the roof's shape, size and orientation influence what is mounted there. The authors used a machine learning approach (Support Vector Machine, SMV, classification) to classify 10,085 building roofs on how much solar energy they would receive in Geneva, Switzerland. To do this the SMV was able to identify the roof shape in 66% of the cases. Those are flat & shed, gable, gambrel & mansard, cross/corner gable & hop, and complex roofs. The authors were interested in how much area was available for receiving solar energy. They claim that for most buildings the ratio between to useful roof area for PV and the footprint of the building is close to one (except for the gable case which has a ratio of 1.18), so the footprint could be used to estimate the maximum area of PV. It is claimed that solar roof-shape provides a great deal of information on efficient solar integration to roof, retrofitting intervention on building roofs, but mainly basic information for designing new buildings.

The authors of an article that discusses assessing the embodied energy in a building [Nizam et al. 2018] suggest that this can be done with building information modeling (BIM) based tools. It is noted that BIM with life cycle analysis (LCA) capability can handle these types of problems. Some previous work has utilized BIM software, either where data were exported from the main BIM authoring tool and then auxiliary tools were used to evaluate the model, causing a disconnect between the model and the analysis, or it has ignored the importance of retaining the LCA results within the BIM environment. This paper presents a way to estimate the embodied energy aspects within the BIM software. An example of the computational approach has been used to show how the system will work.

This paper by Moosavi et al. [2018] deals with a naturally ventilated building equipped with a solar chimney and hybrid evaporative cooling which is investigated for its cooling performance. The paper also reports data gathered during a fourteen-day site test in an occupied office building with an atrium. Quantifying the effects of these building aspects on efficiency for enhancing the indoor air movement and thermal conditions was defined and determined. Among other factors, the wind and stack-induced airflow rates are indicative that the solar chimney reduces the atrium (not the adjacent space) air temperature if a larger exhaust opening is used. It does extend the hours of comfort significantly, and it reduces the cooling requirement somewhat. When the wind speed exceeds 1.4 m/s in a favorable direction it increases the hot air removal. Several other results are noted.

If properly calibrated, energy models can do quite a good job predicting energy use in buildings [Kim et al. 2017]. This paper is written to outline a method that can be used to calibrate models. This method is based upon occupancy and plug load records with the latter being able to be inferred from metered electrical use data if need be. The authors also propose a procedure for situations when building sub-load data are not available. The calculated occupancy rates and plug load data provided higher accuracy than default or averaged data did. An indicator of errors decreased with the use of this system from 21% to 12% compared to the models that used the default data. Since many of the campus buildings where this technique was applied do not have submetering systems to indicate the plug loads the technique was able to use processed electrical load data to estimate occupancy rates. Some data showed for two campus buildings showed that the errors in prediction decreased from 156% to 16% for one and from 128% to 31% for another. The method used seems to be a good way to analyze buildings that do not have plug loads monitored.

A new energy management system is receiving more and more attention. One paper that deals with this topic applies a microgrid to the consideration [Lešić; et al. 2017]. This paper proposes a modular coordination mechanism between building zones comfort control and building microgrid energy flows control based on model predictive control. This approach opens considerations to technologically coordinate modularly heterogeneous building subsystems for economically-optimal operation under user comfort constraints. The system is made to account for both the energy requirements and the microgrid energy price profiles. A simple interface for exchanging building consumption and microgrid energy price profiles is the modularity upon what the approach is based. The proposed coordination mechanism is presented in a realistic case study of maintaining comfort in an office building with the integrated microgrid. Results are presented about yearly cost-effectiveness of the considered configurations. These are suitable for deployment either for long term or short term zero energy business investments.

A study of changes in building energy consumption and its relationships to climate can be used to study the basis for energy saving and carbon emission reduction [Li, M., et al. 2015]. This study of heating and cooling energy was made in several buildings in Tianjin City, China, during the period of 1981-2010 using the software TRNSYS. Results showed that required extreme winter heating decreased over this period, but the required extreme summer cooling increased over the same period. Apparently large buildings experienced less of this effect, but smaller buildings and residences did have it. The authors found that the extreme cooling condition was primarily related to the wet bulb temperature. Four variables affected the extreme daily heating energy but the dry bulb temperature had the main effect. It was noted that there was a 1-3 hour delay in peak heating requirement in the buildings but there was no notable delay for the cooling energy requirement.

The estimation of energy consumption during the building design stage is described in a paper by Zhao et al. [2015] by use of the Building Information Software package. In the paper that is noted, the authors outline how they accomplished this. Several subprograms can be used to do this, and the Revit platform was used in this case. The ability for doing this is demonstrated in the application to the design of an office building. In designing a building in this way, the software can evaluate whether or not the designed building meets the energy efficiency requirements. This can shorten the design review time.

A paper from France, where there are high winter electrical heating requirements, has been recently published [Robillart et al. 2018]. As would be expected, this results in a high winter peak electrical demand, which in turn, causes concerns over the production-consumption balance. Advanced control schemes, like the Model Predictive Control frame work, are examined to shift the heating load while maintaining indoor comfort and taking advantage of indoor thermal mass. Some optimization techniques and other methods were used to develop a real-time control of the electrical heating. The novelty of what was done is to introduce reduced models that have been derived from advanced building software. A simulation is discussed that demonstrates the controller performance for a single-zone test building. It is found that the controller exhibits excellent performance and can easily be applied in real time.

A recent article has appeared on energy modeling relative to building retrofitting [Sanhudo et al. 2018]. It is well known that extensive retrofitting of buildings, although conceptually straightforward, can be, in fact, quite complicated. Part of the reason for this may be that not much is known about the building inner details which are slated for retrofitting as well as how this should be modeled. In this paper, Building Information Modeling (BIM), a concept that is becoming quite popular in the building design area, is applied to retrofits, and a process of doing this is described in this paper. A review of energy conservation is given with a particular orientation to topics relevant to retrofits. How to acquire data about the as-built configuration is also given, using such techniques as laser scanning and infrared thermography. The leading BIM software is examined for its possible applicability to this type of function. Finally, the paper postulates on future innovations in this topic as well as challenges the industry will face related to this.

A paper outlining an effort for designing a lightweight construction house that has many goals has been presented by Block et al. [2017]. Solar has only a minor role in this effort, but it will be emphasized in this brief description. While the paper focused on the building industry challenges, emphases are given to improvement of energy efficiency, onsite power generation, and reduction of materials of construction. Four innovations are discussed in the paper, and only one is noted here because of space limits: A soft actuated, adaptive solar façade. This façade consists of movable modular elements that are mounted curtain-like in from the façade, employing novel actuators. Both solar energy generation as well as shading are the result of this design. Unlike many other papers, this one is highly illustrated.

Chwieduk [2017] (note that there are two papers by this author in this review) has written a paper that has addressed general energy conservation aspects in residential buildings. Modern methods of building energy conservation are discussed. The author differentiates between concepts that might be applied to existing buildings and those that are more appropriate for new buildings. In new buildings, this could involve various aspects of shape and form of the building, materials of construction and renewable energy systems. For new buildings, the use of a solar buffer space is advocated. New technologies are, by and large, reworked versions of older concepts. The author predicts that new buildings will include renewable (primarily solar) technologies. Results of the simulation studies could be used for making new regulations on thermal energy use in buildings.

The writing of the energy consumption of a building as a function of several variables has been done in the past, but seldom has much been done with this interpretation of the energy use variation. In this paper [Yoon and Moon 2018] that approach is used, but the authors use this information to pursue further insights regarding the energy use variation from it. They are particularly concerned about the energy use of the tenants. So, in this paper there is a focus on the tenants, which requires insights about occupant schedule, operation, and equipment. They used a ranking of the importance of the various factors using the Random Forest algorithm and verified the energy use in the various types of accommodations using a Gaussian process regression model. The focus is on the defining of the energy use of each tenant, and, as anticipated, it is shown that the energy use varies with the type of accommodation.

Following a similar theme as the above paper, quantifying the effects of the human presence on the energy consumption in buildings is the focus of this paper by Yan Zhang et al. [2018] also. A quote from the paper is "the understanding of occupant behavior and its role in building energy performance remains vague, confusing and inconsistent." The authors attack this issue by first presenting a review of the literature. This literature review has several areas of focus, including: human behaviors related to window opening, lighting control, and heating/cooling. They then discuss ways of collecting data relative to human performance in the control of building heating and cooling as well as ways of modeling this behavior, as published ways of saving energy. After doing this they estimate the possible savings in energy related to occupant behavior might be 10-25% for residential buildings and 5-30% for commercial buildings. The authors then point out four gaps of knowledge relative to understanding human behavior: understanding it in a systematic framework, furthering this understanding to a larger city scale, for linking this to occupant socio-economic and policy variables, and evaluating the occupant behavior to all of this.

For a number of reasons the estimation of the energy use of a building is extremely important. A paper by Delzendeh et al. [2017] gives a review of a great deal of the literature that is in this topic. Even so, there is widespread recognition that the difference between the predicted and the actual energy use in building can be quite large (up to 300% difference.) Impact of the inhabitants, for example, has been essentially overlooked. This reported research has identified some of the areas that need more work, including larger scale analysis (e.g. urban analysis); interior design, in terms of space layout; psychological cognitive behavioral methods; and the integration of quantitative and qualitative research findings in energy simulation tools to name but a few. In general, though, more work needs to be done on the impact of humans on the energy balance of a building.

Two grid independent zero energy buildings are used to demonstrate the effect of climate on these types energy systems [Leonard and Michaelides 2018]. Grid independence greatly impacts the energy required from the grid, and this characteristic is preferred. One of these houses is assumed to be in the northern part of the US where there is dominance in heating energy required. The other is assumed to be in the southwest part of the country where air conditioning energy needs dominate. Both houses have both electrical energy storage and hydrogen generated storage. A great deal of detail is given to the types of energy outputs and needs that are determined.

In a book chapter, the authors [Gönülol and Tokuç 2018] outline various ways that improvements can be made in the design of residential buildings. This is accomplished using the US DOE Solar Decathlon competition as the basis. Although this competition is undergoing some changes in rules, it has been the case that 20 schools from around the world were chosen to participate, and they compete for prize money by designing, building, and bringing their house to a central location to be judged. The chapter gives a good analysis of the competition and what kinds of things it accomplishes. People who are not familiar with Solar Decathlon competition should read of this fairly concise summary of what has transpired in this particular event over the years.

While many countries have building codes, and many others are developing them, implementation of the codes in all cases is of concern. Hence the method and consistency of how they are brought to active use is an aspect for study, and this is what is addressed in the paper by Evans et al. [2017]. The research covered here identifies elements and practices of codes of 22 countries which amounts to about 70% of the codes that have been developed world wide. These elements and practices include data like type, age, size and location of the building, the type of inspections that may be required and documentation of the building materials used. Some countries have introduced compliance evaluation studies. The article also provides details of practices that provide assistance with implementation of building energy codes.

A paper has appeared [Fan and Xia 2018] that outlines the work that South Africa has accomplished to move to a green building sector. In doing this they have developed an Energy Performance Certificate (EPC) standard for a retrofit. To be able to participate in the green building program the retrofitter has to achieve over a minimum rating on the EPC. The paper describes an optimization model that can help the retrofitter achieve a desired outcome. The model determines optimal retrofit plans for a whole building in a systematic manner, taking into account both the envelope components and the indoor facilities. All plans include a roof-top PV system. The economic facets of the project are dealt with in some detail in the paper. The paper includes a case study.

Chel and Kaushit [2017] have written about the use of renewable energy technologies for sustainable development for energy efficient buildings. They have defined the four main steps in the design and construction of such buildings. The first step, carried out before construction begins, is the nearly zero energy design for the building. This includes elements like passive design for heating and cooling as well as daylighting. Second, it is important to use low embodied energy construction materials. The third element is the use of energy efficient equipment in the building that will cut down on the operational energy use. Lastly, incorporate renewable technologies where ever possible, like electrical generation and water heating. The authors discuss the economics and environmental impacts of these aspects in the paper.

Buildings often fail to perform as they were hoped to do, and often this is because of insufficient energy efficiency, inadequate indoor air quality, or general occupant dissatisfaction. The authors of this paper [Borgstein et al. 2018] have performed detailed energy audits on 33 office buildings in Brazil and categorized the ones that involved errors in building design, or problems with operations and maintenance. There were 333 failures that comprised 55 different modes. Many of the same failure modes occur in small and medium buildings while large buildings usually present a separate set of issues. Problems of operations and management far outnumber situations where the building is poorly designed.

Energy retrofits for large buildings can offer more of a challenge than those for small buildings. Luddeni et al. [2018] have added the application to the former by considering a study of retrofitting a large Italian office building. They used a simulation based approach that combines energy modeling and optimization techniques to see what kinds of energy savings might result for various retrofitting options along a path to achieving net zero energy conditions. To do this they assessed the cost of various retrofit options. The approach they used indicated that they achieve a potential primary energy savings of 54% when outfitted with the energy efficiency measures (EEMs) and rooftop PV. The EEMs that resulted are very depend upon the building characteristics and its local climate. Zero energy predictions were not found in any of this group's studies.

It has been generally recognized that it is important to decrease energy use in buildings, and there are some large-scale efforts underway in this realm. However, the authors of this paper [Hamilton et al. 2017] make the case on what is needed is to support the development and implementation of an on-going evaluation of energy demand policy. As the world moves toward a low carbon built environment, this will require a number of some prerequisite changes that are not necessarily available or are inaccessible or incomplete. This paper describes the basis of the IEA ERC Annex 70: Building Energy Epidemiology, as this draws upon health science epidemiology analogy as a whole group of varied background people to be involved in cross-disciplinary studies to accomplish this. To illustrate this process, the authors present an example from the UK on the type of process that is needed for residential buildings.

Although building designs in a location like Brunei are not terribly important to this review, a paper that is related to those is included here [Shabunko et al. 2018]. The reason is not necessarily because of the importance of the designs, but definitely because it discusses a way of analyzing categories of buildings. In this paper EnergyPlus is used to analyze 400 residential buildings that represent three basic types. The EnergyPlus results were then compared to the actual use data as well as results from a door-to-door survey. This paper shows the way that models can be used to produce benchmarking.

A paper by Rathore et al. [2018] makes the case about why rooftop solar energy should be developed in India. One alternative seems to be fossil-fueled plants. But these cause environmental degradation and new plants of this type should not be developed. Another possibility is large scale solar. While there are some plants of this type being developed, to rely on these would require large pieces of land to install them, and this is something quite scarce in India. A final alternative is rooftop solar. The area for its installation is certainly available, and it has all the positive characteristics of other types of PV. It is a proven technology, provides environmental benefits, and saves using new land. This is the alternative that is clearly the best.

1.4.4 Greater Than Single Building Scale

A paper in the category of urban-scale building modeling, where building modeling is done specifically to see how a group of buildings perform, has been put forth by Chen and Hong [2018]. This does not use the detailed approach like the analysis of a single building might require but rather uses higher level building information. Three zoning methods were used: One Zone (uses a one thermal zone per floor within the target building's footprint), Autozone (splits the building's footprint into perimeter and core zones) and Prototype (uses US DOE's reference building prototype shapes). It was shown that variations exist between the output of the various models, and it is recommended that the AutoZone model be used.

District level energy modeling approaches and including associated software tools are the topic of a recent paper [Allegrini et al. 2015]. The authors claim that it is no longer appropriate to model a building with assumed climatic condition or its converse: modeling the climate with an assumed form for a building impacted. Consideration is given to new tools and models pertinent to district sized areas. The following aspects are included: district energy systems, renewable energy generation systems and the urban microclimate considerations. A number of cross disciplinary tools are discussed. They end with an evaluation of district level types of analysis.

A paper that describes an analysis of energy flows in a large city setting [Ahmed et al. 2017] is one of several recent ones by González and coworkers. To accomplish this, they use high resolution weather data for New York City interfaced with a single building energy model. They typically consider some extreme weather condition period for analysis. Among other data they give hourly energy consumption. This plus locational information is then available from the analysis. As they point out, the information from these kinds of analyses could be of value for policy makers targeting localized energy efficiency and GHG mitigation measures.

When solar energy systems are located in a dense network like there is Belgium, information from these kinds of systems can be used to derive global horizontal solar data [Bertrand et al. 2018]. This paper incorporates hourly PV power generation information from roughly 1500 residential systems for a one-year period. A network of radiometric stations is used to verify the data developed. Several factors influence the results of this approach including the PV system efficiency, the weather conditions, and the density of the PV systems. As might be expected, the quality of the data suffers as the information about the orientation and tilt becomes less accurate. The system accuracy also suffers at lower sun angles.

This paper [Xingxing Zhang et al. 2018] puts forth the argument for clustering buildings in urban areas. Doing this would allow easier access to renewable energy sources and these could be coupled with improving building performance and distributed energy management, resulting in great improvement in the energy use in buildings. All three of the potential energy applications could be used, namely solar electric power, solar thermal, and energy efficiency. Since solar plays such an important role in this, application possibilities are reviewed in the paper. Building density is also considered. Modeling techniques that incorporate analysis for all of these elements are considered. Key factors to incorporate in this analysis approach are outlined.

Decreases of building energy consumption and greenhouse gas emissions seem to be an almost universal desire. However, the actual location where the action takes place is usually in the cities. This is because of the role the cities play in dealing with options in this aspect of energy conservation. A wide range of modelling suggestions has been made in the past about how to deal with this situation. Much of this information about city modeling approaches is reviewed in this paper [W. Li et al. 2017], where categories of the models are listed. Because there are significant differences in the noted models, the strengths and weaknesses of each are listed. The paper ends with a discussion of the challenges related to the models that exist.

Extreme hot weather can tax the local grid operation. This is particularly important in large population areas, where imposing large demands on the grid can have a negative effect on many electrical users. If these potentially acute weather conditions may be predicted, then it may be possible to forecast building energy demands. In turn it may be possible to make short term adjustments, or, more definitely, long term adjustments to handle these loads in a better fashion. This is the focus of a paper by Ortiz et al. [2017]. The authors have consulted with a variety of resources to develop insights to building energy requirements for New York City. As a result, it is possible to compare predictions to actual energy use in buildings. An additional element to the paper is that use of while roofs as a means of decreasing heat loads has been examined.

GIS is being shown to be extremely valuable in a number of applications. In the paper by Groppi et al. [2018] it has been applied in a study to assist in the assessment of the decrease of nonrenewable of thermal and electrical consumptions of homes on a city scale. This is done by inferring the amount of renewable energy being used and thus leading to a decrease in use of conventional energy. The approach was used on a small-size city in Italy. It was found that the in the installation in newer buildings that the addition of PV saved more conventional energy that did solar thermal while in older buildings the two applications saved similar energy. Solar thermal installations saved more thermal energy than PV systems save electrical energy. Historic buildings most likely have solar thermal installations rather than PV installations, usually because the roof space is more limited on historic buildings.

A study of 90 residential buildings related to their life energy costs has been reported by Chastas et al. [2016]. It turned out this effort was complicated by a number of factors, because of some problems in the definitions and other factors that arose. While there were seemingly clear definitions of the boundaries to consider as well as other stipulations, there were some uncertainties derived from the life cycle inventory methods used. Operating energy seemed to dominate the life cycle energy in the past. But that has now changed such that embodied energy has become the dominant form in the newer, low energy buildings in the current situation. For example, the use of renewable energy aspects in a passive house for the generation of electricity classifies it in the range of embodied energy of a nearly zero energy building. However, the authors point out the uncertainties in many of the classification aspects.

Phoenix, Arizona, has a program called Energize Phoenix, and this is the basis of a paper related to a research project that used data from this program. Retrofit data covering 201 residential buildings and 636 commercial buildings from January 2008 to April 2013 were used. Retrofits for residential buildings were divided into five types, while those for the commercial buildings were divided into 6 types. Monthly energy savings ranged from 12% for commercial buildings to 8% for residential buildings but these savings were 30-50% lower than the predicted values. They also considered energy savings that were related to the building attributes. Learning effects were also observed in this program. So there a possibility that improved energy performance can result.

Broberg and Egüez [2018] have investigated whether or not the type of ownership of multi-dwelling building matters in the building's energy efficiency. They question, for example, if rental apartment buildings are less energy efficient than cooperative apartment buildings and whether public ownership has a negative impact on energy efficiency. The authors give a number of reasons why there could be differences, and this is what has led to the study. They find that cooperatively owned buildings are a great deal more energy efficient than are rental buildings, while publically owned buildings have somewhat better energy efficiency.

Several actively controlled building façade technologies offer a variety of interesting and mostly function in ways to manage building's energy and comfort, including on-site generation of electricity. In the paper by Powell et al. [2018], a novel addition to this technology is proposed. This is a concept the uses adaptive reflective panels on the building exterior. This approach will allow sharing of sunlight in ways that are generally impossible. The authors present the details about how the technique models the operation and assesses the performance of such a system. This involves a resistive-capacitive technique that is coupled to a Monte Carlo ray tracing simulation. It functions in a way to contribute to energy savings. It has been shown to function in a variety of important ways.

1.4.5 Meteorological Affects

A paper by Buffat et al. [2018] considers the estimation of solar irradiation over geographically large areas. To do this the authors have applied insights about long term space and time variations of the flux, as well as the typical orientations of the surfaces where PV might be installed. They recognize that the flux components have uncertainties that are affected in different ways because of the variations in way they are modeled. To accomplish this, they use long term (22 years) satellite data with a temporal variation of 30 minutes and a spatial resolution between 3.8 and 5.6 m. The possible building effects include the types of roof tops (handled with the use of Digital Surface Models) as well as the building footprints. Definition of slope and orientation of rooftops have a resolution of 0.5 m. To address the potential electricity that might be generated, they used models of the characteristics of different types of existing PV plants for monthly periods. Validation of their results showed an accuracy of 0.28% in August and 28.08% in December. No significant variations related to spatial and technological variations were found.

Authors of this next paper [Dey et al. 2018] use a solar tree to perform appropriate location/application of tuning of the solar power generation curve. They start with the optimization of single panel orientation using a data driven approach. Latitude *and* azimuth angles both need to be considered over a large latitude range to define optimal orientations. This is particularly true for locations that have an asymmetric daily insolation pattern. They showed the feasibility of this approach by designing a 1 kW solar tree at four locations. Orientation of the solar panels was accomplished to minimize shadow losses using a generic algorithm. Ray optic simulations for solar trees were validated for San Francisco and Paris, indicating less than 2% shading losses which increased their power generation by 2.04% and 7.38% respectively compared to latitude tilt. Generally, the solar tree simulation showed excellent matches and be easily extended to other locations and capacities.

Machete et al. [2018] state that is very important to know the solar energy incident on a building's roof and facades to exploit good applications of various renewable energy analyses. They have used a 3D GIS tool to handle both 2.5D and 3D modelling approaches to go along with the use of three solar radiation tools, and applied this to a city block in Lisbon for both the winter and summer solstices for different levels of detail for the surround area. It was found that both the surrounding built environment and the topographic relief have a great deal of impact on the simulations, compared to situations when this was not included. The authors extol the use of a 3D tool for this kind of a study.

Peronato et al. [2018] have indicated governmental authorities have supported the development of detailed maps (cadastres) to show the solar potential of existing buildings. These maps usually only provide some simple evaluations, and usually neglect many of the details that would show solar impacts and environmental conditions. However, the authors argue that this kind of information would be very valuable and demonstrate this in this paper. Uncertainty in solar generation potential of existing buildings is shown using a multi-scale approach. An analysis that presents a visualization technique is described. Using multiple simulation scenarios, insights to setting confidence intervals due to two typical uncertainty factors (vegetation and weather) are shown. The technique is quite versatile and can be run at various scales using space- and time-aggregated results. These should be of great value to decision-makers.

1.4.6 Miscellaneous Topics

The Life Cycle Costing Assessment (LCCA) is used quite frequently in evaluation of the Sustainability Life Cycle in the Solar Energy sector. As the primary component for assessing the environmental and social areas, it has been quite effective. For example, it is valuable in choosing between PV installation versus a solar thermal system. It can also help with clarifying their component evaluations of electrical energy storage vs. thermal energy storage. The authors [Naves et al. 2018] discuss a wide range applications using LCCA, and the reader is referred their paper to consider all of them.

Verclas [2018] has put forth a possible application to energy use in low- and moderate-income communities that are dependent on transportation. This concept, an outgrowth of two programs that among them provide low cost solar power, combines solar rooftop systems with electric vehicle incentives. Among other aspects, this concept addresses the fact that many times these types of communities do not have sufficient transportation options. As the concept is described, it provides capital to facilitate the development of the system. Details of the system design is described in detail in the paper. It is concluded that broadening the deployment of this system could reduce costs and increase the access to renewable energy for vulnerable families.

A method of energy analysis for buildings using exergy is described in the paper by Zhou [2018]. It probably is more accurate to refer to it as a separate method for accounting for the different types of energy in determining a balance. Energy is a universal term that accounts for all forms (e.g. heat and work) as the same importance in the determining of a balance. However, as was put forth many years after energy analyses were in use, exergy distinguishes between heat and work to put those kinds of terms a different basis. Exergy is based upon work, and work is a higher quality form of energy than energy. This is because to convert heat into work, some type of conversion process needs to take place where the energy content is decreased. There are many rules related to these kinds of analyses, and Zhou goes through many of them in a building application. So, for example, hot water cannot be converted 100% to work (like the driving of a piston). On the other hand, something like electricity, a manufactured form of energy, can be considered like work...it can directly turn a motor, for example. The author calls out renewable energy factors specifically in this discussion.

Illinois residential electricity consumers will be able to participate in a greatly expanded renewable energy market, and this situation is discussed in a paper [Borovik and Albers 2018]. Source of this opportunity is a recently enacted the Future Energy Jobs Act. Some of the aspects available include proactive options which might involve installing one's own system, to passive ones. The latter include a program in which most Illinois residential consumers will automatically be able to participate, just by being a customer of the public utility.

This review of institutional cooking using solar energy [Indora and Kandpal 2018] begins with estimates of the energy requirements for cooking. It also gives an overview of cooking technologies as well as their policies and promotional programs. There is an emphasis on concentrating solar sources along with the design, construction and operational details pertinent to both direct and indirect systems are given. Case studies (mainly from India) give feedback on the efficacies of institutional large-scale systems. A few systems of each type are described in the detail, as well as major findings and observations about each type.

It is interesting to include a relatively current paper on infrared thermography in this general building-related review. While it is not something one would usually find mounted in a home, for example, it is a tool that can be quite useful in diagnosing heat leaks in a building. Lucchi [2018] has written an excellent state of the art consideration of this valuable tool. Among other aspects of the article, it contains a list of 148 publications relative to the instrument and its use. In its range of topics covered are details of its historical growth, general procedures for its use, a discussion of the range of types of cameras currently in use, various technical issues and limitations to its use. The passive application is by far the most frequently used. This can pinpoint heat leaks as well as provide a map of temperatures using the camera type device. An overview of its use in energy audits in given.

1.5 LITERATURE CITED

This Section contains the literature cited for all four main sections with the two sections of part 2, as well as 3, and 4. In other words there are two sets of bibliographic information for the second section (2 parts) given as A parts i) and ii), while the other two main sections are given as parts B and C.

1.5.1 Solar Thermal Power Generation References

Abdulhamed, Ali Jaber, Adam, Nor Mariah, Ab-Kadir, Mohd Zainal Abidin, Hairuddin, Abdul Aziz, 2018. Review of solar parabolic-trough collector geometrical and thermal analyses, performance, and applications, *Renewable and Sustainable Energy Reviews*, **91**, 822-831.4r

Abutayeh, Mohammad, Alazzam, Anas, 2017. Adapting steady-state solar power models to include transients, *Journal of Solar Energy Engineering*, **139**, 7 pages.

Aichmayer, L., Garrido, J., and Laumert, B., 2018. Scaling effects of a novel solar receiver for a micro gas-turbine based solar dish system, *Solar Energy*, **162**, 248-264.

Al-Kayiem, H. H., and Aja, O. C., 2016. Historic and recent progress in solar chimney power plant enhancing technologies, *Renewable and Sustainable Energy Reviews*, **58**, 1269-1292.

Bauer, T., Pfleger, N., Laing, D., Steinmann, W.-D., Eck, M., Kaesche, S., 2013. Chapter 20. High-Temperature Molten Salts for Solar Power Application, in *Molten Salts Chemistry.* Elsevier, 415-438.

Behr, Omar, 2018. Solar thermal power plants—A review of configurations and performance comparison, *Renewable and Sustainable Energy Reviews*, **92**, 608-627.

Bittencourt de Sá, A., Filho, V., Tadrist, L., Passos, J., 2018. Direct steam generation in linear solar concentration: Experimental and modeling investigation — A review, *Renewable and Sustainable Energy Reviews*, **90**, 910-936.

Breeze, Paul, 2016. Chap. 11: Solar Integration and the Environmental Impact of Solar Power, in *Solar Power Generation*, Elsevier Science & Technology.

Buck, R., Schwarbözl, P., 2018. Chapter 4.17 Solar Tower Systems in *Comprehensive Energy Systems* (Editor in Chief, Ibrahim Dincer), Elsevier, 692-732.

Bukhary S., Ahmad, S., and Batista, J., 2018. Analyzing land and water requirements for solar deployment in the Southwestern United States, *Renewable and Sustainable Energy Reviews*, **82**, 3288-3305.

Butti, K., Perlin, J., 1980. *A Golden Thread, 2500 Years of Solar Architecture and Technology*, Chesire Books.

Calise, Francesco, d'Accadia, Massimo Dentice, Libertini, Luigi, Vicidomini, Maria, 2018. Thermoeconomic analysis of an integrated solar combined cycle power plant, *Energy Conversion and Management*, **171**, 1038-1051.

Chaanaoui, M., Vaudreuil, S., and Bounahmidi, T., 2016. Benchmark of concentrating solar power plants: historical, current and future technical and economic development, *Procedia Computer Science*, **83**, 782-789.

Costa, S. C. S., Diniz, A. S. A. C., Kazmerski, L. L., 2018. Solar energy dust and soiling R & D progress: Literature review update for 2016, *Renewable and Sustainable Reviews*, **82**, 2504-2536.

DeAngelis, F., Seyf, H. R., Berman, R., Schmidt, G., Moore, D., Henry, A., 2018. Design of a high temperature (1350 °C) solar receiver based on a liquid metal heat transfer fluid: Sensitivity analysis, *Solar Energy,* **164**, 200-209.

Dincer, Ibrahim, Bicer, Yusuf, 2018. 4.23 Solar Thermochemical Energy Conversion, Comprehensive Energy Systems, Vol. 4, Elsevier, 895-946.

Du, Eshun, Zhang, Ning, Hodge, Bri-Mathias, Kang, Chongqing, Kroposki, Benjamin, Xia, Qing, 2018. Economic justification of concentrating solar power in high renewable energy penetrated power systems, *Applied Energy,* **222**, 649-661.

Dunham, M. T., Iverson, B. D., 2014. High-efficiency thermodynamic power cycles for concentrated solar power systems, *Renewable and Sustainable Energy Reviews*, **30**, 758-770.

Elmaihy, Ali, El Weteedy, Ahmed, 2017. Short-duration performance acceptance test of parabolic trough solar field: A case study, *Journal of Solar Energy Engineering,* **139**, 9 pages.

Enríquez, Luis Coco, Munoz-Anton, Javier, Martínez-Val Penalosa, José María, 2017. Thermodynamic optimization of supercritical CO_2 Brayton power cycles coupled to line-focusing solar fields, *Journal of Solar Energy Engineering*, **139**, 8 pages.

Fernández-González, Daniel, Ruiz-Bustinza, I., González-Gasca, Carmen, Noval, Juan Piñuela, Mochón-Castaños, Javier, Sancho-Gorostiaga, José, Verdeja, Luis Felipe, 2018. Concentrated solar energy applications in materials science and metallurgy, *Solar Energy*, **170**, 520-540.

Freedman, Justin P., Wang, Hao, Prasher, Ravi S., 2018. Analysis of nanofluid-based parabolic trough collectors for solar thermal applications, *Journal of Solar Energy Engineering,* **140**, 8 pages.

Ge, Y. T., Li, L., Luo, X., Tassou, S. A., 2118. Performance evaluation of a low-grade power generation system with $C\,O_2$ transcritical cycle, *Applied Energy*, **227**, 220-230.

González-Roubaud, E., Pérez-Osorio, D., Prieto, C., 2017. Review of commercial thermal energy storage in concentrated solar power plants: Steam vs. molten salts, *Renewable and Sustainable Energy Reviews*, **80**, 133-148.

Guney, Mukrimin Sevket, 2016. Solar power and application methods, *Renewable and Sustainable Energy Reviews*, **57**, 776-785.

Guo, Penghua, Wang, Yunfeng, Li, Jingyin, and Wang, Yuan, 2016. Thermodynamic analysis of a solar chimney power plant system with soil heat storage, *Applied Thermal Engineering,* **100**, 1076-1084.

Hafez, A. Z., Soliman, A., El-Metwally, K. A., Ismail, I. M., 2017. Design analysis factors and specifications of solar dish technologies for different systems and applications, *Renewable and Sustainable Energy Reviews*, **67**, 1019-1036.

Hafez, A. Z., Attia, A. M., Eltwab, H. S., ElKousy, A. O., Afifi, A. A., AbdElhamid, A. G., AbdElqader, A. N., Fateen, S-E. K., El-Metwally, K. A., Soliman, A., Ismail, I. M., 2018. Design analysis of solar parabolic trough thermal collectors, *Renewable and Sustainable Energy Reviews*, **82**, 1215-1260.

Hairat, Manish Kumar, Ghosh, Sajal, 2017. 100 GW solar power in India by 2022 – A critical review, *Renewable and Sustainable Energy Reviews*, **73**, 1041-1050.

Helman, Udi, 2017. Chapter 26: Economic and Reliability Benefits of Solar Plants, *Renewable Energy Integration, 2nd Edition,* (Editor: Lawrence Jones), Elsevier, 351-372.

Hernandez, R. R., Easter, S. B., Murphy-Mariscal, M. L., Maestre, F. T., Tavassoli, M., Allen, E. B., Barrows, C. W., Belnap, J., Ochoa-Hueso, R., Ravi, S., Allen, M. F., 2014. Environmental impacts of utility-scale solar energy, *Renewable and Sustainable Energy Reviews*, **29**, 766-779.

Hess, David J., 2016. The politics of niche-regime conflicts: Distributed solar energy in the United States, *Environmental Innovation and Societal Transitions,* **19**, 42-50.

Ho, Clifford, 2016. A review of high-temperature particle receivers for concentrating solar power, *Applied Thermal Engineering,* **109**, 958-969.

Hussain, Faisal M., Al-Sulaiman, Fahad A., 2018. Performance analysis of a solar chimney power plant design aided with reflectors, *Energy Conversion and Management*, **177**, 30-42.

Islam, Md Tasbirul, Huda, Naxmul, Abdullah, A. B., R. Saidur, 2018. A comprehensive review of state-of-the-art concentrating solar power (CSP) technologies: Current status and research trends, *Renewable and Sustainable Energy Reviews*, **91**, 987-1018.

Ju, Xing, Xu, Chao, Hu, Yangqing, Han, Xue, Wei, Gaosheng, Du, Xiaoze, 2017. A review on the development of photovoltaic/concentrated solar power (PVCSP) hybrid systems, *Solar Energy Materials & Solar Cells*, **161**, 305-327.

Kabir, E., Kumar, P., Kumar, S., Adelodun, A. A., Kim, K.-H., 2018. Solar energy: Potential and future prospects, *Renewable and Sustainable Energy Reviews*, **82**, 894-900.

Kalogirou, Soteris A., 2013. Chap. 10: Solar Thermal Power Systems in *Solar Energy Engineering: Processes and Systems*, Elsevier Science & Technology, 541-581

Karas, D. E., Byun, Jongmin, Moon, Jaeyun, Jose, Cilla, 2018. Copper-oxide spinel absorber coatings for high-temperature concentrated solar power systems, *Solar Energy Materials and Solar Cells*, **182**, 321-330.

Kassem, A., Al-Haddad, K., and Komljenovic, D., 2017. Concentrated solar thermal power in Saudi Arabia: Definition and simulation of alternative scenarios, *Renewable and Sustainable Energy Reviews,* **80**, 75-91.

Kedare, S. B., Desai, N. B., 2017. Solar Thermal Process Heat, *Volume 4, Encyclopedia of Sustainable Technologies*, Elsevier, 367-376.

Khakrah, Hamidreza, Shamloo, Amir, Hannani, Siamak Kazemzadeh, 2017. Determination of parabolic trough solar collector efficiency using nanofluid: A comprehensive numerical study, *Journal of Solar Energy Engineering,* **139**, 11 pages.

Khan, Jibran, and Arsalan, Mudassar H., 2016. Solar power technologies for sustainable electricity generation—A review, *Renewable and Sustainable Energy Reviews*, **55**, 414-425.

Kumar, Apurv, Kim, Jin-Soo, Lipinski, Wojciech, 2017. Radiation absorption in a particle curtain exposed to direct high-flux solar irradiation, *Journal of Solar Energy Engineering*, **140**, 17 pages.

Kumar, V., Shrivastava, R. L., Untawale, S. P., 2015. Fresnel lens: A promising alternative of reflectors in concentrated solar power, *Renewable and Sustainable Energy Reviews*, **44**, 376-390.

Kuravi, S., Trahan, J., Goswami, D. Y., Rahman, M. M., Stefanakos, E. K., 2013. Thermal energy storage technologies and systems for concentrating solar power plants, *Energy and Combustion Science*, **39**, 285-319.

Kylili, Angeliki, Fokaides, Paris A., Ioannides, Andreas, Kalogirou, Soteris, 2018. Environmental assessment of solar thermal systems for the industrial sector, *Journal of Cleaner Production*, **176**, 99-109.

Leonard, Matthew D., Michaelides, Efstathios E., Michaelides, Dimitrios N., 2018. Substitution of coal power plants with renewable energy sources – Shift of the power demand and energy storage, *Energy Conversion and Management*, **164**, 27-35.

Letcher, Trevor M., 2018. Why solar energy? Chapter 1 in *A Comprehensive Guide to Solar Energy Systems*,

Lilliestam, Johan, Barradi, Touria, Caldés, Natalia, Gomez, Marta, Hanger, Susanne, Kern, Jürgen, Komendantova, Nadejda, Mehos, Mark, Hong, Wai Mun, Wang, Zhifeng, Patt, Anthony, 2018. Policies to keep and expand the option of concentrating solar power for dispatchable renewable electricity, *Energy Policy*, **116**, 193 197.

Liqreina, A., Qoaider, L., 2014. Dry cooling of concentrating solar power (CSP) plants, an economic competitive option for the desert regions of the MENA region, *Solar Energy*,**103**, 417-424.

Liu, M., Tay, N. H. S., Bell, S., Belusko, M., Jacob, R., Will, G., Saman, W., Bruno, F., 2016. Review on concentrating solar power plants and new developments in high temperature thermal energy storage technologies, *Renewable and Sustainable Energy Reviews,* **53**, 1411-1432.

Malali, P. D., Chaturvedi, S. K., and Abdel-Salam, T., 2017. Performance optimization of a regenerative Brayton heat engine coupled with a parabolic dish solar collector, *Energy Conversion and Management*, **143**, 85-95.

Milani, Dia, Luu, Minh Tri, McNaughton, Robbie, Abbas, Ali, 2017. Optimizing an advanced hybrid of solar-assisted supercritical CO_2 Brayton cycle: A vital transition for low-carbon power generation industry, *Energy Conversion and Management*, **148**, 1317-1331.

Merino, Luis Moreno, Fernández, Núria Imbern, Valsero, Juan José Duran, Aguilera, Héctor, 2018. Concentrating solar power plants versus groundwater resources in Mediterranean areas of Spain: The environmental dilemma, *Journal of Environmental Management*, **206**, 409-417.

Nsengiyumva, Walter, Chen, Shi Guo, Hu, Lihua, Chen, Xueyong, 2018. Recent advancements and challenges in Solar Tracking Systems (STS): A review. *Renewable and Sustainable Energy Reviews,* **81**, 250-279.

Okonkwo, Eric C., Abid, Muhammad, Ratlamwala, Tahir A. H., 2018. Numerical analysis of heat transfer enhancement in a parabolic trough collector based on geometry modifications and working fluid usage, *Journal of Solar Energy Engineering,* **140**, 11 pages.

Okoye, C. O., Solyali, O., Taylan, O., 2016. A new economic feasibility approach for solar chimney power plant design, *Energy Conversion and Management*, **126**, 1013-1027.

Ophoff, Cédric, Ozalp, Nesrin, 2017. 2017. A novel iris mechanism for solar thermal receivers, *Journal of Solar Energy Engineering*, **139**, 10 pages.

Pandey, A. K., Hossain, M. S., Tyagi, V. V., Rahim, N. A., Selvaraj, J. A./L., Sari, A., 2018. Novel approaches and recent developments on potential applications of phase change materials in solar energy, *Renewable and Sustainable Energy Reviews*, **82**, 281-323.

Perlin, J., 2013. *Let It Shine, the 6,000-Year Story of Solar Energy*, New World Library.

Pietzcker, Robert Carl, Stetter, Daniel, Manger, Susanne, Luderer, Gunnar, 2014. Using the sun to decarbonize the power sector: The economic potential of photovoltaics and concentrating solar power, *Applied Energy*, **135**, 704-720.

Poživil, Peter, Steinfeld, Aldo, 2017. Integration of a pressurized-air solar receiver array to a gas turbine power cycle for solar tower applications, *Journal of Solar Energy Engineering,* **139**, 8 pages.

Pramanik, Santanu, Ravikrishna, R. V., 2017. A review of concentrated solar power hybrid technologies, *Applied Thermal Engineering,* **127**, 602-637.

Prieto, Cristina, Cooper, Patrick, Fernández, A. Inés, Cabeza, Luisa F., 2016. Review of technology: Thermochemical energy storage for concentrated solar power plants, *Renewable and Sustainable Energy Reviews,* **60**, 909-929.

Rea, Jonathan E., Oshman, Christopher J., Olsen, Michele L., Hardin, Corey L., Glatzmaier, Greg C., Siegel, Nathan P., Parilla, Philip A., Ginley, David S., Toberer, Eric S., 2018. Performance modeling and techno-economic analysis of a modular concentrated solar power tower with latent heat storage, *Applied Energy*, **217**, 143-152.

Sangi, Roozbeh, Amidpour, Majid, Hosseinizadeh, Behzad, 2011. Modeling and numerical simulation of solar chimney power plants, *Solar Energy*, **85**, 829-838.

Sansaniwal, Sunil Kumar, Sharma, Vashimant, Mathur, Jyotirmay, 2018. Energy and exergy analyses of various typical solar energy applications: A comprehensive review, *Renewable and Sustainable Energy Reviews,* **82**, 1576-1601.

Santos, J. J. C. S., Palacio, J. C. E., Reyes, A. M. M., Carvalho, M., Freire, A. J. R., Barone, M. A., 2018. Chapter 12: Concentrating Solar Power, *Advances in Renewable Energies and Power Technologies,* 373-402.

Shahsavari, Amir, Akbari, Morteza, 2018. Potential of solar energy in developing countries for reducing energy-related emissions, *Renewable and Sustainable Energy Reviews*, **90**, 275-291.

Shoai-Tehrani, Bianka, Akimoto, Keigo, Sano, Fuminori, 2018. Low-carbon investments from the perspective of electric utilities: The burden of the past. *Utilities Policy*, **51**, 18-32.

Stein, W. H., Buck, R., 2017. Advanced power cycles for concentrated solar power, *Solar Energy*, **152**, 91-105.

Suman, Siddharth, Khan, Mohd. Kaleem, Pathak, Manabendra, 2015. Performance enhancement of solar collectors–A review, *Renewable and Sustainable Energy Reviews*, **49**, 192-210.

Tehrani, S. Saeed Mostafavi, Taylor, Robert A., Nithyanandam, Karthik, Ghazani, Ardalan Shafiei, 2017. Annual comparative performance and cost analysis of high temperature, sensible thermal energy storage systems integrated with a concentrated solar power plant, *Solar Energy*, **153**, 153-172.

Toro, Claudia, Lior, Noam, 2017. Analysis and comparison of solar-heat driven Stirling, Brayton and Rankine cycles for space power generation, *Energy*, **120**, 549-564.

Trabelsi, S. E., Qoaider, L., Guizani, A., 2018. Investigation of using molten salt as heat transfer fluid for dry cooled solar parabolic trough power plants under desert conditions, *Energy Conversion and Management*, 156, 253-263.

Trainer, Ted, 2014. The limits to solar thermal electricity, *Energy Policy*, **73**, 57-64.

Trainer, Ted, 2017a. 'Can renewables meet total Australian energy demand: A "disaggregated" approach, *Energy Policy,* **109**, 539-544.

Trainer, Ted, 2017b. Some problems in storing renewable energy, *Energy Policy,* **110**, 386-393.

Trainer, Ted, 2018. Estimating the EROI of whole systems for 100% renewable electricity supply capable of dealing with intermittency, *Energy Policy*, **119**, 684-653

Turchi, C. S., Stekli, J., Bueno, P. C., 2017. Chapter 11: Concentrating Solar Power in *Fundamentals and Applications of Supercritical Carbon Dioxide (sCO_2) Based Power Cycles,* Elsevier Ltd., 2017, 269-292.

Turchi, C. S., Vidal, J., Bauer, M., 2018. Molten salt power towers operating at 600–650 °C: Salt selection and cost, *Solar Energy*, **164**, 38-46.

Wang, Ruilin, Sun, Jie, Hong, Hui, Liu, Qibin, 2017. An on-site test method for thermal and optical performances of parabolic-trough loop for utility-scale concentrating solar power plant, *Solar Energy*, **153**, 142-152.

Wenger, Erez, Epstein, Michael, Kribus, Abraham, 2017. Thermo-electro-chemical storage (TECS) of solar energy, *Applied Energy*, **190**, 788-799.

Xu, Ben, Li, Peiwen, Chan, Cholik, 2015. Application of phase change materials for thermal energy storage in concentrated solar thermal power plants: A review to recent developments, *Applied Energy*, **160**, 286-307.

Xu, Xiankun, Wang, Xiaoxin, Li, Peiwen, Li, Yuanyuan, Hao, Qing, Xiao, Bo, Elsentriecy, Hassan, Gervasio, Dominic, 2018. Experimental Test of Properties of KCl-$MgCl_2$ Eutectic Molten Salt for Heat Transfer and Thermal Storage Fluid in Concentrated Solar Power Systems, *Journal of Solar Energy Engineering,* **140**, 9 pages.

Xu, Xinhai, Vignarooban, K., Xu, Ben, Hsu, K., Kannan, A.M., 2016. Prospects and problems of concentrating solar power technologies for power generation in the desert regions, *Renewable and Sustainable Energy Reviews*, **53**, 1106–1131.

Zeitouny, Joya, Lalau, Noémie, Gordon, Jeffry M., Katz, Eugene A., Flamant, Gilles, Dollet, Alain, Vossier, Alexis, 2018. Assessing high-temperature photovoltaic performance for solar hybrid power plants, *Solar Energy Materials and Solar Cells*, **182**, 61-67.

Zhou, Xinping, Xu, Yangyang, Hou, Yaxiong, 2017. Effect of flow area to fluid power and turbine pressure drop factor of solar chimney power plants, *Journal of Solar Energy Engineering*, **139**, 9 pages.

1.5.2 Photovoltaic Power Generation References

Abuella, Mohamed, Chowdhury, Badrul, 2018. Forecasting of solar power ramp events: A post-processing approach, *Renewable Energy*, **133.** 1380-1392.

Akinyele, D. O., 2017. Environmental performance evaluation of a grid-independent solar photovoltaic power generation (SPPG) plant, *Energy*, **130**, 515-529.

AL-Rousan, Nadia, Isa, Nor, Ashidi Mat, Desa, Mohd Khairunaz Mat, 2018. Advances in solar photovoltaic tracking systems: A review, *Renewable and Sustainable Energy Reviews*, **82**, 2548–2569.

Al-Shetwi, Ali Q., Sujod, Muhamad Zahim, 2018. Modeling and control of grid-connected photovoltaic power plant with fault ride-through capability, *Journal of Solar Energy Engineering,* **140**, 1-8.

Antonanzas, J., Osorio, N., Escobar, R., Urraca, R., Martinez-de-Pison, F. J., Antonanzas-Torres, F., 2016. Review of photovoltaic power forecasting, *Solar Energy*, **136**, 78-111.

Ashby, Michael F., 2016. Chapter 12: Solar PV, in *Materials and Sustainable Development*, Elsevier Ltd., 181-196.

Behera, Manoja Kumar, Majumder, Irani, Nayak, Niranjan, 2018. Solar photovoltaic power forecasting using optimized modified extreme learning machine technique, *Engineering Science and Technology, an International Journal*, **21**, 428-438.

Breeze, Paul, 2016. Chapter 11: Solar Integration and the Environmental Impact of Solar Power, in *Solar Power Generation*, Elsevier Science & Technology, 81-87.

Bright, Jamie M., Killinger, Sven, Lingfors, David, Engerer, Nicholas A., 2018. Improved satellite-derived PV power nowcasting using real-time power data from reference PV systems, *Solar Energy*, **168**, 118-139.

Cabrera-Tobar, Ana, Bullich-Massagué, Eduard, Aragüés-Peñalba, Mònica, Gomis-Bellmunt, Oriol, 2016. Topologies for large scale photovoltaic power plants, *Renewable and Sustainable Energy Reviews*, **59**, 309-319.

Capdevila, Hugo, Naidoo, Vijandren, Graeber, Max, 2016. Soiling forecast and measurements for large PV power generation project in desert environments, *IEEE 43rd PV Specialists Conference*, 2071-2075.

Chaudhary, Priyanka, Rizwan, M., 2018. Energy management supporting high penetration of solar photovoltaic generation for smart grid using solar forecasts and pumped hydro storage system, *Renewable Energy*, **118**, 928-946.

Chang, Rui, Shen, Yanbo, Luo, Yong, Wang, Bo, Yang, Zhenbin, Guo, Peng, 2018. Observed surface radiation and temperature impacts from the large-scale deployment of photovoltaics in the barren area of Gonghe, China, *Renewable Energy*, **118**, 131-137.

Chiacchio, Ferdinando, D'Urso, Diego, Famoso, Fabio, Brusca, Sebastian, Aizpurua, Jose Ignacio, Catterson, Victoria M., 2018. On the use of dynamic reliability for an accurate modelling of renewable power plants, *Energy*, **151**, 605-621.

Comello, Stephen, Reichelstein, Stefan, Sahoo, Anshuman, 2018. The road ahead for solar PV power, *Renewable and Sustainable Energy Reviews*, **92**, 744-756.

Crago, Christine L., Koegler, Eric, 2018. Drivers of growth in commercial-scale solar PV capacity, *Energy Policy*, **120**, 481-491.

Creayla, Christine May C., Garcia, Felan Carlo C., Macabebe, Erees Queen B., 2017. **Next** day power forecast model using smart hybrid energy monitoring system and meteorological data, *Procedia Computer Science,* **105**, 256-263.

Dincer, Ibrahim, Bicer, Yusuf, 2018. 4.19 PV-Based Energy Conversion Systems, in *Comprehensive Energy Systems*, Vol. 4, (Editor Ibrahim Dincer), Elsevier Inc., 762-793.

Ding, Yulong, Li, Yongliang, Liu, Chuanping, Sun, Ze, 2015. Chapter 2: Solar Electrical Energy Storage in *Solar Energy Storage* (Brent Sorensen, Editor), Elsevier Science and Technology, 7-22.

Dougherty, Brian, Boyd, Matthew, 2017. Model (at least) twice, build once: Experiences with the Design-Bid-Build process for solar photovoltaic arrays, *Journal of Solar Energy Engineering*, **139**, 1-8.

Durusu, Ali, Erduman, Ali, 2018. An improved methodology to design large-scale photovoltaic power plant, *Journal of Solar Energy Engineering*, **140**, 11 pages.

Gandoman, Foad H., Abdel Aleem, Shady H. E., Omar, Noshin, Ahmadi, Abdollah, Alenezi, Faisal Q., 2018. Short-term solar power forecasting considering cloud coverage and ambient temperature variation effects, *Renewable Energy*, **123**, 793-805.

Hacke, Peter, Lokanath, Sumanth, Williams, Paul, Vasan, Arvind, Sochord, Paul, TamizhMani, GovindaSamy, Shinohara, Hirofumi, Kurtz, Sarah, 2018. A status review of photovoltaic power conversion equipment reliability, safety, and quality assurance protocols, *Renewable and Sustainable Energy Reviews*, **82**, 1097–1112.

Helman, Udi, 2017. Chapter 26 Economic and reliability benefits of solar plants, in *Renewable Energy Integration* (Lawrence E. Jones, Editor), Elsevier, 351-372.

Hernandez, R. R., Easter, S. B., Murphy-Mariscal, M. L., Maestre, F. T., Tavassoli, M., Allen, E. B., Barrows, C. W., Belnap, J., Ochoa-Hueso, R., Ravi, S., Allen, M. F., 2014. Environmental impacts of utility-scale solar energy, *Renewable and Sustainable Energy Reviews*, **29**, 766-779.

Honrubia-Escribano, Andrés, Ramirez, F. Javier, Gómez-Lázaro, Emilio, Garcia-Villaverde, Pedro M., Ruiz-Ortega, Maria J., Parra-Requena, Gloria, 2018. Influence of solar technology in the economic performance of PV power plants in Europe. A comprehensive analysis, *Renewable and Sustainable Energy Reviews*, **82**, 488-501.

Islam, Syed M., Nayar, Chem V., Abu-Siada, Ahmed, Hasan, Md Mubashwar, 2018. Chapter 25: Power Electronics for Renewable Energy Sources, in *Power Electronics Handbook, 4th Ed*, (Muhammed Rascid, Ed.), Elsevier Incorporated, 783-825.

Jamali, Siamak, Yari, Mortaza, Mahmoudi, S. M. S., 2018. Enhanced power generation through cooling a semi-transparent PV power plant with a solar chimney, *Energy Conversion and Management*, **175**, 227-235.

Jiang E., Yu, Fletcher, John, Burr, Patrick, Hall, Charles, Zheng, Bowen, Wang, Da-Wei, Ouyang, Zi, Lennon, Alison, 2018. Suitability of representative electrochemical energy storage technologies for ramp-rate control of photovoltaic power, *Journal of Power Sources*, **384**, 396-407.

Lai, Chun Sing, Jia, Youwei, Lai, Loi Lei, Xu, Zhao, McCulloch, Malcolm D., Wong, Kit Po, 2017. A comprehensive review on large-scale photovoltaic system with applications of electrical energy storage, *Renewable and Sustainable Energy Reviews*, **78**, 439-451.

Leonard, Matthew D., Michaelides, Efstathios E., Michaelides, Dimitrios N., 2018. Substitution of coal plants with renewable energy sources–Shift of the power demand and energy storage, *Energy Conversion and Management*, **164**, 27-35.

Letcher, Trevor M., 2018. Why solar energy? Chapter 1 in *A Comprehensive Guide to Solar Energy Systems with a Special Focus on Photovoltaic Systems* (Editors: Trevor Letcher and Vasilis M. Fthenakis), Academic Press, 3-15.

Li, Guiqiang, Shittu, Samson, Diallo, Thierno M. O., Yu, Min, Zhao, Xudong, Ji, Jie, 2018. A review of solar photovoltaic-thermoelectric hybrid system for electricity generation, *Energy*, **158**, 41-58.

Ludin, Norasikin Ahmad, Mustafa, Nur Ifthitah, Hanafiah, Marlia M., Ibrahim, Mohd Adib, Teridi, Mohd Asri Mat, Sepeai, Suhaila, Zaharim, Azami, Sopian, Kamaruzzaman, 2018. Prospects of life cycle assessment of renewable energy from solar photovoltaic technologies: A review, *Renewable and Sustainable Energy Reviews*, **96**, 11-28.

Majumdar, Debaleena, Pasqualetti, Martin J., 2018. Analysis of land availability for utility-scale power plants and assessment of solar photovoltaic development in the state of Arizona, USA, *Renewable Energy*, **134**, 1213-1231.

May, Geoffrey J., Davidson, Alistair, Monahov, Boris, 2018. Lead batteries for utility energy storage: A review, *Journal of Energy Storage*, **15**, 145-157.

Munshi, Amit H., Sasidharan, Nikhil, Pinkayan, Subin, Barth, Kurt L., Sampath, W. S., 2018. Thin-film CdTe photovoltaics – The technology for utility scale sustainable energy generation, *Solar Energy*, **173**, 511-516.

Nghitevelekwa, K., Bansal, R. C., 2018. A review of generation dispatch with large-scale photovoltaic systems, *Renewable and Sustainable Energy Reviews*, **81**, 615-624.

Nikolaidis, Pavlos, Poullikkas, Andreas, 2018. Cost metrics of electrical energy storage technologies in potential power system operations, *Sustainable Energy Technologies and Assessments*, **25**, 43-59.

Ouai, Atallah, Mokrani, Lakhdar, Machmoum, Mohamed, Houari, Azeddine, 2018. Control and energy management of a large scale grid-connected PV system for power quality improvement, *Solar Energy*, **171**, 893-906.

Prehoda, Emily W., Pearce, Joshua M., 2017. Potential lives saved by replacing coal with solar photovoltaic electricity production in the U.S., *Renewable and Sustainable Energy Reviews*, **80**, 710-715.

Ramalingam, Kakkan, Indulkar, Chandrasen, 2017. Chapter 3, Solar Energy and Photovoltaic Technology, in *Distributed Generation Systems, Design, Operation and Grid Integration* (G. B. Gharehpetian and S. M. M. Agah, editors), Butterworth-Heinemann, 69-147.

Raza, Muhammad Qamar, Mithulananthan, N., Summerfield, Alex, 2018. Solar output power forecast using an ensemble framework with neural predictors and Bayesian adaptive combination, *Solar Energy*, **166**, 226-241.

Rodríguez, Fermín, Fleetwood, Alice, Galarza, Ainhoa, Fontán, Luis, 2018. Predicting solar energy generation through artificial neural networks using weather forecasts for microgrid control, *Renewable Energy*, **126**, 855-864.

Sánchez-Pantoja, Núria, Vidal, Rosario, Pastor, M. Carmen, 2018. Aesthetic impact of solar energy systems, *Renewable and Sustainable Energy Reviews*, **98**, 227-238.

Semeraro, Teodoro, Pomes, Alessandro, Del Giudice, Cecilia, Negro, Danilo, Aretano, Roberta, 2018. Planning ground based utility scale solar energy as green infrastructure to enhance ecosystem services, *Energy Policy*, **117**, 218-227.

Shahsavari, Amir, Akbari, Morteza, 2018. Potential of solar energy in developing countries for reducing energy-related emissions, *Renewable and Sustainable Energy Reviews*, **90**, 275-291.

Sivaneasan, Balakrishnan, Kandasamy, Nandha Kumar, Lim, May Lin, Goh, Kwang Ping, 2018. A new demand response algorithm for solar PV intermittency management, *Applied Energy*, **218**, 36-45.

Späth, Leonhard, 2018. Large-scale photovoltaics? Yes please, but not like this! Insights on different perspectives underlying the trade-off between land use and renewable electricity development, *Energy Policy*, **122**, 429-437.

Srinivasan, Sunderasan, Kottam, Vamshi Krishna Reddy, 2018. Solar photovoltaic module production: Environmental footprint, management horizons and investor goodwill, *Renewable and Sustainable Energy Reviews*, **81**, 874-882.

Sueyoshi, Toshiyuki, Wang, Derek, 2017. Measuring scale efficiency and returns to scale on large commercial rooftop photovoltaic systems in California, *Energy Economics*, **65**, 389-398.

Trainer, Ted, 2018. Estimating the EROI of whole systems for 100% renewable electricity supply capable of dealing with intermittency, *Energy Policy*, **119**, 648-653.

Ulsrud, Kirsten, Rohracher, Harald, Winther, Tanja, Muchunku, Charles, Palit, Debajit, 2018. Pathways to electricity for all: What makes village-scale solar power successful? *Energy Research & Social Science*, **44**, 32-40.

Vasel, Ahmad, Iakovidis, Frantzis, 2017. The effect of wind direction on the performance of solar PV plants, *Energy Conversion and Management,* **153**, 455-461.

Verzijlbergh, Remco A., Heijnen, Petra W., de Roode, Stephan R., Los, Alexander, Jonker, Harm J. J., 2015. Improved model output statistics of numerical weather prediction based irradiance forecasts for solar power applications, *Solar Energy*, 118, 634-645.

Waqas, Adeel, Jie, Ji, 2018. Effectiveness of phase change material for cooling of photovoltaic panel for hot climate, *Journal of Solar Energy Engineering*, 1-10.

Webb, Jeremy, Wilson, Clevo, Steinberg, Theodore, Stein, Wes, 2017. Chapter 19: Solar Grid Parity and its Impact on the Grid, in *Innovation and Disruption at the Grid's Edge: How Distributed Energy Resources Are Disrupting the Utility Business Model*, (F. P. Sioshansi ed.), Elsevier Science and Technology, 389-408.

Wolfe, Philip R., 2018. Chapter II-5-B: Utility-Scale Solar Power, in *McEvoy's Handbook of Photovoltaics: Fundamentals and Applications, 3rd Edition* (edited by Soteris A. Kalogirou), Academic Press, 1073-1093.

Zhang, Chao, Wei, Yi-Li, Cao, Peng-Fei, Lin, Meng-Chang, 2018. Energy storage system: Current studies on batteries and power condition system, *Renewable and Sustainable Energy Reviews*, **82**, 3091-3106.

1.5.3 Hydrogen Energy Development References

Aasadnia, Majid, Mehrpooya, Mehdi, 2018. Large-scale liquid hydrogen production methods and approaches: A review, *Applied Energy*, **212**, 57-83.

Abad, A. Velazquez, Dodds, P. E., 2017. Production of hydrogen in *Encyclopedia of Sustainable Technologies, Vol. 3,* (Martin Abraham, Editor), Elsevier Inc., 293-304.

Abdalla, Abdalla M., Hossain, Shahzad, Nisfindy, Ozzan B., Azad, Atia T., Dawood, Mohamed, Azad, Abul K., 2018. Hydrogen production, storage, transportation and key challenges with applications: A review, *Energy Conversion and Management*, **165**,602-627.

Acar, Canan, Dincer, Ibrahim, 2018. 4.24 Hydrogen energy conversion systems in *Comprehensive Energy Systems* (Editor in Chief, Ibrahim Dincer), Elsevier, Inc., 947-983.

Al-Zareer, Maan, Dincer, Ibrahim, Rosen, Marc A., 2018. Analysis and assessment of a hydrogen production plant consisting of coal gasification, thermochemical water decomposition and hydrogen compression systems, *Energy Conversion and Management*, **157**. 600-618.

Arsalis, Alexandros, Alexandrou, Andreas N., Georghiou, George E., 2018. Thermoeconomic modeling of a completely autonomous, zero-emission photovoltaic system with hydrogen storage for residential application, *Renewable Energy*, **126**, 354-369.

Assaf, Jihane, Shabani, Bahman, 2018a. Experimental study of a novel hybrid solar-thermal/PV-hydrogen system: Towards 100% renewable heat and power supply to standalone applications, *Energy*, **157**, 862-876.

Assaf, Jihane, Shabani, Bahman, 2018b. Multi-objective sizing optimisation of a solar-thermal system integrated with a solar-hydrogen combined heat and power system, using genetic algorithm, *Energy Conversion and Management*, **164**, 518-532.

Attia, Yasser, Samer, Mohamed, 2017. Metal clusters: New era of hydrogen production, *Renewable and Sustainable Energy Reviews*, **79**, 878-892.

Badea, Gheorghe, Naghiu, George Sebastian, Giurca, Ioan, Aşchilean, Ioan, Megyesi, Emanuel, 2017. Hydrogen production using solar energy - technical analysis, *Energy Procedia,* **112**, 418-425.

Baetcke, Lars, Kaltschmitt, Martin, 2018. Hydrogen Storage for Mobile Application: Technologies and Their Assessment, Chapter 5 in HYDROGEN SUPPLY CHAIN, Elsevier, 167-206.

Baniasadi, Ehsan, 2017. Concurrent hydrogen and water production from brine water based on solar spectrum splitting: Process design and thermoeconomic analysis, *Renewable Energy*, **102**, 50-64.

Baykara, Sema Z., 2018. Hydrogen: A brief overview on its sources, production and environmental impact, *International Journal of Hydrogen Energy*, **43**, 10605-10614.

Bicer, Yusuf, Dincer, Ibrahim, 2017. Experimental investigation of a PV-coupled photoelectrochemical hydrogen production system, *International Journal of Hydrogen Energy,* **42**, 2512-2521.

Bicer, Yusuf, Dincer, Ibrahim, 2018. Clean fuel options with hydrogen for sea transportation: A life cycle approach, *International Journal of Hydrogen Energy*, **43**, 1179-1193.

Brennan, Sile, Molkov, Vladimir, 2018. Pressure peaking phenomenon for indoor hydrogen releases, *International Journal of Hydrogen Energy*, **43**, 18530-18541.

Brown, Daryl, Tegrotenhuis, Ward, Wegeng, Robert, Mankins, John, 2014. Solar powered steam-methane reformer economics, *Energy Procedia*, **49**, 1916-1921.

Burhan, Muhammad, Shahzad, Muhammad Wakil, Ng, Kim Choon, 2018. Hydrogen at the rooftop: Compact CPV-hydrogen system to convert sunlight to hydrogen, *Applied Thermal Engineering*, **132**, 154-164.

Cao, Fei, Wei, Qingyu, Liu, Huan, Lu, Na, Zhao, Liang, Guo, Liejin, 2018. Development of the direct solar photocatalytic water splitting system for hydrogen production in Northwest China:

Design and evaluation of photoreactor, *Renewable Energy*, **121**, 153-163.

Çelik, Dicle, Yildiz, Meltem, 2017. Investigation of hydrogen production methods in accordance with green chemistry principles, *International Journal of Hydrogen Energy,* **42**, 23395-23401.

Chakik, Fatima ezzahra, Kaddami, Mohammed, Mikou, Mohammed, 2017. Effect of operating parameters on hydrogen production by electrolysis of water, *International Journal of Hydrogen Energy,* **42**, 25550-25557.

Chi, Jun, Yu, Hongmei, 2018. Water electrolysis based on renewable energy for hydrogen production, *Chinese Journal of Catalysis*, **39**, 390-394.

Colón, Gerardo, 2016. Towards the hydrogen production by photocatalysis, *Applied Catalysis Part A: General*, **518**, 48–59.

Corumlu, Vahit, Ozsoy, Ahmet, Ozturk, Murat, 2018. Thermodynamic studies of a novel heat pipe evacuated tube solar collectors based integrated process for hydrogen production, *International Journal of Hydrogen Energy,* **43**, 1060-1070.

Coutanceau, Christophe, Baranton, Stève, Audichon, Thomas, 2017. Hydrogen electrochemical production, *Hydrogen and Fuel Cells Primers*, (series editor, Bruno Pollet), Elsevier Publishers. 1-6.

Coutanceau, Christophe, 2018. Hydrogen production from thermal reforming, Chapter 1 in *Hydrogen Electrochemical Production* (Bruno Pollet, Editor), Academic Press, 7-15.

Dagdougui, Hanane, 2018. Ch 2 Hydrogen production and current technologies, in *Hydrogen Infrastructure for Energy Applications: Production, Storage, Distribution and Safety,* 7-21.

Demir, Murat Emre, Dincer, Ibrahim, 2018. Cost assessment and evaluation of various hydrogen delivery scenarios, *International Journal of Hydrogen Energy*, **43**, 10420-10430.

Dincer, Ibrahim, Acar, Canan, 2017. Innovation in hydrogen production, *International Journal of Hydrogen Energy*, **42**, 14843-14864.

Dincer, Ibrahim, Acar, Canan, 2018. Smart energy solutions with hydrogen options, *International Journal of Hydrogen Energy*, **43**, 8579-8599.

Dincer, Ibrahim, Bicer, Yusuf, 2018. 4.22 Electrochemical energy conversion, in *Comprehensive Energy Systems* (Editor in Chief Ibrahim Dincer), Elsevier, 856-893.

Edelia, Erfika M., Winkler, Roy, Sengupta, Debalina, El-Halwagi, Mahmoud M., Mannan, M. Sam, 2018. A computational fluid dynamics evaluation of unconfined hydrogen explosions in high pressure applications, *International Journal of Hydrogen Energy*, **43**, 16411-16420.

Fereidooni, Mojaba, Mostafaeipour, Ali, Kalantar, Vali, Goudarzi, Hossein, 2018. A comprehensive evaluation of hydrogen production from photovoltaic power station, *Renewable and Sustainable Energy Reviews*, **82**, 415-423.

Gökçek, Murat, Kale, Cihangir, 2018. Optimal design of a Hydrogen Refueling Station (HRFS) powered by Hybrid Power System, *Energy Conversion and Management*, **161**, 215-224.

He, Yumin, Wang, Dunwei, 2018. Toward practical solar hydrogen production, *Chem*, **4**, 405-408.

Hwangbo, Soonho, Lee, Seungchul, Yoo, Changkyoo, 2017. Optimal network design of hydrogen production by integrated utility and biogas supply networks, *Applied Energy*, **208**, 195-209.

Jarunglumlert, Teeraya, Prommuak, Chattip, Putmai, Namtip, Pavasant, Prasert, 2018. Scaling-up bio-hydrogen production from food waste: Feasibilities and challenges, *International Journal of Hydrogen Energy,* **43**, 634-648.

Karapekmez, Aras, Dincer, Ibrahim, 2018. Modelling of hydrogen production from hydrogen sulfide in geothermal power plants, *International Journal of Hydrogen Energy*, **43**, 10569-10579.

Khalid, Farrukh, Dincer, Ibrahim, Rosen, Marc A., 2018. Model development and analysis of a novel high-temperature electrolyser for gas phase electrolysis of hydrogen chloride for hydrogen production, *International Journal of Hydrogen Energy*, **43**, 9112-9118.

Khetkorn, Wanthanee, Rastogi, Rajesh P., Incharoensakdi, Aran, Lindblad, Peter, Madamwar, Datta, Pandey, Ashok, Larroche, Christian, 2017. Microalgal hydrogen production – A review, *Bioresource Technology*, **243**, 1194-1206.

Leonardi, Salvatore Gianluca, Bonavita, Anna, Donato, Nicola, Neri, Giovanni, 2018. Development of a hydrogen dual sensor for fuel cell applications, *International Journal of Hydrogen Energy*, **43**, 11896-11902.

Likkasit, Chonnawee, Maroufmashat, Azadeh, Elkamel, Ali, Ku, Hong-ming, Fowler, Michael, 2018. Solar-aided hydrogen production methods for the integration of renewable energies into oil & gas industries, *Energy Conversion and Management*, **168**, 395-406.

Liu, Taixiu, Liu, Qibin, Lei, Jing, Sui, Jun, Jin, Hongguang, 2018. Solar-clean fuel distributed energy system with solar thermochemistry and chemical recuperation, *Applied Energy*, **225**, 380-391.

Luo, Ming, Yi, Yang, Wang, Shuzhong, Wang, Zhuliang, Du, Min, Pan, Jianfeng, Wang, Qian, 2018. Review of hydrogen production using chemical-looping technology, *Renewable and Sustainable Energy Reviews*, **81**, 3186-3214.

Minuto, F. D., Balderas-Xicohténcatl, R., Policiccho, A., Hirscher, M., Agostino, R. G., 2018. Assessment methodology of promising porous materials for near ambient temperature hydrogen storage applications, *International Journal of Hydrogen Energy*, **43**, 14550-14556.

Mohsin, M., Rasheed, A.K., Saidur, R., 2018. Economic viability and production capacity of wind generated renewable hydrogen, *International Journal of Hydrogen Energy,* **43**, 2621-2630.

Monnerie, N., von Storch, H., Houaijia, A., Roeb, M., Sattler, C., 2017. Hydrogen production by coupling pressurized high temperature electrolyser with solar tower technology, *International Journal of Hydrogen Energy,* **42**, 13498-13509.

Nikolaidis, Palos, Poullikkas, Andreas, 2017. A comparative overview of hydrogen production processes, *Renewable and Sustainable Energy Reviews*, **67**, 597-611.

Ohaeri, Enyinnaya, Eduok, Ubong, Szpunar, Jerzy, 2018. Hydrogen related degradation in pipeline steel: A review, *International Journal of Hydrogen Energy*, **43**, 14584-14617.

Okemoto, Atsushi, Tanaka, Kazuhito, Kudo, Yumika, Gohda, Shin, Koshiba, Yasuko, Ishida, Kenji, Horie, Takafumi, Taniya, Keita, Ichihashia, Yuichi, Nishiyama, Satoru, 2018. Hydrogen production for photocatalytic decomposition of water with urea as a reducing agent, *Catalysis Today*, **307**, 231-236.

Omar, Muhammed Arslan, Altinişik, Kemal, 2016. Simulation of hydrogen production system with hybrid solar collector, *International Journal of Hydrogen Energy,* **41**, 12836-12841.

Parkinson, Brett, Tabatabaei, Mojgan, Upham, David C., Ballinger, Benjamin, Greig, Chris, Smart, Simon, McFarland, Eric,

2018. Hydrogen production using methane: Technoeconomics of decarbonizing fuels and chemicals, *International Journal of Hydrogen Energy,* **43**, 2540-2555.

Sakr, I. M., Abdelsalam, Ali M., El-Askary, W. A., 2017. Effect of electrodes separator-type on hydrogen production using solar energy, *Energy*, **140**, 625-632.

Salkuyeh, Yaser Khojasteh, Saville, Bradley A., MacLean, Heather L., 2018. Techno-economic analysis and life cycle assessment of hydrogen production from different biomass gasification processes, *International Journal of Hydrogen Energy,* **43**, 9514-9528.

Schweitzer, Daniel, Albrecht, Friedemann Georg, Schmid, Max, Beirow, Marcel, Spörl, Reinhold, Dietrich, Ralph-Uwe, Seitz, Antje, 2018. Process simulation and techno-economic assessment of SER steam gasification for hydrogen production, *International Journal of Hydrogen Energy,* **43**, 569-579.

Sengodan, Sivaprakash, Lan, Rong, Humphreys, John, Du, Dongwei, Xu, Wie, Wang, Huanting, Tao, Shanwen, 2018. Advances in reforming and partial oxidation of hydrocarbons for hydrogen production and fuel cell applications, *Renewable and Sustainable Energy Reviews*, **82**, 761-780.

Seyitoglu, S. S., Dincer, I., Kilicarslan, A., 2017. Energy and exergy analyses of hydrogen production by coal gasification, *International Journal of Hydrogen Energy,* **42**, 2592-2600.

Sharma, Preeti, Melkania, Uma, 2018. Effect of bioaugmentation on hydrogen production from organic fraction of municipal solid waste, *International Journal of Hydrogen Energy,* **43**, 7270-7298.

Sheikhbahaei, V., Baniasadi, E., Naterer, G. F., 2018. Experimental investigation of solar assisted hydrogen production from water and aluminum, *International Journal of Hydrogen Energy,* **43**, 9181-9191.

Show, Kuan-Yeow, Yan, Yuegen, Ling, Ming, Ye, Guoxiang, Li, Ting, Lee, Duu-Jong, 2018. Hydrogen production from algal biomass – Advances, challenges and prospects, *Bioresource Technology,* **257**, 290-300.

Siddiqui, Osamah, Dincer, Ibrahim, 2018. Examination of a new solar-based integrated system for desalination, electricity generation and hydrogen production, *Solar Energy*, **163**, 224-234.

Sołowski, Gaweł, Shalaby, Marwa. S., Abdallah, Heba, Shaban, Ahmed. M., Cenian, Adam, 2018. Production of hydrogen from biomass and its separation using membrane technology, *Renewable and Sustainable Energy Reviews*, **82**, 3152-3167.

Sorgulu, F., Dincer, I., 2018a. A renewable source based hydrogen energy system for residential applications, *International Journal of Hydrogen Energy*, **43**, 5842-5851.

Sorgulu, F., Dincer, I., 2018b, Cost evaluation of two potential nuclear power plants for hydrogen production, *International Journal of Hydrogen Energy,* **43**, 10522-10529.

Sreedhar, I., Kamani, Krutarth M., Kamani, Bansi M., Reddy, Benjaram M., Venugopal, A., 2018. A Bird's Eye view of the process and engineering aspects of hydrogen storage, *Renewable and Sustainable Energy Reviews*, **91**, 838-830.

Stamatakis, Emmanuel, Yiotis, Andreas, Ginanissi, Stella, Tolias, Ilias, Stubos, Athanassios, 2018. Modeling and simulation supporting the application of fuel cell & hydrogen technologies, *Journal of Computational Science*, **27**, 10-20.

Stern, Alvin G., 2018. A new sustainable hydrogen clean energy paradigm, *International Journal of Hydrogen Energy*, **43**, 4244-4255.

Suleman, F., Dincer, I., Agelin-Chaab, M., 2015. Environmental impact assessment and comparison of some hydrogen production options, *International Journal of Hydrogen Energy*, **40**, 6976-6987.

Sun, Bing, Zhao, Xiaotong, Xin, Yanbin, Zhu, Xiaomei, 2017. Large capacity hydrogen production by microwave discharge plasma in liquid fuels ethanol, *International Journal of Hydrogen Energy,* **42**, 24947-24054.

Tebibel, H., Khellaf, A., Menia, S., Nouicer I., 2017. Design, modelling and optimal power and hydrogen management strategy of an off grid PV system for hydrogen production using methanol electrolysis, *International Journal of Hydrogen Energy,* **42**, 14950-14967.

Tolias, I. C., Giannissi, S. G., Venetsanos, A. G., Keenan, J., Shentsov, V., Makarov, D., Coldrick, S., Kotchourko, A., Ren, K., Jedicke, O., Melideo, D., Baraldi, D., Slater, S., Duclos, A., Verbecke, F., Molkov, V., 2018. Best practice guidelines in numerical simulations and CFD benchmarking for hydrogen safety applications, *International Journal of Hydrogen Energy*, **44**, 9050-9062.

Trompoukis, Christos, Abass, Aimi, Schüttauf, Jan-Willem, Bosserez, Tom, Rongé, Jan, Lauwaert, Johan, Martens, Johan A., Baets, Roel, 2018. Porous multi-junction thin-film silicon solar cells for scalable water splitting, *Solar Energy Materials and Solar Cells*, **182**, 196-203.

Wang, Dai, Muratori, Matteo, Eichman, Joshua, Wei, Max, Saxena, Samveg, Zhang, Cong, 2018. Quantifying the flexibility of hydrogen production systems to support large-scale renewable energy integration, *Journal of Power Sources*, **399**, 383-391.

Wang, Jianlong, Yin, Yanan, 2018. Fermentative hydrogen production using various biomass-based materials as feedstock, *Renewable and Sustainable Energy Reviews*, **92**, 284-306.

Wang, Xian, Meng, Qinglei, Gao, Liqin, Jin, Zhao, Ge, Junjie, Liu, Changpeng, Xing, Wei, 2018. Recent progress in hydrogen production from formic acid decomposition, *International Journal of Hydrogen Energy,* **43**, 7055-7071.

Wulf, Christina, Reuß, Markus, Grube, Thomas, Zapp, Petra, Robinius, Martin, Hake, Jürgen-Hake, Stolten, Detlef, 2018. Life Cycle Assessment of hydrogen transport and distribution options, *Journal of Cleaner Production*, **199**, 431-443.

Xu, Jiuping, Li, Qiulin, Xie, Heping, Ni, Ting, Ouyang, Chi, 2018. Tech-integrated paradigm based approaches towards carbon-free hydrogen production, *Renewable and Sustainable Energy Reviews*, **82**, 4279-4295.

Yan, Fuwu, Xu, Lei, Wang, Yu, 2018. Application of hydrogen enriched natural gas in spark ignition IC engines: from fundamental fuel properties to engine performances and emissions, *Renewable and Sustainable Energy Reviews*, **82**, 1457-1488.

Yılmaz, Fatih, Balta, M. Tolga, 2017. Energy and exergy analyses of hydrogen production step in boron based thermochemical cycle for hydrogen production, *International Journal of Hydrogen Energy,* **42**, 2485-2491.

Yüksel, Yunus Emre, 2018. Thermodynamic assessment of modified Organic Rankine Cycle integrated with parabolic trough collector for hydrogen production, *International Journal of Hydrogen Energy,* **43**, 5832-5841.

Yuksel, Yunus Emre, Ozturk, Murat, Dincer, Ibrahim, 2018. Analysis and performance assessment of a combined geothermal power-based hydrogen production and liquefaction system, *International Journal of Hydrogen Energy*, **43**, 10268-10280.

1.5.4 Building Applications of Solar References

Ahmed, Krarti, Ortiz, Luis E., González, J. E., 2017. On the spatio-temporal end-user energy demands of a dense urban environment, *Journal of Solar Energy Engineering*, **139**, 11 pages.

Ali, Ahmed Hamza H., 2017. Performance-cost and global warming assessments of two residential scale solar cooling systems versus a conventional one in hot arid areas, *Sustainable Energy Technologies and Assessments*, **20**, 1-8.

Allegrini, Jonas, Orehounig, Kristina, Mavromatidis, Georgios, Ruesch, Florian, Dorer, Viktor, Evins, Ralph, 2015. A review of modelling approaches and tools for the simulation of district-scale energy systems, *Renewable and Sustainable Energy Reviews*, **52**, 1391-1404.

Bavaresco, Mateus Vinícius, Ghisi, Enedir, 2018. Influence of user interaction with internal blinds on the energy efficiency of office buildings, *Energy & Buildings*, **166**, 538-549.

Bertrand, Cédric, Housmans, Caroline, Leloux, Jonathon, Journée, Michel, 2018. Solar irradiation from the energy production of residential PV systems, *Renewable Energy*, **125**, 306-318.

Block, P., Schlueter, A., Veenendaal, D., Bakker, J., Begle, M., Hischier, I., Hofer, J., Jayathissa, P., Maxwell, I., Echenagucia, T. Méndez, Nagy, Z., Pigram, D., Svetozarevic, B., Torsing, R.,

Verbeek, J., Willmann, A., Lydon, G.P., 2017. NEST HiLo: Investigating lightweight construction and adaptive energy, *Journal of Building Engineering*, **12**, 332-341.

Borgstein, E. H., Lamberts, R., Hensen, J. L. M., 2018. Mapping failures in energy and environmental performance of buildings, *Energy and Buildings*, **158**, 476-485. Blame it on the owner — Ownership and energy performance of multi-dwelling buildings, *Energy Economics*, **72**, 108-119.

Borovik, Michael R., Albers, John D., 2018. Participation in the Illinois solar renewable energy market, *The Electricity Journal*, **31**, 33-39.

Broberg, Thomas, Egüez, Alejandro, 2018. Blame it on the owner — Ownership and energy performance of multi-dwelling buildings, *Energy Economics*, **72**, 108-119.

Buffat, René, Grassi, Stefano, Raubal, Martin, 2018. A scalable method for estimating rooftop solar irradiation potential over large regions, *Applied Energy*, **216**, 389-401.

Chastas, Panagiotis, Theodosiou, Theodoros, Bikas, Dimitrios, 2016. Embodied energy in residential buildings-towards the nearly zero energy building: A literature review, *Building and Environment*, **105**, 267-282.

Chel, Arvind, Kaushik, Geetanjali, 2017. Renewable energy technologies for sustainable development of energy efficient building, *Alexandria Engineering Journal*, **57**, 655-669.

Chemisana, D., Zacharopoulos, Aggelos, 2015. Building-integration of high-concentration photovoltaic systems, *High Concentrator Photovoltaics* P. Pérez-Higueras and E.F. Fernández, eds, Green Energy Technology, 353-376.

Chen, Yixing, Hong, Tianzhen, 2018. Impacts of building geometry modeling methods on the simulation results of urban building energy models, *Applied Energy*, **215**, 717-735.

Chwieduk, Dorota,2017. Towards modern options of energy conservation in buildings, *Renewable Energy*, **101**, 1194-1202.

Chwieduk, Dorota, 2018. Impact of solar energy on the energy balance of attic rooms in high latitude countries, *Applied Thermal Engineering*, **136**, 548-559.

Comello, Stephen, Reichelstein, Stefan, 2017. Cost competitiveness of residential solar PV: The impact of net metering restrictions, *Renewable and Sustainable Energy Reviews*, **75**, 46-57.

Del Pero, Claudio, Aste, Niccolò, Paksoy, Halime, Haghighat, Fariborz, Grillo, Samuele, Leonforte, Fabrizio, 2018. Energy storage key performance indicators for building application, *Sustainable Cities and Society*, **40**, 54-65.

Delzendeh, Elham, Wu, Song, Lee, Angela, Zhou, Ying, 2017. The impact of occupants' behaviors on building energy analysis: A research review, *Renewable and Sustainable Energy Reviews*, **80**, 1061-1071.

Dey, Sumon, Lakshmanan, Madan Kumar, Pesala, Bala, 2018. Optimal solar tree design for increased flexibility in seasonal energy extraction, *Renewable Energy*, **125**, 1038-1048.

Dincer, Ibrahim, Bicer, Yusuf, 2018. 4.19 PV-Based Energy Conversion Systems, in *Comprehensive Energy Systems*, Elsevier Inc., 762-793.

Elzeyadi, Ihab, 2017. The impacts of dynamic façade shading typologies on building energy performance and occupant's multi-comfort, *Architectural Science Review*, **60**, 316-324.

Evans, Meredydd, Roshchanka, Volha, Graham, Peter, 2017. An international survey of building energy codes and their implementation, *Journal of Cleaner Production*, **158**, 382-389.

Fan, Yuling, Xia, Xiaohua, 2018. Energy-efficiency building retrofit planning for green building compliance, *Building and Environment*, **135**, 312-321.

Fanney, A. Hunter, Healy, William, Payne, Vance, Kneifel, Joshua, Ng, Lisa, Doughety, Brian, Ullah, Tania, Omar, Farhad, 2017. Small changes yield large results at NIST's Net-Zero Energy Residential Test Facility, *Journal of Solar Energy Engineering*, **139**, 11 pages.

Ge, T. S., Wang, R. Z., Xu, Z. Y., Pan, Q. W., Du, S., Chen, X. M., Ma, T., Wu, X. N., Sun, X. L., Chen, J. F., 2018. Solar heating and cooling: Present and future development, *Renewable Energy*, **126**, 1126-1140.

Gönülol, Okay, Tokuç, Ayça, 2018. Net zero residential building architecture in the future, Chap. 1.2 in Exergetic, Energetic and Environmental Dimensions (Editors: Ibrahim Dincer, Can Ozgur Colpan, Onder Kizilkan), Academic Press, 39-53.

Gounni, Ayoub, El Alami, Mustapha, Mabouk, Mohamed Tahar, Kheiri, Abdelhamid, 2018. Experimental study of thermal performance of a reduced scale cavity equipped with phase change material: Study of the optimal phase change material location, *Journal of Solar Energy Engineering*, **140**, 1-6.

Granqvist, Claes G., Niklasson, Gunnar A., 2018. Solar energy materials for thermal applications: A primer, *Solar Energy Materials and Solar Cells*, **180**, 213-226.

Groppi, Daniele, de Santoli, Livio, Cumo, Fabrizio, Garcia, Davide Astiaso, 2018. A GIS-based model to assess buildings energy consumption and usable solar energy potential in urban areas, *Sustainable Cities and Society*, **40**, 546-558.

Hamilton, Ian, Summerfield, Alex, Oreszczyn, Tadj, Ruyssevelt, Paul, 2017. Using epidemiological methods in energy and buildings research to achieve carbon emission targets, *Energy and Buildings*, **154**, 188-197.

Hsieh, Chun-Ming, Li, Juan-Juan, Zhang, Liman, Schwegler, Ben, 2018. Effects of tree shading and transpiration on building cooling energy use, *Energy and Buildings*, **150**, 283-397.

Indora, Sunil, Kandpal, Tara C., 2018. Institutional cooking with solar energy: A review, *Renewable and Sustainable Energy Reviews*, **84**, 131-154.

Jakica, Nebojsa, 2018. State-of-the-art review of solar design tools and methods for assessing daylighting and solar potential for building-integrated photovoltaics, *Renewable and Sustainable Energy Reviews*, **81**, 1296–1328.

Jani, D. B., Mishra, Manish, Sahoo, Pradeep Kumar, 2018. A critical review on application of solar energy as renewable regeneration heat source in solid desiccant – vapor compression hybrid cooling system, *Journal of Building Engineering*, **18**, 107-124.

Jones, Byron W., Powell, Robert, 2015. Evaluation of distributed building thermal energy storage in conjunction with wind and solar electric power generation, *Renewable Energy*, **74**, 699-707.

Kang, Jing, Wang, Shengwei, Gang, Wenjie, 2017. Performance of distributed energy systems in buildings in cooling dominated regions and the impacts of energy policies, *Applied Thermal Engineering*, **127**, 281-291.

Kheiri, Farshad, 2018. A review on optimization methods applied in energy-efficient building geometry and envelope design, *Renewable and Sustainable Energy Reviews*, **92**, 897-920.

Kim, Yang-Seon, Heidarinejad, Mohammad, Dahlhausen, Matthew, Srebric, Jelena, 2017. Building energy model calibration with schedules derived from electricity use data, *Applied Energy*, **190**, 997-1007.

Krarti, Moncef, 2018. Chapter 2: Advanced building energy efficiency systems, in Optimal Design and Retrofit of Energy Efficient Buildings, Communities, and Urban Centers (Moncef Krarti, Editor), Elsevier, 45-114.

Krese, Gorazd, Koželj, Rok, Butala, Vincenc, Stritih, Uroš, 2018. Thermochemical seasonal solar energy storage for heating and cooling of buildings, *Energy and Buildings*, **164**, 239-253.

Leonard, Matthew D., Michaelides, Efstathios E., 2018. Grid-independent residential buildings with renewable energy sources, *Energy*, **148**, 448-460.

Lešić, Vinko, Martinčević, Anita, Vašak, Mario, 2017. Modular energy cost optimization for buildings with integrated microgrid, *Applied Energy*, **197**, 14-28.

Li, M., Shi, J., Guo, J., Cao, J., Niu, J., Xiong, M., 2015. Climate impacts on extreme energy consumption of different types of buildings, *PLOS ONE*, **10**, 12 pages.

Li, W., Zhou, Y., Cetin, K., Eom, J., Wang, Y., Chen, G., Zhang, X., 2017. Modeling urban building energy use: A review of modeling approaches and procedures, *Energy*, **141**, 2445-2457.

Liang, Jing, Qiu, Yueming, James, Timothy, Ruddell, Benjamin L., Dalrymple, Michael, Earl, Stevan, Castelazo, Alex, 2018. Do energy retrofits work? Evidence from commercial and residential buildings in Phoenix, *Journal of Environmental Economics and Management*, **92**, 726-743.

Lizana, Jesús, Chacartegui, Ricardo, Barrios-Padura, Angela, Valverde, José Manuel, 2017. Advances in thermal energy storage materials and their applications towards zero energy buildings: A critical review, *Applied Energy*, **203**, 219-239.

Lucchi, Elena, 2018. Applications of the infrared thermography in the energy audit of buildings: A review, *Renewable and Sustainable Energy Reviews*, **82**, 3077-3090.

Luddeni, Giacomo, Krarti, Moncef, Pernigotto, Giovanni, Gasparella, Andrea, 2018. An analysis methodology for large-scale deep energy retrofits of existing building stocks: Case study of the Italian office building, *Sustainable Cities and Society*, **41**, 296-311.

Machete, Rita, Falcão, Ana Paula, Gomes, M. Glória, Rodrigues, A. Moret, 2018. The use of 3D GIS to analyze the influence of urban context on buildings' solar energy potential, *Energy & Buildings*, **177**, 290-302.

Martinopoulos, Georgis, 2018. Life Cycle Assessment of solar energy conversion systems in energetic retrofitted buildings, *Journal of Building Engineering*, **20**, 256-263.

May, Geoffrey J., Davidson, Alistair, Monahov, Boris, 2018. Lead batteries for utility energy storage: A review, *Journal of Energy Storage*, **15**, 145-157.

Mohajeri, Nahid, Assouline, Dan, Guiboud, Berenice, Bill, Andreas, Gudmundsson, Agust, Scartezzini, Jean-Louis, 2018. A city-scale roof shape classification using machine learning for solar energy applications, *Renewable Energy*, **121**, 81-93.

Moosavi, Leila, Zandi, Majid, Bidi, Mokhtar, 2018. Experimental study on the cooling performance of solar-assisted natural ventilation in a large building in a warm and humid climate, *Journal of Building Engineering*, **19**, 228-241.

Nizam, Raja Shahmir, Zhang, Cheng, Tian, Lu, 2018. A BIM based tool for assessing embodied energy for buildings, *Energy & Buildings*, **170**, 1-14.

Ortiz, Luis E., Gonzalez, Jorge E., Gutierrez, Estatio, Arend, Mark, 2017. Forecasting building energy demands with a coupled weather-building energy model in a dense urban environment, *Journal of Solar Energy Engineering*, **139**, 8 pages.

Peronato, G., Rastogi, P., Rey, E., Andersen, M., 2018. A toolkit for multi-scale mapping of the solar energy-generation potential of buildings in urban environments under uncertainty, *Solar Energy*, **173**, 861-874.

Powell, Daniel, Hischier, Illias, Jayathissa, Prageeth, Svetozarevic, Bratislav, Schlüter, Arno, 2018. A reflective adaptive solar façade for multi-building energy and comfort management, *Energy & Buildings*, **177**, 303-315.

Rathore, Pushpendra Kumar Singh, Chauhan, Durg Singh, Singh, Rudra Pratap, 2018. Decentralized solar rooftop photovoltaic in India: On the path of sustainable energy security, *Renewable Energy*, **131**, 297-307.

Reed, A. L., Novelli, A. P., Doran, K. L., Ge, S., Lu, N., McCartney, J. S., 2018. Solar district heating with underground thermal energy storage: Pathways to commercial viability in North America, *Renewable Energy*, **126**, 1-13.

Robillart, M., Schalbart, P., Chaplais, F., Peuportier, B., 2018. Model reduction and model predictive control of energy-efficient buildings for electrical heating load shifting, *Journal of Process Control*, **74**, 23-34.

Saffari, Mohammad, de Gracia, Alvaro, Fernández, Cèsar, Cabeza, Luisa F., 2017. Simulation-based optimization of PCM melting temperature to improve the energy performance in buildings, *Applied Energy*, **202**, 420–434.

Sanhudo, Luís, Ramos, Nuno M.M., Martins, João Poças, Almeida, Ricardo M.S.F., Barreira, Eva, Simões, M. Lurdes, Cardoso, Vítor, 2018. Building information modeling for energy retrofitting – A review, *Renewable and Sustainable Energy Reviews*, **89**, 249-260.

Saxena, Rajat, Biplab, Kumar, Rakshit, Dibakar, 2018. Quantitative assessment of phase change material load abatement in composite climatic condition, *Journal of Solar Energy Engineering*, **140**, 15 pages.

Shabunko, V., Lim, C. M., Mathew, S., 2018. EnergyPlus models for the benchmarking of residential buildings in Brunei Darussalam, *Energy & Buildings*, **169**, 507-516.

Shirazi, Ali, Taylor, Robert A., Morrison, Graham L., White, Stephen D., 2018. Solar-powered absorption chillers: A comprehensive and critical review, *Energy Conversion and Management*, **171**, 59-81.

Stritih, U., Tyagi, V. V., Stropnik, R., Paksoy, H., Haghighat, F., Joybari, M. Mastani, 2018. Integration of passive PCM technologies for Net-Zero Energy Buildings, *Sustainable Cities and Society*, **41**, 286-295.

Üçtug, Fehmi Görkem, Azapagic, Adisa, 2018. Environmental impacts of small-scale hybrid energy systems: Coupling solar photovoltaics and lithium-ion batteries, *Science of the Total Environment*, **643**, 1579-1589.

Vanaga, Ruta, Blumberga, Andra, Freimanis, Ritvars, Mols, Toms, Blumberga, Dagnija, 2018. Solar façade module for nearly zero energy building, *Energy*, **157**, 1025-1034.

Verclas, Kirsten, 2018. Electric vehicle and solar energy pilot: Opportunity to address suburban energy challenges, *The Electricity Journal*, **31**, 48-56.

Vieira, Filomeno M., Moura, Pedro S., de Almeida, Aníbal T. 2017. Energy storage system for self-consumption of photovoltaic energy in residential zero energy buildings, *Renewable Energy*, **103**, 308-320.

Wu, Yunpeng, Krishnan, Padmaja, Zhang, Min-Hong, Yu, Liya E, 2018. Using photocatalytic coating to maintain solar reflectance and lower cooling energy consumption of buildings, *Energy & Buildings*, **164**, 176-186.

Yoon, Young Ran, Moon, Hyeun Jun, 2018. Energy consumption model with energy use factors of tenants in commercial buildings using Gaussian process regression, *Energy & Buildings*, **168**, 215-224.

Zhang, Xingxing, Lovati, Marco, Vigna, Ilaria, Widén, Joakim, Han, Mengjie, Gal, Csilla, Feng, Tao, 2018. A review of urban energy systems at building cluster level incorporating renewable-energy-source (RES) envelope solutions, *Applied Energy*, **230**, 1034-1056.

Zhang, Yan, Bai, Xuemei, Mills, Franklin P., Pezzey, John C. V., 2018. Rethinking the role of occupant behavior in building energy performance: A review, *Energy & Buildings*, **172**, 279-294.

Zhao, Dongliang, Lu, Xing, Fan, Tianzhu, Wu, Yuen Shing, Lou, Lun, Wang, Qiuwang, Fan, Jintu, Yang, Ronggui, 2018. Personal thermal management using portable thermoelectrics for potential building energy saving, *Applied Energy*, **218**, 282-291.

Zhao, Qin, Chen, Zeyu, Zhu, Yiyun, Li, Zhe, Hei, Xinhong, 2016. Toward automatic review of building energy efficiency based on Building Information Modeling, *12th International Conference on Computational Intelligence and Security*, The IEEE Computer Society.

Zhou, Yan, 2018. Evaluation of renewable energy utilization efficiency in buildings with exergy analysis, *Applied Thermal Engineering*, **137**, 430-439.

1.6 SOME SOLAR RELATED TECHNOLOGIES AND THEIR APPLICATIONS[1]

1.6.1 Solar Power Conversion, Utility Scale

1.6.1.1 Introduction The use of solar energy for generating power is a concept that has been around for a long time. Some early examples of this, to be dealt with in the sections that follow, include dishes and troughs that were developed in a flurry of creative activity between the late 1800s and the early 1900s. There was somewhat of a hiatus from much development in first half of the 20th century until work again resumed after the Second World War. This was driven by interest in exploring space as well as other thrusts. An excellent summary of the early history of solar applications has been given by Butti and Perlin [1980].

Momentum to examine renewable energy generally picked up considerably with the oil embargo of the 1970s. At that time, President Jimmy Carter outlined the challenge the US particularly, and the world more generally, faced in terms of dealing with energy issues. He warned that the situation was "the moral equivalent of war" in a 1977 speech. In this same year, he was responsible for forming the US Department of Energy, which combined some existing agencies, including the US Energy Research and Development Administration. He also initiated the Solar Energy Research Institute, in Golden, Colorado, that later became the National Renewable Energy Laboratory.

Following that time, interest again waned as a result of falling energy prices. President Ronald Reagan removed the solar panels that were placed on the White House under President Carter. Oil again became the primary focus in energy. Various starts and stops to solar development followed, including the start of the large development of the Luz systems in the Mohave Desert area of California, which was later stopped when existing tax credits went away.

Although many countries of the world, particularly Israel and Japan, had active programs in solar water heating for many years, solar power generation had taken a low profile existence. About the turn of the 21st century this began to change. Not only was there a heightened awareness of problems with continuing the reliance on Middle East oil, as had been the case all along, but this became more strikingly apparent as various types of strife broke out there, including the wars in Iran and the terroristic activities of many Middle East groups. Also, more and more people came to appreciate the concerns about peak oil. Simultaneously, concern grew about climate change and the impact of burning of fossil fuels on that. Worldwide, there is a growing awareness of the greenhouse gas increases in the atmosphere. As time moved toward the later part the early decade of the twenty first century, there was a rebirth of solar power generation activity. Some of Europe was reacting with lucrative feed-in tariffs

[1] This is an update of "Chapter 1: Some Solar Related Technologies and Their Applications" by Robert Boehm from ASME Publication "Energy and Power Generation Hand Book: Established and Emerging Technologies" New York, 2011. There may be changes to the names of some of the companies mentioned. Generally, the original name is used.

that have encouraged development, particularly in Spain and Germany. In the US, much of the encouragement was being given by individual states. This aspect was quite different than the wave of development under Carter, which was primarily driven from the Federal level, and thus was easily changed by a new president with a difference in philosophy. This time found mandates, tax incentives, and rebate programs interwoven in a great deal of society, including at the state and local levels.

Solar development really flourished in this period. Some states, but certainly not all, moved quite decisively into solar development. The Federal government also found solar research and development growing. This period ran until about 2016, and then the Federal government began to be more restrained in its development. This trend has held until the publishing of this book.

Generally, new power sources require some types of financial incentives when bringing them to market. There is usually a need for some type of government support. Nuclear power has received massive subsidies, but these are masked somewhat by the fact that both power applications and weapons' development were at play at the same time. Solar development has been plagued by the fact that the costs of related equipment were high because the market has been small, and the market has been small because the costs are high. In the photovoltaic arena, one where a great deal of progress has been made with generally more financial support from the government than other types of solar power generation, historically a 20% reduction in cost has been realized for each doubling of production. However, world competition as well as increase manufacturing led to prices greatly decreasing for PV. Toward the middle of the second decade of the twenty-first century prices for PV began to level off. However, it was the case that PV was the lowest cost power generation in some locations.

The development story is complicated by the overlaying economic situations as well as political reasons. Many solar installations tend to be high capital investment items because, in a sense, the "fuel" is being paid for upon construction. In good economic times, this high cost, while still a concern, can be covered by loans. During economic hard times, with the usual tightening of credit during those periods, the solar market can become severely constrained. Currently the situation is that PV manufacturing has greatly increased, and corresponding prices are at their lowest (ever) prices. However, the US has placed a tariff on all foreign manufactured PV panels, and this has changed the market conditions.

Two major ways of producing power are either by thermal means, where high temperatures from solar heating can be used in some type of heat engine, or by photovoltaic approaches. Because these are two quite different approaches, each will be touched upon separately in what follows.

1.6.1.2 The Use of Concentration Concentration opens a number of opportunities for solar power systems. In thermal systems, concentration allows much higher temperatures to be generated than would be possible with simple flat plate systems. This, in turn, results in higher temperatures within the associated thermal engine. As is well known from Second Law of Thermodynamics ideas, higher temperatures result in higher engine thermal efficiencies. Another benefit of concentration that goes hand-in-hand with this one is that the profile of the heated element is relatively smaller than flat plate systems (comparing energy harvested per unit area of collector). This allows special attention to be given to minimizing the heat loss from the receiver elements.

In PV systems, concentration permits tradeoffs between solar cells, which are generally of a different type than is used in flat plate systems and more expensive, against tracking mechanisms and support structure including lenses or mirrors, which may be less expensive in comparison to the PV elements. Concentration allows for each PV cell to produce proportionally more power per unit, and concentration may also increase the cell's efficiency. One possible concern is that some types of cells (silicon particularly) can have a significant negative temperature effect on efficiency compared to others. High ambient temperatures can result in noticeable decreases in production efficiency. Another concern that has come into play more recently, is that flat panels generally have less overall expense compared to concentrating systems.

Direct beam radiation is required for concentrating systems to be effective. Diffuse components of the radiation are not of value in these approaches unless the concentration is small. In addition, large amounts of flux are preferred to make the capital investment in these systems more profitable. Several locations in the world meet this requirement. For example, solar radiation resource in the US is shown in Figure 1.1. This resource is enormous, as shown in Figure 1.2 which is a representation that a square area of land approximately 100 miles (161 km) on a side could furnish with current technology all of the US's electricity needs. Of course, the transmission of this generated power to the whole country poses a significant problem. Storage is another with current technology.

However, several issues come into play when the choices of locations for concentrating solar plants are selected. For these plants, only the beam radiation is to be considered. Also, rugged landscape is to be shunned. Generally, a relatively flat area is preferred. When these factors are considered, the areas where solar plants can be located are far less numerous. See Figure 1.3. Here all land areas with greater than a 3% grade have been eliminated. Another factor that comes into play that is depicted here is that the plant has to be near a major power grid corridor. Construction of long lengths of major power lines can be extremely expensive.

Concentration can be accomplished in a variety of ways in two general categories: single axis (line focus) and dual axis (point focus) tracking. There are many kinds of single axis approaches because of a fixed axis (which can be fixed in a variety of ways) in one dimension and a moveable one in the other. In all of the concentrating systems, the beam can be developed either by refracting or reflecting the incident energy. Generally applications of refracting systems are less frequently used in practice, and these are usually for PV systems.

Another issue in concentration is the way this takes place in terms of the imaging aspects of it [Welford 1983]. This involves the concept of imaging optics vs. non-imaging optics. In general, the latter has a great deal to offer in the design of concentrating systems over the former. For one thing, the amount of light concentrated in non-imaging optical systems can be much greater than the comparable situation in imaging systems. Secondly, non-imaging approaches allow much better control of radiation, giving, say, more uniform irradiance or controlling the distribution better.

Last, but certainly not least, cleaning is required for the concentrating elements of all CSP systems. Some geographic areas are more prone to lens or mirror soiling than others, but cleaning is a necessity. Usually water is used for this purpose, so some water

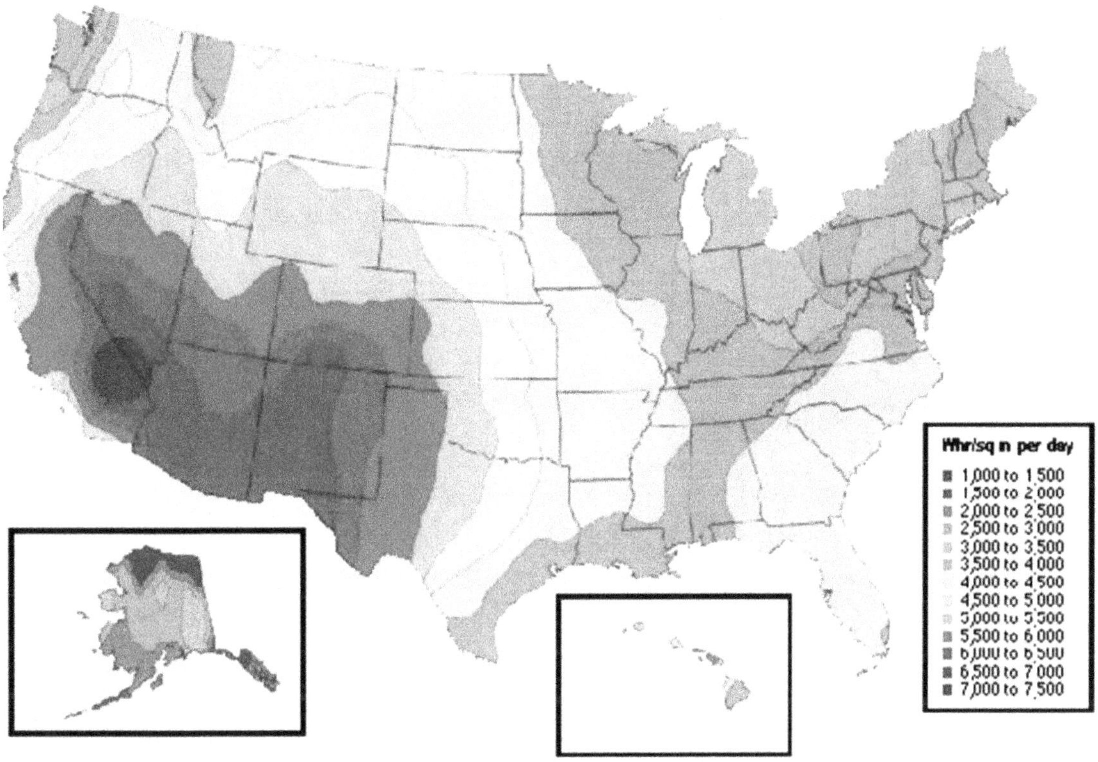

FIG. 1.1: SHOWN IS THE MAP OF THE US INDICATING THE AVERAGE SOLAR RADIATION PER YEAR IN UNITS OF W-HR/SQ M AREA/DAY. (*Source:* US Department of Energy)

FIG. 1.2: IN THIS REPRESENTATION FROM THE NATIONAL RENEWABLE ENERGY LABORATORY, THE FACT IS DEPICTED THAT THE TOTAL ELECTRICITY REQUIREMENTS OF THE US COULD BE FURNISHED FROM SOLAR ENERGY FALLING AND CONVERTED ON A SQUARE 100 MILES X 100 MILES (161 KM X 161 KM) AREA OF THE STATE OF NEVADA.

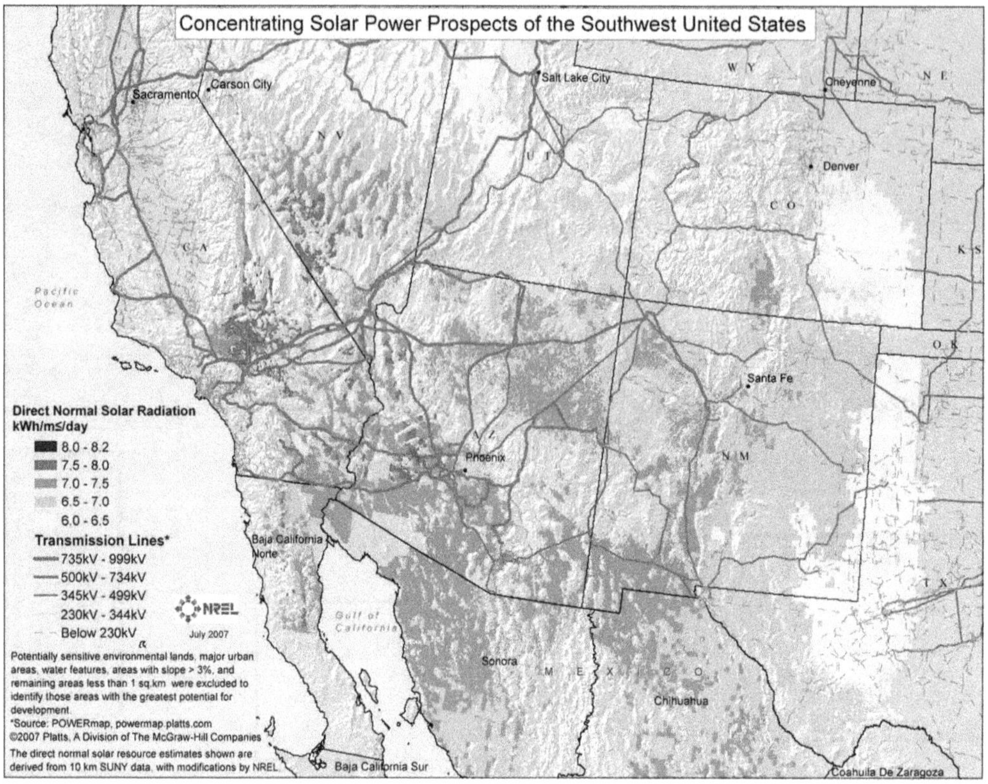

FIG. 1.3: THIS PLOT OF THE SOUTHWESTERN US AND NORTHWESTERN MEXICO SHOWS THE DIRECT NORMAL SOLAR FLUX (UNITS OF KW-HR/SQ M AREA/DAY) ONLY ON LAND THAT HAS LESS THAN A 3% GRADE. MAJOR TRANSMISSION LINES ARE ALSO SHOWN. (*Source:* NREL)

must be available for these kinds of systems even if it is not used for cooling purposes. Some treatments are available that facilitate rain running off surfaces quickly if they are oriented in a direction that gravity can be of assistance. It is desirable to have surfaces that stow facing downward during periods of darkness and inclement weather to minimize the cleaning requirements, but this may complicate the design requirements.

1.6.1.3 Thermal Power Generation Approaches

1.6.1.3.1 Rankine Cycles Many solar thermal power generation schemes (particularly the trough and tower systems to be discussed below) involve Rankine cycles to generate power from the heat collected by the solar field. There is both good and bad news about these kinds of plants.

Comparisons of the resulting operational levels of thermal power plants when using concentration are given in Table 1.1 [Zachary 2010]. Here the performance characteristics of trough and tower plants are compared. Flat plate performance, which is not really a viable approach in utility scale solar thermal power plants, is shown also for completeness.

On the plus side, the power block of these plants is in many respects common to conventional gas-fired or coal-fired units, although there may be design differences due to the transient nature of the solar source. This commonality greatly simplifies their design and operation. These plants can utilize conventional steam plant components, from the turbine to the condenser. Any major power plant component manufacturer can supply the power block for these plants.

The other side of the coin is that both the high and low temperatures of these types of plants may be limited. For example, solar plants typically operate at high temperatures that are lower than those of conventional plants. See Table 1.1 for typical values for solar plants. This poses implications for the condenser end. While the condensing temperature affects the operation of all Rankine plants, when the high temperature is moderate, the condenser temperature has a relatively higher impact on plant efficiency.

TABLE 1.1

CYCLE PERFORMANCE COMPARISON [ZACHARY 2010]

Type of CSP	Maximum Steam Temperature Expected, °C	Potential Cycle Efficiency Ranges, %*
Flat Panels	110	7-10
Fresnel Collector	310	20-25
Parabolic Trough	395	28-32
Solar Tower	550	35-42

*Depends upon many factors of plant design.

In general, wet cooling offers lower condensing temperatures than does dry cooling. Since most solar thermal power plants will be located in arid regions of the world, scarce water for cooling could be a significant limitation.

1.6.1.3.2 Trough Technology This approach uses a fluid circulating through a field of collectors where energy is reflected from a very large dimension parabolic trough (often aluminized or mirrored in some way on the inner side) to a small tubular arrangement along the axis. The small tube is often a metallic absorber where the fluid flows and a glass tube (usually evacuated) is concentrically located around it to increase the resistance to heat loss. A significant array of piping is used to connect all of the troughs together to bring the heated fluid to the main power plant and to return the cooled fluid back to the field.

The concentrator elements in the plant can offer a large wind profile. It is the case that plants of this type, like most tracking kinds of plants, have some arrangement for stowing their solar collection elements during high wind to give as low of a wind profile as possible.

Between 1984 and 1990 the Israeli company Luz International built several plants using this technology in the Mojave Desert region of California. An interesting analysis of this development is given in the book by Berger [1997]. In all, a total of 364 MWe of these plants was constructed. Each of these series of plants was over 15% more cost efficient than its predecessor [Ricker 2008].

Solargenix (now Acciona Solar Power), recently made further improvements to this technology and built plants that included a 1 MWe organic Rankine cycle plant in Arizona and a 64 MWe (75 MWe maximum capacity) plant in Southern Nevada. The latter is shown in Figure 1.4. This plant includes 760 parabolic troughs with more than 180,000 mirrors, 18,240 4-m long tubes, and heats the field fluid to 390°C (735°F).

It is usually the case that a synthetic oil circulates through the field returning hot to a heat exchanger where the heat is transferred to steam. The steam then traverses a fairly conventional Rankine cycle engine.

Several other companies have begun to develop potentially large solar thermal plants that also use a line focus approach. One of the approaches that could have been used to replace the parabolic trough includes a segmented set of nearly flat mirror facets that can individually follow the sun. This is denoted as Fresnel collector in Table 1.1. One such type like this was manufactured by Ausra Power (Palo Alto, California) that used a technology that was developed in Australia. See Figure 1.5. The benefits from this approach were claimed in the field design. The facets were lower to the ground and offered a smaller profile to the wind than do parabolic troughs. Also, the receiver tubes are stationary, so no flexible connections were required between trough segments as is the case with the parabolic troughs, and thus the new design was claimed to be able to handle high-pressure steam in the field. No heat exchanger would then be required to remove heat from the field fluid for the Rankine cycle fluid for these types of designs.

1.6.1.3.3 Power Tower (Central Receiver) Systems In the power tower system, a large number of ground-mounted mirrors are made to follow the sun on a nearly instantaneous fashion, reflecting the sunlight to the top of a centrally located tower where a receiver section is located. The receiver design could take a variety of forms, but one involves a series of parallel tubes that cover the outside of the structure or are located in the back of a cavity-type receiver. Depending upon the type of fluid circulating through the receiver (both steam and molten salt have been used in various experiments that have been performed previously), there may or may not need to be a heat exchanger to transfer the collected heat into steam for the conventional power block. Heat rejection systems for these types of units are either wet or dry cooling towers. Of course, the latter use no water for cooling (ideal in desert areas), but operate at a slight deficiency in engine efficiency compared to wet cooled systems.

These kinds of systems can be large. The original unit was Solar One, located just outside of Barstow, California, which is also in the Mohave Desert region. See Figure 1.5. The tall tower was illuminated by a surround field of 1818 heliostats. Each of the heliostats had an area of 40 m^2. The unit produced 10 MWe when the receiver fluid was steam. A later modification to the plant enlarged the field somewhat and used molten salt mixture of 60% sodium nitrate and 40% potassium nitrate as the receiver fluid, with a heat

FIG. 1.4: NEVADA SOLAR ONE, A 64 MWE (75 MWE MAXIMUM) PLANT BUILT BY ACCIONA SOLAR POWER BEGAN ELECTRICAL PRODUCTION NEAR LAS VEGAS DURING 2007. (*Courtesy* Acciona Solar Power)

FIG. 1.5: A CONCEPT OF A FRESNEL (LINE FOCUS) SOLAR THERMAL PLANT HAS BEEN PUT FORTH BY AUSRA POWER. IN THIS APPROACH, LONG AND TRANSVERSELY NEARLY FLAT FACETS EACH ROTATE ON THE GROUND TO FOCUS THE DIRECT BEAM RADIATION ONTO A FIXED RECEIVER OVERHEAD. (*Courtesy* of Ausra Power).

FIG. 1.6: SHOWN IS A SOLAR CENTRAL RECEIVER PLANT THAT OPERATED DURING THE 1980S AND 1990S NEAR BARSTOW, CALIFORNIA. COURTESY OF SANDIA NATIONAL LABORATORIES.

exchanger to transfer the heat to steam for the power block. Some degree of thermal storage was incorporated into both configurations. When steam was the receiver fluid, an organic heat transfer fluid was used as the storage medium. When molten salt was used as the receiver fluid, it served as the storage medium.

Abengoa Solar had two tower projects in Spain. See Figure 1.7. One, PS10, was a tower that generated 11 MW of electricity. To do this, it used 624 heliostats that were each 120 m^2 in area. Unlike Solar One in California, this plant does not use a surround field. Rather it has a north-side field, which Abengoa Solar felt was more efficient for their location. Next to PS10 is PS 20 that is designed to generate 20 MWe, and it started producing power in 2009. The field consisted of 1,255 mirrored heliostats, each with the same area as those on PS10. PS20's tower was 531 feet tall.

More recently two systems were built in the Mojave Desert area of the Southwest US. One of these called the Ivanpah Solar Power Facility is located just west of Primm, Nevada, in California. The three-tower steam plant has a gross power output of 392 MW. It uses dry cooling and does not have any storage capability. Another solar thermal plant near Tonopah, Nevada, consists of one tower and has a rated output of 110 MW. Its working fluid is a molten salt which is also used as a thermal storage medium, giving it the ability to generate power nearly 10 hours after the sun goes down.

1.6.1.3.4 Dish Stirling Systems These units are usually found on a dish arrangement where both the mirror facets and the engine/generator follow the sun throughout the day. Unlike the trough and the power tower arrangement which can be constructed for large power ratings from a single engine, this approach requires many smaller units (typically around 25 kW each) to be constructed.

The basis of this approach is the Stirling cycle, which is known to be one of the more efficient thermodynamic engines because it has nearly constant temperature heat transfer processes. In these devices, the working fluid (a gas) is contained in the engine housing and is usually either hydrogen or helium. A low molecular weight fluid is preferred for this application because of its good heat transfer characteristics and low-pressure drop behavior. Ambient air is used for heat rejection purposes. The dish maintains a high solar flux on the heat input area of the engine, and air is circulated over a heat rejection area, usually through one or more compact heat exchangers.

Two different designs of dish-Stirling systems are shown in Figure 1.8. The one in the right foreground is a design developed by McDonald-Douglas in the 1980s that later sold the terrestrial application rights to Stirling Energy Systems (SES). For many years, this type of system held the highest demonstrated efficiency conversion rates of sunlight to electricity for any type of system of around 30%. The SES unit used fixed mirror facets and an engine that was developed by the Scandinavian firm Kockums. SAIC International developed the unit shown on the left side, and it used stretched membrane facets that could be focused by exerting a slight vacuum on the chamber behind the membrane in each facet. It used an engine developed by Stirling Thermal Motors (STM). Both units produced an output of nominally 25 kW. When installing these kinds of systems, a significant pedestal has to be constructed that penetrates the ground a distance appropriate to counteract wind loads, but these types of units are designed to move to stow position in very high winds.

FIG. 1.7: THIS PHOTO SHOWS PS10 IN OPERATION IN THE BACKGROUND AND PS20 UNDER CONSTRUCTION IN THE FOREGROUND. PS20 STARTED POWER PRODUCTION IN 2009. (From Abengoa Solar.)

FIG. 1.8: TWO DIFFERENT DESIGNS OF DISH-STIRLING SOLAR DEVICES ARE SHOWN ON THE UNLV CAMPUS IN THE EARLY 2000S. THE UNIT ON THE RIGHT IS A DESIGN OWNED BY SES AND ON THE LEFT IS A DESIGN FROM SAIC INTERNATIONAL. (*Courtesy* of UNLV.)

Later SES in conjunction with Sandia National Laboratories redesigned several aspects of its system that they call the Suncatcher -;. They moved toward a more production-oriented design, and several of these units are shown in Figure 1.9. Even more recently a 1MWe solar field of this design was installed outside Phoenix, Arizona. The exclusive development sister company of SES is Tessera. The system operated for a short time and was sold to a foreign developer who moved it to China.

As is the case with both the trough and power tower technologies, several companies were said to be developing dish systems of this design. Although much interest in this approach was expressed, the author is not aware of any systems of this type in actual commercial production at the present time.

1.6.1.3.5 Supercritical CO_2 Engines As the second decade of the 21st century is coming to an end, these types of engines are being considered for a variety of applications. These systems can be driven by a range of heat sources, including various combustion systems, nuclear systems and solar power. Overall, they offer many promising aspects. For example, they are able to operate with higher temperature sources (the main limitation currently is related to the ability of materials to handle these temperatures), theoretically resulting higher operating thermal efficiencies. And the physical unit of a given rating can be somewhat smaller than some of the solar thermal plants described above.

1.6.1.3.6 Thermal Storage It is often said that solar systems cannot be used to furnish a constant rate of power generation because the sun shines only a fraction of the hours in a day. While this is obviously true about the sun, it doesn't have to be true about the power plant. The last big frontier in solar power generation is the use of effective means of storage. This would allow solar energy collected during sunny periods to be used in periods when the solar flux is low or zero in magnitude. While battery storage can be used in electrical systems, thermal storage of the best for large-scale solar systems and will be discussed in what follows.

When the solar thermal system operates on a Rankine cycle, the solution is clearly a thermal storage element in the plant design. A larger field than is needed to drive the plant during the day would be used, storing some of the energy to be used at a later time when it is needed. Traditionally storage has not been used to any great extent in solar thermal power plants because there were not incentives for this to happen. However, with the possibilities of solar being called upon to furnish 24-hour-a-day power, the issue becomes quite important.

Two approaches have been used in the past with solar thermal plants. As noted previously in this presentation, earlier designs have used either molten nitrate salts or high temperature heat transfer fluids. When the latter is used, the storage tanks may have a filler inserted, such as rock or something that is of lower cost than the fluid being used. Molten salt has also been used for storage applications. In molten salt flowing systems, this fluid cannot be allowed to solidify (cool below the freezing/melting temperature, but still greatly above ambient temperature) unless extensive heat tracing is incorporated in the design. Some newer designs have other fluids adding or removing heat from nonflowing salt systems.

FIG. 1.9: UNITS OF A MORE RECENT DESIGN OF THE SES DISH-STIRLING SYSTEM (THE SUNCATCHER™;) ARE SHOWN HERE AT SANDIA LABORATORIES UNDERGOING EVALUATION. (From Sandia Laboratories)

These kinds of systems may be able to take advantage of phase change phenomena (solid/liquid) in the salt, offering additional storage capacity per unit volume of salt.

Designers often favor the two-tank approach in flowing systems. One tank is for hot fluid and the other is for cold fluid. A typical evening will begin with the hot tank fully charged and the cold tank relatively depleted. As the hot fluid is used during the night to generate steam, it is returned in a cooled condition to the cold tank. Efforts have examined the use of stratified tanks with hot fluid on the top and cold fluid on the bottom, with a moving boundary between them (thermocline approach).

For dish-Stirling systems, as well as all PV systems discussed below, storage presents a more difficult problem. Perhaps the future will bring some viable possibilities either including or beyond storage batteries, but only time will tell.

1.6.1.4 Photovoltaic Approaches to Utility Scale Power Generation

1.6.1.4.1 Overview Comments The photovoltaic effect has been known since Becquerel identified it in 1839. Many incremental developments have occurred at various time intervals after that. One was the discovery of a method of monocrystalline silicon production by Czochralski in 1918. However, it was not until 1941 that the first monocrystalline cell was constructed. In the 1950s photovoltaic cells received a great deal more attention where applications were apparent to the growing space exploration interests. In 1955 Hoffman Electronics-Semiconductor Division introduced a commercial product that demonstrated 2% efficiency with 14 mW peak power for \$25/cell. Both the types of cells and their associated efficiencies have increased with time. This is illustrated in Figure 1.10. Now there are many companies producing a wide range of PV products for both utility applications as well as residential installations.

Silicon has been the mainstay of the industry for over a half century, has proven to be quite a durable product, and has shown steady increases in efficiency and decreases in cost. In addition to the monocrystalline form that has all of these good characteristics, silicon cells are also produced in an amorphous form that, although it has lower efficiency, can be applied in thin film form with correspondingly lower cost than the monocrystalline variety. A good review of solar cell development through the date of publication is the book by Perlin [1999].

One type of cell given special attention to here is CdTe. While cadmium is toxic, the amount of tellurium on the earth's surface is not well known but may be limited and the cell efficiency is lower than many other types, this cell is still one to be kept in mind because of its relatively low cost. It is made from several thin layers of materials. The active layers are less than the diameter of a hair thick. This cell is less toxic than the cadmium is by itself. Cost for the raw materials of this cell is a negligible part of the overall cost of the cell. Estimates indicate that tellurium, although quite small in quantity in the earth's crust, may ultimately be in sufficient quantities in older cells to furnish what is needed by recycling of CdTe.

Another type of cells that will be given brief attention here are those of the multijunction (often triple junction) types that have

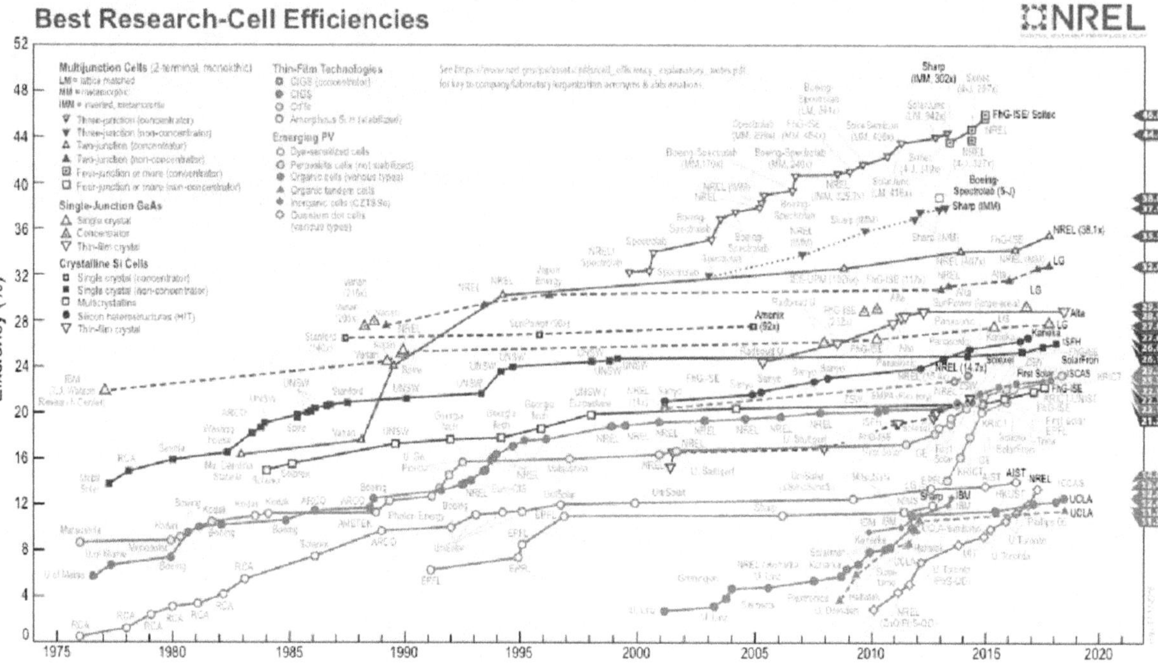

FIG. 1.10: THE EFFICIENCIES AND TYPES OF PHOTOVOLTAIC CELLS HAVE BEEN GROWING YEAR BY YEAR. (From NREL)

currently been used either on spacecraft or high concentration PV systems (see more discussion on the latter below). One of the factors that limits PV cells to the typically low efficiency that most demonstrate is that a given semiconductor material is sensitive to only a limited range of the solar spectrum. By using thin layers of carefully selected, but differing, materials, this sensitivity range is greatly expanded. The result is higher efficiency. From Figure 10 it can be seen that triple junction cells show the highest of all efficiencies indicated there. Another positive aspect of many types of multi junction cells is that their efficiency is less sensitive to increases in operational temperature than are those of the silicon type.

In more recent times, attention has been brought to bear on concentrating systems applied to photovoltaic systems [Swanson 2000]. In the period when his paper was written, CPV systems had not lived up to their promise of large-scale applications, partially because the cost of alternative fuels was relatively low, and flat plate PV approaches were being developed quite rapidly. The latter had the ability to be applied to roof structures, thus opening a large market for them. Concern about tracker reliability was also a factor that held back CPV development. Since the market was small, the companies involved in this technology could not realize the economies of mass production and all of the related benefits that this implies.

This situation is obviously changing. Cell efficiencies have increased substantially over the years as noted above. Swanson [2000] indicated that 10% of the output efficiency goes to pay for everything but the cells, so as efficiency increases, it encourages the application of CPV approaches.

One additional benefit of CPV systems compared to fixed flat plate systems is their ability to capture more of the sun's energy because of their tracking nature, compared to fixed flat plate PV systems. When more efficient cells can be applied to the CPV systems, this can greatly boost the energy generated for this approach.

In all, Swanson [2000] discusses the following benefits for CPV compared to fixed flat plate systems.

- Lower cost
- Superior efficiency
- High annual capacity factor
- Less materials' availability issues
- Less toxic material use (compared to some thin film systems)
- Ease of recycling
- Ease of rapid manufacturing capacity scale-up
- High local manufacturing content

For the purposes of this presentation we will put CPV into two categories with an imprecise division point: low concentration and high concentration. This is a result, in general, of the temperature of operation and the corresponding quality of cell required to appropriately utilize the concentrated beam. Low concentration will use cells that can be one-sun units, and passive cooling could always be used. On the other hand, high concentration units will require improved fill factors and other design features in addition to some well-designed cooling system, either passive or active.

For example, very high concentration systems for PV applications have been reported rendering up to 10^4 suns [Gordon et al. 2004]. These are obviously not production systems, but rather special experimental systems. At the other end of the concentration range, JX Crystals is producing a 3x system [Fraas et al. 2007]. Hence a wide range of concentrations have been available for PV systems.

The cells used in these kinds of systems can range from conventional one-sun types up to multi-junction (mj), very high efficiency, types. In between lie high efficiency single junction cells. It had been argued that high concentration systems were more appropriate than mj cells [Slade et al. 2005], but that is changing in favor of the mj cells as the latter become more developed for this market.

Unlike concentrating thermal systems, where virtually all concentrators are of the reflecting type, CPV systems can have refracting or reflecting configurations. This is due to the possibility that a given lens element can focus onto a single cell individually. Typically, Fresnel lenses are used for the former and mirror-like reflectors are used for the latter.

Concentration not only boosts the cell incident flux, it also generates more heat within the cell compared to one-sun situations. When concentrations are sought at very high levels, enhanced heat removal from the cell is almost always called for.

1.6.1.4.2 High Concentration Systems High concentration systems might be defined as using concentrations above 10, but this is not a precise definition. Many of the systems now being marketed have design concentrations in the range of 100-500x, but 1000x doesn't seem too far from reality.

One of the companies that was quite active in this field was Amonix (no longer in business), which was located in the Los Angeles, California, area. For quite some time, Amonix was marketing a 25 kW system and a 35 kW system with a 250x concentration. They then moved to 500x systems with multi-junction cells that had nearly doubled their units' output. Multi-junction cells were then being included in designs from several manufacturers [Zubi et al. 2009]. Cooling of the cells was performed passively. As is the case with any high concentration system, a two-axis tracker is used to follow the sun across the sky. A photo of several Amonix high concentration PV (HCPV) systems is shown in Figure 1.11.

Another approach to HCPV systems is the use of a dish. Some investigation of this type of system had taken place in the past (see Figure 1.12). In the figure a design that was developed by SAIC is shown. The nominal output of this device was also 25 kW, using fixed facet mirrors.

Two distinct differences are present between the two HCPV systems shown here. First, in the dish system the flux comes from a large number of facets to a particular point on the PV array. This simplifies to the concentration assembly, but it also can cause hot spots to exist. A similar kind of behavior is sometimes demonstrated on dish Stirling units. A second issue that differs between the Amonix unit and the SAIC unit is the cooling approaches used. In the first case the cooling is accomplished in a passive manner, while the latter is performed actively using the system denoted in Figure 1.13 [Mahderekal et al. 2006]. There are plusses and minuses for each of the approaches, but presumably the active system could offer better control in varying conditions.

1.6.1.4.3 Low Concentration PV Systems The other end of the range of concentration involves the use of low concentration designs. As was noted earlier this is not a precisely defined category, but it might include concentration ratios of less than 10. For these kinds of systems, the precision of the tracking can be reduced significantly. Also, the heat that needs to be removed from the cells is greatly reduced and may all be done by passive interaction with the ambient conditions. Finally, many of these systems use one-sun cells. In one such design by JX Crystals, a

FIG. 1.11: AN ARRAY OF 25-KW AMONIX DESIGNS IN SPAIN AT THE GUASCOR FOTON INSTALLATION, LICENSE PARTNER AMONIX INC. NEWER UNITS HAD MUCH HIGHER POWER OUTPUT BECAUSE OF THE USE OF MULTI-JUNCTION CELLS. (From Amonix.)

FIG. 1.12: A DISH-PV CONCENTRATING SYSTEM IS SHOWN OPERATING ON SUN. THE CONCENTRATION WAS APPROXIMATELY 250X. (From UNLV.)

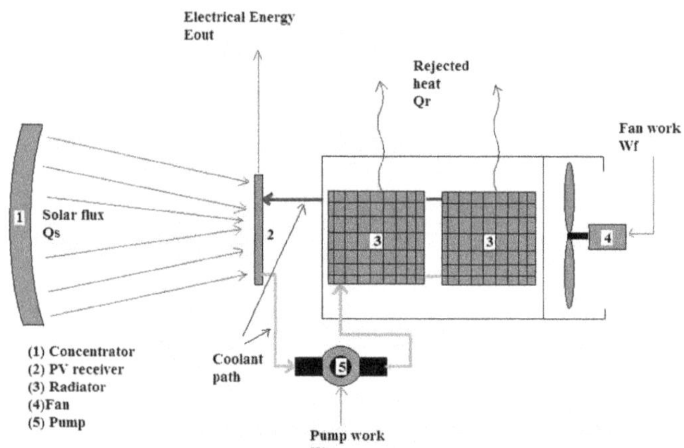

FIG. 1.13: THE SCHEMATIC DESIGN OF THE ASSEMBLY AT THE SNOUT OF THE DISH ARRAY SHOWN IN FIG.12 IS DEPICTED HERE. (After Mahderekal et al. 2006)

3x concentration ratio was used. As was stated by the developer, the bulk of the PV cells were replaced by reflecting aluminum that was on the order of 20 times less expensive than the cells [Frass et al. 2007]. This design used aluminum "valleys" with conventional cells mounted at the base of the valley. One axis tracking was possible with a carousel type of arrangement in this application. See Figure 1.14. This unit was installed flat on a roof without penetrations.

1.6.1.4.4 Tradeoffs between One-Sun, Low Concentration, and High Concentration Systems

- One-sun systems, particularly fixed, flat plate assemblies, are simple to install and maintain.
- One-sun systems take up more ground area for the same power out than do concentrating types of systems, in general. This is particularly true if cells like the lower cost but lower efficiency CdTe type are used.
- One-axis tracking on flat plate systems can be adapted that increase the power output but do not increase significantly the initial cost and maintenance expenses.
- Tracking capabilities for high concentration systems must be more precise than are needed in low concentration systems.
- Low concentration systems can use one-sun cells and passive cooling designs.
- Some low concentration systems can be mounted flat on the ground without penetrating the latter.
- Each cell in a high concentration system produces significantly more power than does each in a low concentration or one-sun system.
- Cell cooling is a key concern in high concentration systems.
- High efficiency cells that are of high cost will be more cost effective in high concentration compared to low concentration systems.
- To minimize the costs of the tracking system per unit of power produced, high concentration systems must be on the order of 25-50 kW each to minimize the cost of the tracking system in the overall cost of power. These types of systems must have some careful design to minimize wind effects on the stability of the installation and dissipate heat generated in the cell.

1.6.2 HYDROGEN PRODUCTION AND USE

1.6.2.1 Background Hydrogen is the most commonly-occurring element in the universe. It can be used for a variety of applications including manufacturing, electricity generation, and motive power. In addition, it can be produced from solar and wind energy systems, and it makes an excellent storage medium for these renewable resources. NASA and the oil refining business have used hydrogen for years in achieving their goals with few hydrogen accidents. In spite of all of these good qualities, why has it not been used more as a core fuel in the world? For one thing, it is not a fuel in the sense as, say, natural gas and gasoline are fuels. Those two examples can be found in reservoirs at various places around the globe. Hydrogen, on the other hand, is an energy carrier. It has to be produced, much like electricity has to be generated.

Another characteristic of hydrogen is that it is the smallest atom in existence. The result of this is that it typically takes up a large volume to store, and it is prone to leaking unless the containing system is extremely well sealed. It can cause corrosion stress damage to some common materials that might be used in contact with it. Finally, it has a wide flammability range compared to most common fuels.

None of these drawbacks is insurmountable. What is clear, though, is that the infrastructure for generating, distributing, and utilizing hydrogen on a country-wide scale is not in place. However, it remains an intriguing option for future developments.

FIG. 1.14: THE JX CRYSTALS ONE-AXIS TRACKING 3X CONCENTRATION PV SYSTEM IS SHOWN. (*Courtesy* UNLV).

Hydrogen has been used to some extent over a large number of years. It is not our purpose here to cover all aspects of its history. Instead we will give some insights on how this can be generated from solar energy, and how it might be utilized in the modern world.

1.6.2.2 Generation of Hydrogen

Most of the world's supply of hydrogen is generated from a process called steam-methane reforming (SMR). In this technique, methane (normally as natural gas) and steam are brought together at elevated temperatures (in the range of 700°C-1100°C). A major hydrogen distribution system is located on the Gulf coast area of the US, and this is a significant part of a US hydrogen industry that produced 10.6 million metric tons of hydrogen in 2006, most of it generated using SMR [EIA 2008]. This is enough hydrogen to furnish 20-30 million cars with fuel or for energy uses in 5-8 million homes. SMR is currently the least expensive way to produce hydrogen. Its drawbacks, however, from a sustainability point of view include that it uses what is essentially a nonrenewable energy supply (natural gas) and it generates greenhouse gases (primarily CO_2).

A cleaner approach, both environmentally as well as in hydrogen purity, can be electrolysis where water is broken into its molecular components using electricity. This is the second largest source of hydrogen, but the amount generated from this is much smaller than from SMR. How environmentally clean of a generating approach this is depends upon the source of electricity that is used. If the latter is generated from dirty fossil-fired plants, that fact becomes part of the hydrogen's carbon footprint. On the other hand, electrolysis produces very pure hydrogen. So, when very high purity of hydrogen is needed, electrolysis is the desired source.

Because hydrogen can be made from electricity, if the electricity is made from renewably-generated power, then the hydrogen becomes renewable also. This approach is a good addition to a time-variable renewable power generation source like solar or wind. Like "making hay while the sun shines," it is possible to save solar energy by the use of electrolytic hydrogen that is generated from it. Certainly, one of the drawbacks to this approach currently is the high cost of generating hydrogen. When the cost of the photovoltaic, or other means of power generation from the sun, is factored in with the cost of the electrolysis unit, this does equate to higher cost hydrogen than that available from SMR approaches. However, PV systems have been decreasing markedly in cost, and there are new approaches called photoelectrochemical that could bring the costs down significantly. An example of this has been reported by Wang et al. [2008].

Several solar powered hydrogen generation stations were developed in the early part of this century in the US. One of these types of hydrogen generation and filling stations was built by UNLV, Proton Energy, Air Products, and the Southern Nevada Water Authority at the Las Vegas Valley Water District headquarters in Las Vegas. This system used four one-axis tracking flat plate PV units for power. This, or grid power, can then furnish the electricity to an electrolyzer. Water is converted there to hydrogen and oxygen, where the hydrogen is piped to a compressor and the oxygen

FIG. 1.15: THE UNLV/LAS VEGAS VALLEY WATER DISTRICT SOLAR POWERED HYDROGEN GENERATION AND FILLING STATION IS SHOWN. TO THE RIGHT IN THE FIG. IS THE ELECTROLYZER SYSTEM, TO THE LEFT ARE THE COMPRESSOR, STORAGE AND DISPENSING STATION, AND IN THE REAR TWO OF THE FOUR PV PANELS ARE SHOWN THAT POWER THE SYSTEM.

is vented. Storage of the hydrogen takes place at 6000 psi for refueling of hydrogen vehicles. A photo of the filling station with two of its four PV panels showing is given in Figure 1.15.

One of the big issues in developing a hydrogen infrastructure for vehicle powering is how to transport the hydrogen from where it is generated to where it is used. Various options like using tanker trucks or pipelines have obvious drawbacks. One possibility is to have dispersed generation filling stations that make hydrogen on site. In locations where the solar resource is very good (like the Southwest portion of the US), this could be used. In other locations that might have substantial wind availability, that could be used. Consideration of local renewable resources to generate hydrogen on the sites where it is dispensed could furnish hydrogen for a significant amount of the US land area.

1.6.2.3 Utilization of Hydrogen Hydrogen can be used in the various ways any fuels are used: combustion processes, engine fuels, electricity generation in fuel cells, and others. Here a focus on transportation applications will be discussed. Not all transportation applications will be covered, however. It is well known that hydrogen can be used to power rockets as has been shown over the years. It can also be used to fuel aircraft, although this application has not been demonstrated to any large extent to date. In what follows, we will discuss road vehicle applications. Generally, the motive power for hydrogen road vehicles will be internal compression ignition or spark ignition combustion engines and fuel cells.

First consider using some type of internal combustion engine (ICE) for motive power. In this application, the ICE is a machine that has been around significantly over a century so a great deal of infrastructure has been developed for both the manufacture as well as the maintenance of these devices. If they can be used for hydrogen applications, then one piece of the required hydrogen utilization infrastructure would be pretty much in place. What is required to convert a conventional (say gasoline powered) engine into a hydrogen power engine? It is actually quite simple. For one thing, the spark timing must be changed quite significantly because of the ease of ignition of hydrogen as well as the resulting flame speed being so high. Also, the fuel injection/air-fuel-ratio needs to be controlled quite carefully. For all of these reasons, the engine computer control must be appropriately calibrated. One of the bigger concerns using hydrogen in ICE's is to minimize the NOx formed. This can be done with high air fuel ratio. This, however, causes reduced power with normally aspirated engines.

Of course, almost all modern engines use some type of fuel injection, and some of them (in addition to Diesel engines) use direct cylinder injection. Adaptation of direct injectors to hydrogen engines is important for a number of reasons. One of them is that the flame speed in hydrogen is extremely fast, much faster than typical hydrocarbon fuels. Also, even more critically than HC fueled engines, the combustion process may need to be controlled to minimize NOx formation. Direct in-cylinder fuel injection can offer a significant improvement over carburetion or port injection. However, hydrogen fuel injectors, whether they be for port or direct cylinder injection of hydrogen need to be designed more carefully because of the drying nature of hydrogen fuel compared to hydrocarbon fuels. This is because hydrocarbon fuels have lubricity, so sliding metal surfaces work well in that environment. Hydrogen is a drying fuel, and sliding metal surfaces are to be shunned in hydrogen systems. Work has been ongoing related to special in-cylinder fuel injection systems at several locations. One such design that combines an in-cylinder fuel injector integral with a spark plug, and which can be used to replace the conventional spark plug is shown in Figure 1.16 [Wilson 2010]. Something like this greatly simplifies the conversion of a hydrocarbon spark ignition engine to a hydrogen in-cylinder fuel injected engine.

Another way that hydrogen can be used in vehicles is in conjunction with fuel cells. Good insights into the development of fuel cells have been given by Andujar and Segura [2009]. These devices are essentially the opposite process of electrolyzers. In the latter, water and electricity are used to make hydrogen and oxygen.

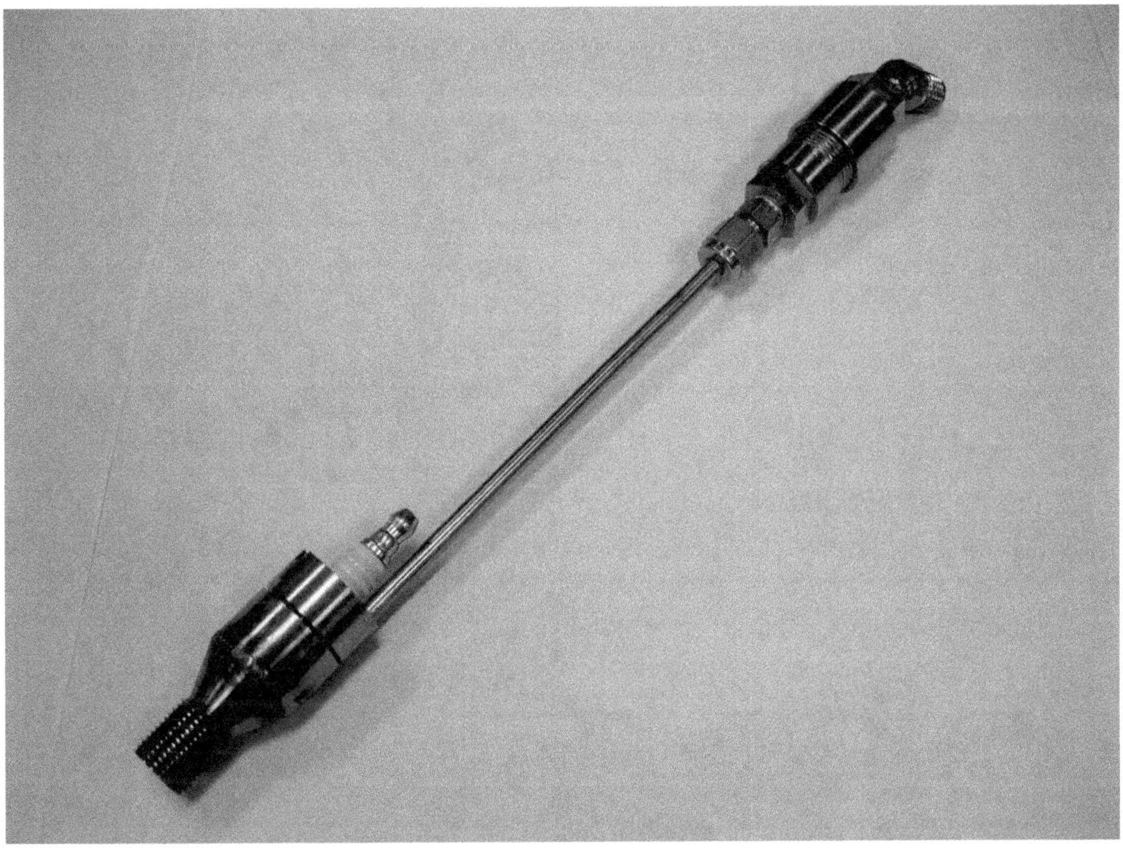

FIG. 1.16: AN INTEGRAL IN-CYLINDER FUEL INJECTOR AND SPARK PLUG ASSEMBLY IS SHOWN. THIS ASSEMBLY REPLACES THE CONVENTIONAL SPARK PLUG OF A CONVENTIONAL SPARK IGNITION ENGINE [WILSON 2010].

In fuel cells, hydrogen and oxygen (the latter usually from atmospheric air) combine to make water and electricity. Hence fuel cell vehicles are really electric vehicles that use hydrogen fuel to generate the electricity. These devices have a number of very desirable characteristics, including quiet operation and higher efficiency than ICE's. Down sides include that they are currently expensive and their membranes can be poisoned with impurities in the air.

The most typical type of fuel cell to use for vehicular propulsion is of the proton exchange membrane type, and most automobile manufactures have cars with this type of propulsion. These cells operate at relatively low temperatures (about 175 °F or 80 °C), have high power density, can vary their output rather quickly (but not as fast as automobile engines) to meet shifts in power demand, and are suited for applications -- such as in automobiles --a where quick startup is required. The proton exchange membrane is a semi-permeable membrane that allows hydrogen ions to pass through it. The membrane is coated on both sides with highly dispersed metal alloy particles (mostly platinum) that are active catalysts. This type of fuel cell is sensitive to fuel impurities. Cell outputs generally range from 50 to 250 kW, although they are also available in smaller sizes.

Depending upon the function of the vehicle, it could be a straightforward task to convert an all-electric vehicle to a fuel cell hybrid design that uses hydrogen fuel. One such example is shown in Figure 1.17. This Taylor-Dunn electric utility vehicle was converted to be a hybrid hydrogen fuel cell electric vehicle. In doing this, approximately 50% of the original batteries were removed, and the space they had taken was used to install a 5.5 kW fuel cell and two high-pressure hydrogen storage tanks. Necessary power control aspects were added to allow the system to operate off of a base power from the fuel cell and to tap the storage battery reserve power if needed for high load periods. When the fuel cell is actually producing more power than the vehicle needs for its motion, the extra energy generated can be used to recharge the batteries. A block diagram of the system used for this conversion is shown in Figure 1.18. In the companion curve (Figure 1.19), the green denotes the amount of power directly from the fuel cell, while the red indicates the power that is from energy stored in the battery.

Whatever is used as the prime mover, one significant issue that needs to be addressed is how the hydrogen is stored on board. As noted earlier, because hydrogen has the lowest molecular weight of all molecules, considerable space is needed to store it. It is not unusual for gaseous hydrogen storage tanks to be designed to hold 5000-6000 psi, and in spite of this, the tanks have to be large for reasonable vehicle range. Other means of storage have also been applied. BMW, for example, has used liquefied hydrogen storage, which is more efficient space-wise than gaseous hydrogen storage. However, the liquid has to be stored at very low temperatures. Some boil-off of hydrogen due to heat transfer through the insulated tank walls will occur. This might cause substantial loss of hydrogen if the vehicle is not used for a long period of time.

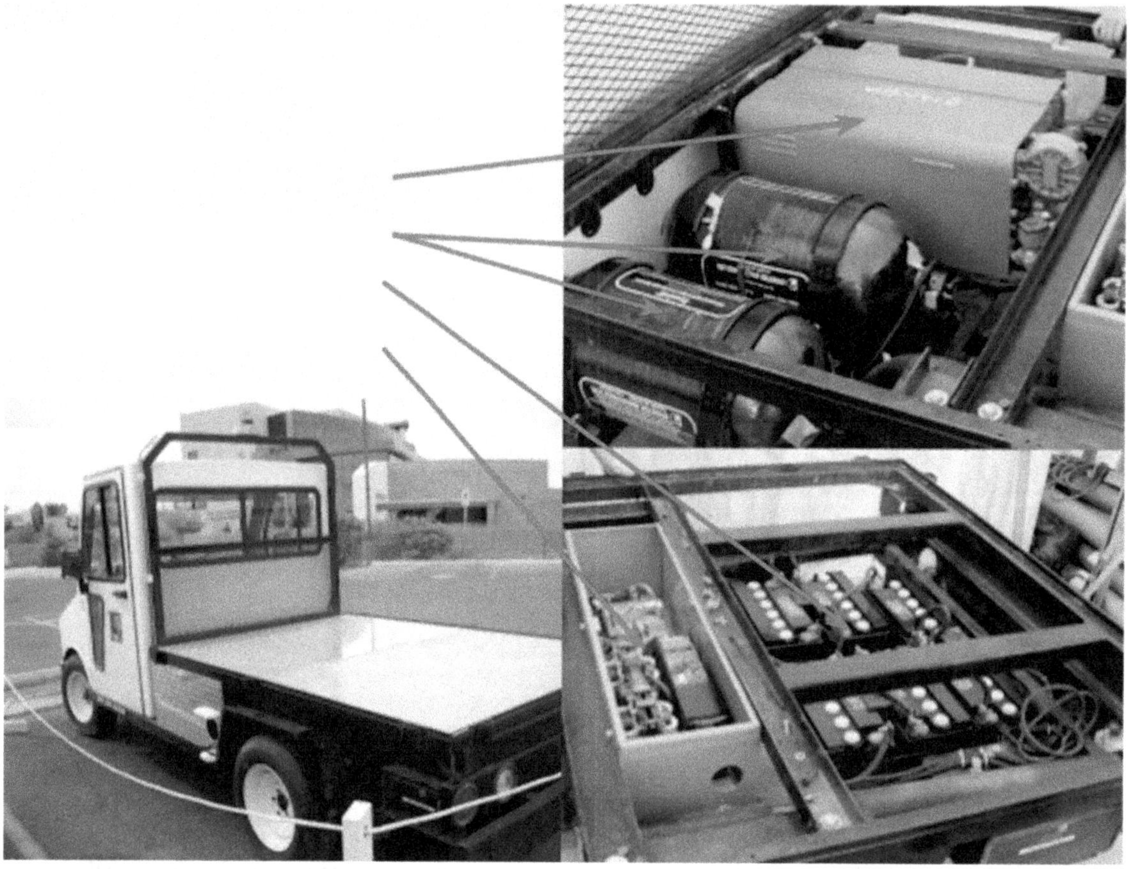

FIG. 1.17: AN ELECTRIC UTILITY VEHICLE THAT HAD BEEN CONVERTED TO A FUEL-CELL-ELECTRIC HYBRID CONFIGURATION IS SHOWN. THE SPACE ORIGINALLY USED FOR HALF OF THE ORIGINAL BATTERY PACK WAS USED FOR A FUEL CELL UNIT AND HYDROGEN STORAGE TANKS. (*Courtesy* of UNLV).

Another way to store hydrogen is with metal hydride beds. These are like a metal sponge to hydrogen, soaking up gas when the bed is cooled, and releasing the hydrogen when the bed is heated. The downside is that the beds are heavy, and this approach has typically been used only in large vehicles like buses and trucks. The upside is that hydrogen is stored in these beds at relatively low pressure (say 100-200 psi). An approximate map of the storage characteristics of hydrogen is shown compared to gasoline in Figure 1.20.

A key element in designing with hydrogen in any application is to allow venting to the atmosphere in case of some type of malfunction. Since the gas is so light, it will rapidly go straight up in the atmosphere in the area of the hydrogen leak as long as it has a clear passage upward.

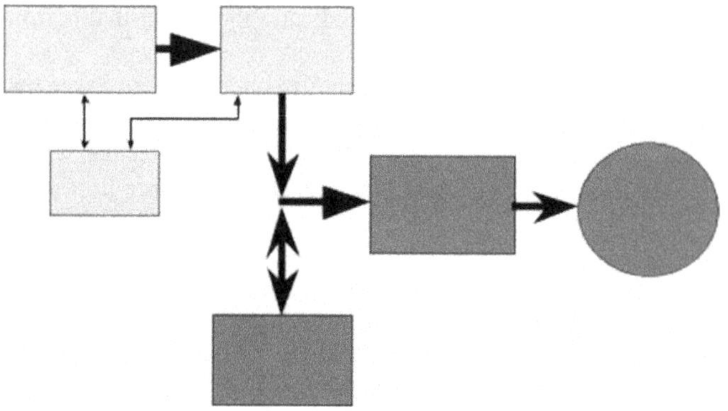

FIG. 1.18: A THE VARIOUS MAJOR COMPONENTS OF THE POWER SYSTEM DEPICTED FOR THE VEHICLE CONVERSION.

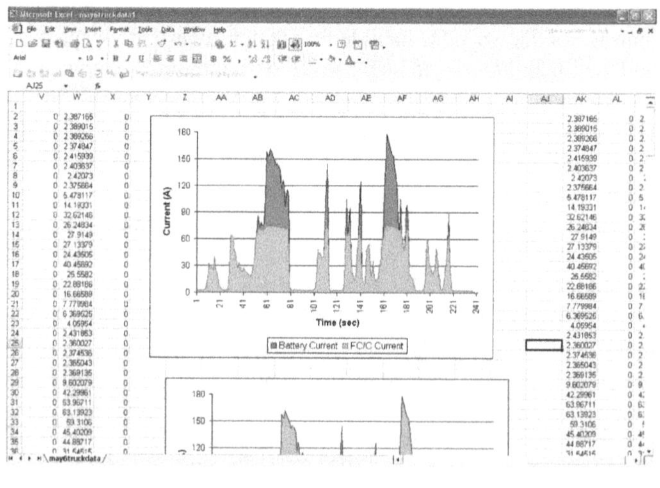

FIG. 1.19: SOME EXAMPLES OF THE POWER DRAW DURING A DRIVING CYCLE ARE SHOWN DENOTING WHAT AMOUNT COMES FROM THE FUEL CELL AND WHAT COMES FROM THE ENERGY STORED IN THE BATTERIES. (*Courtesy* of UNLV).

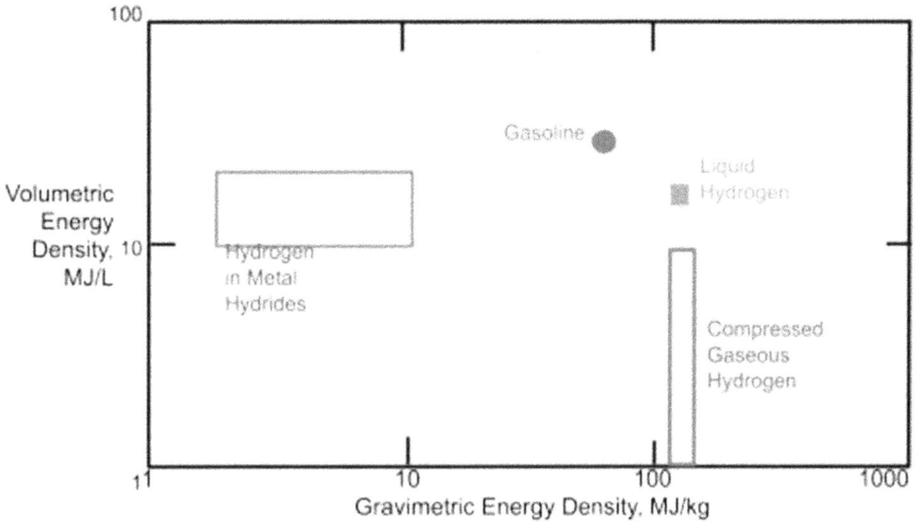

FIG. 1.20: AN APPROXIMATE COMPARISON OF THE THREE WAYS OF STORING HYDROGEN IN TERMS OF THE ENERGY STORED PER UNIT OF WEIGHT AND PER UNIT OF VOLUME. THESE ARE CONTRASTED TO GASOLINE STORAGE. NOTE THAT LIQUEFIED HYDROGEN, ALTHOUGH REQUIRING VERY LOW TEMPERATURES FOR STORAGE, ACHIEVES CHARACTERISTICS ON THIS MAP THAT ARE NOT TOO DIFFERENT THAN GASOLINE.

1.6.3 Building Applications of Solar Energy

1.6.3.1 General Philosophy Knowledge of interaction of solar energy with buildings is very important for the planning for a low energy future. Sometimes these interactions need to be minimized as much as possible, and at other times they need to be accepted to the full extent; but at all times, the building needs to be designed for smart solar interactions. These may include the use of the photons in their basic form (lighting and PV), and at other times the photon conversion to thermal energy is desired (space and water heating).

Of course, the environmental conditions will have an impact on how all this is done. Hot, dry climates will have to be handled differently than temperate climates. It could be that different parts of the building may be designed to achieve different kinds of outcomes. Also, the orientation of the building compared to the sun path in the different seasons of the year should be carefully incorporated in the design.

By far the most important aspect to general building design is cost-effective energy conservation. Forms that this may take vary, but they almost always involve carefully chosen insulation and high performance windows. In addition, high electrical use devices in the house (particularly air conditioners in warm or hot climates) need to be chosen for high efficiency.

A critical tool in good design is the use of building energy simulation (BES). Like many types of simulation software packages, BES codes tend to range between codes that are extremely comprehensive and often hard to master and easier codes that may not be as detailed in their ability to analyze. BES codes that allow "what if" kinds of analyses to examine the changes to performance with change in design are particularly good to use. One such example that has been quite easy to apply is ENERGY-10 [2010]. There may have been some aspects of the design that were not easily accounted for in this code, but the major elements could be. It also allowed changes to the building design within a given sized and oriented structure. Codes can be used with detailed measurements to estimate the cost-effectiveness of conservation components on actual buildings [Zhu et al. 2009].

However, it is also critical that component costs of various options be considered in the analysis, and these are sometimes more difficult to gather without direct access to the local construction industry data. One simple way to handle this (if cost information is available) is to consider a base case design, assess its performance and cost (both capital cost as well as operating costs), and then consider "what if" options. Looking at simple payback may be sufficient to see what the long-term cost implications of the design may be.

After the basic structure has been optimized for performance and cost, then various solar applications can be considered for their impact on the overall building. One of the significant problems in all of this is incorporating good estimates of cost increases of conventional energy. If the latter remain stable (or nearly so) compared to current costs, a different set of conclusions may be drawn than if the conventional energy greatly increases in cost over time. One way to handle this is to consider two cost estimation scenarios. One can be considering no increase in cost. The other can be for cost escalation at the average historical cost rate of increase. Each state's public utility commission, or even the local utility, should be able to furnish the latter information.

Human comfort and possible indoor environmental pollution needs to be considered in energy efficient design. Many of these factors are discussed by Omer [2008].

In the next section a concept that is attracting a great deal of interest, particularly in the US and Europe, is discussed: Zero Energy Buildings. Generally, these buildings have a great deal of energy saving design incorporated in their construction, and they also include some means of alternative energy for generation to make up for the energy used in the house. Following this, we discuss the two most-common types of solar add-ons for buildings. These are solar domestic water heating, and photovoltaics. In both of those applications, a south-facing, unshaded orientation is preferred. Some inclinations that are nearly southerly might offer less solar energy harvest than directly south, but the deficit may be small. Inclinations to be used vary somewhat because of ease of application or desire to achieve some particular goal. For good all year performance, an optimal angle of orientation is to face south with a tilt angle equal to the local latitude. If more winter input is desired, it can be achieved by using a higher angle, but this will be at the expense of summer input. Oftentimes solar units on homes will be mounted at whatever the roof inclination is. Of course, for flat roof installation, any orientation can be achieved with a mounting system.

Cleaning is another issue. Rain, if it occurs periodically like it does in many parts of the nation, will generally do a good job in cleaning solar collection units. Many solar collection systems have not been cleaned during their entire lifetime. If the atmosphere is particularly dirty, and rain is particularly sparse, it may be that a periodic cleaning may improve system performance.

1.6.3.2 Net Zero Energy Buildings This is a concept that has, on one hand, been around for a long time, but on the other hand is only now being more widely adopted in the US. In buildings of this sort, very good energy conserving design is coupled with some means of generating energy from a renewable energy source. If the customer matches the net energy use from the grid by an excess generation over the year, this is often called a net site zero energy building. In some quarters this is referred to as a zero-site energy building. Another distinction sometimes made is to denote a building as being net zero source energy. This is related to the site definition, but also accounts for losses in generating the energy and bringing it to the consumer. A similar definition can work for net zero energy costs. In some service areas, utilities pay local generators for the renewable energy they send to the grid. If the payment to the utility matches the payment from the utility, this is denoted as net zero energy costs. A similar definition can be based upon emissions.

FIG. 1.21: TWO EXAMPLES OF ZERO ENERGY HOMES ARE SHOWN WHERE THE SIZE OF THE PV ARRAY IS MUCH LARGER IN PHYSICAL SIZE COMPARED TO THE SDWH SYSTEM. (*Credits:* Left, NREL #14164; Right, UNLV.)

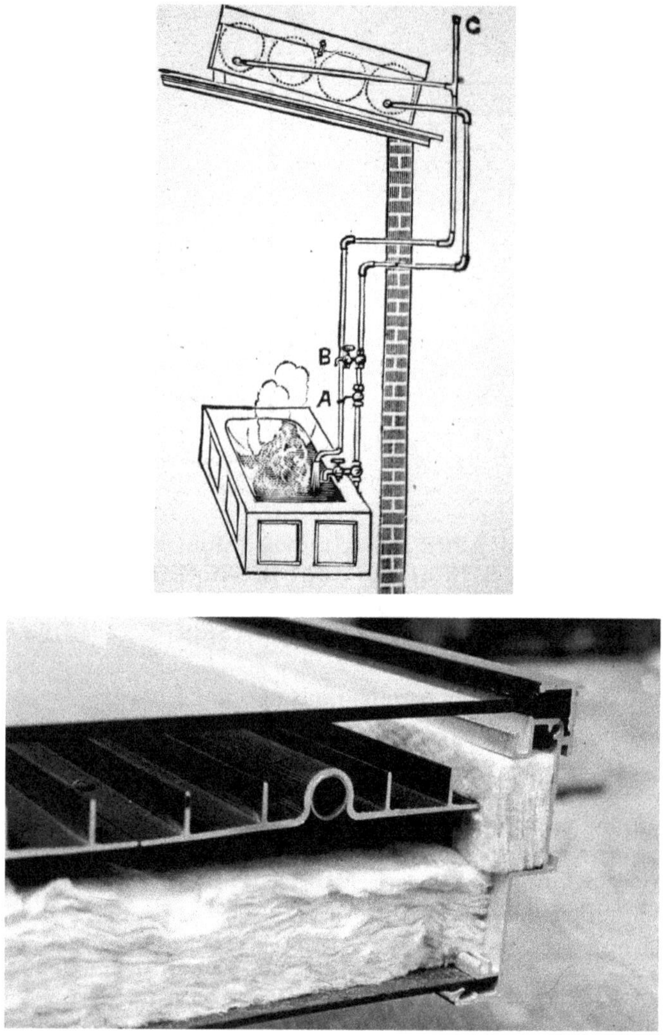

FIG. 1.22: TWO OF THE MANY TYPES OF SDWH SYSTEMS ARE REPRESENTED HERE: A) THE INTEGRATED COLLECTOR-STORAGE SYSTEM AS IT IS CALLED TODAY IS SHOWN IN THIS SKETCH THAT ACCOMPANIED A 1893 US PATENT, AND B) A CUTAWAY VIEW OF A COLLECTOR TYPE THAT MIGHT BE USED WITH A DRAIN-DOWN OR AN ANTIFREEZE PROTECTED SYSTEM.

On net zero energy buildings in a good solar climate, both a solar domestic water heater and a photovoltaic array will often be applied. Each of these topics is addressed separately in sections that follow. Of course, the electricity energy demand for a building will typically be greater than the energy demand for heating of water. As a result, the size of the two collection systems is relatively much larger for the PV and smaller for the SDWH. See Figure 1.21 where some examples of the size of the PV system can be compared to that of the SDWH system. It should be noted that the first of these buildings uses a SDW heating system that has freeze protection, while the second uses an ICS system. Both of these are described in the next section.

1.6.3.3 Solar Domestic Water Heating A great deal of the world has applied solar domestic water heating more than we have in the US. Europe, Asia, and the Middle East are among those areas where this is the case.

In considering solar domestic water heating, several aspects should be considered. Of course, a south or near-south exposure for the solar collection unit is desired. Careful attention to the climatic conditions is also required. Climates that have freezing conditions at some points in the year will require freeze protection that may not be required in more balmy regions. Some types of systems may require a quite-heavy installation on or under the roof, and the basic building design may not allow this without significant strengthening of the structure. Water quality can have an important effect on some types of systems. Also important for cost-effective installations is the cost of the competing conventional energy used for water heating. A rule of thumb that is almost always true is that conventional DWH systems that use electricity for heat input are usually good candidates for SDWH replacement.

Typical elements of a SDWH system include some type of solar collection element, some means of storing the water being heated, and a way of connecting these two together. Finally, of course, it is necessary to merge this between the domestic water source and the hot water system of the building. It is quite common to install the SDWH system between the water input to the building and the conventional water heating system. This allows the latter to serve as "back-up." If significant periods of inclement weather occur minimizing the SDWH output, the conventional system can furnish the necessary heating service.

Two types of SDWH systems will be described here, but it should be remembered that many more options than those described are available on the market. One of the two types of systems considered is the integrated collector-storage (ICS) configuration. The other is a freeze protected type that either allows the collector to drain when heating is not taking place or uses an antifreeze loop. An ICS system and a collector for a freeze-protected system are shown in Figure 1.22.

First consider the ICS system. It includes a combination collection/storage unit typically located on the roof of the building. The collection/storage unit is often constructed of several very large diameter copper tubes capped off at the ends, perhaps 4"-6" in diameter, that are connected with small tubing in a serpentine pattern. The city water supply source for the building is connected directly to the inlet of this unit, and the outlet of the unit is connected to the conventional DWH system of the building. When hot water is used, the DWH system supplies the water, the ICS system supplies the DWH, and the water supply source feeds the ICS unit. Through a day, water use will allow the supply water to flow slowly (because the ICS unit has a very large volume relative to the demand use) through the ICS unit, ultimately heating it to whatever the climatic conditions allow. This results in a system that has very little impact on floor space in the building. It operates in a totally passive manner. That is: no pump or special piping is required other than to connect the unit to the water source and the conventional heater. No second fluid, like antifreeze solution, is required. Two downsides include the fact that the unit is heavy, and the location where it is located, if on the building, may need structural reinforcement. Another concern is that there is no positive freeze protection other than the thermal inertia of the large unit filled with water. If severe freezing conditions occur in the local area, freeze damage would be an issue. Finally, another drawback is that the storage of the hot water occurs out in the weather and the night. This influences the overall efficiency of the system because of heat losses from the water at those times.

Many approaches have been put forth to make a system less susceptible to freezing. One of these uses an antifreeze loop and is thus preferred over the ICS system where severe freezing conditions can occur. The system may use two tanks (if so, the first is called the preheat tank, and the second is called the back-up tank, the latter typically being the conventional water heater), or one tank which serves both purposes. If only one tank is used, a heat exchanger is incorporated, and care should be used to try to achieve stratification in that tank. This is not always easy to accomplish. A distinct loop with a pump, thermostatic control system, expansion tank, and back flow preventer is filled with an antifreeze solution, typically a mixture of water and propylene glycol. When the antifreeze solution in the collector is warmer than the tank, the pump is turned on. The antifreeze solution moves from the collector to the heat exchanger with the result of heating the top of the storage tank. The cooled antifreeze solution circulates back to the collector. The heat exchanger that is used can be one in the tank (perhaps a bayonet unit), or a separate loop that brings water from the tank to an external heat exchanger where it exchanges heat with the antifreeze solution. Obviously, this type of approach is more complicated than the ICS approach, but it does offer positive freeze protection. This can be an extremely important characteristic. Every few years, the antifreeze solution needs to be drained, and the system recharged with a new solution.

1.6.3.4 Photovoltaic Applications on Buildings Most photovoltaic systems on buildings are hooked in parallel with the local electrical grid. This will mean the system must interface to the local utility, and usually the latter will have fairly strict rules about how this takes place. This, in a sense, allows the local grid to serve as "storage" for the local PV system, furnishing power when generation is not sufficient for the building requirements. Of course, if it is not desirable or if it is impossible to hook to a grid because there isn't one locally, then typically storage batteries will be needed. This off-grid type of system will not be explored here, but there are certainly many of them throughout the US. The placement of PV systems on buildings is normally referred to as "distributed generation," and this is in distinction from "utility scale generation" addressed earlier in this review. Both distributed generation issues as well as electrical storage issues have been reviewed in the paper by Toledo et al. [2010].

Usually PV panels are placed on the roof of the building and oriented to the south or nearly south (depending upon the roof orientation). Since PV modules generate DC power, this needs to be converted to AC by use of an inverter. Various meters and disconnect switches are usually required by the local utility for final hook up to the grid. Some locations have what is call "net metering." This means that the power meter for grid power can turn forward (when the building is taking power from the grid) or backwards (when the PV system on the building is making more power than is being used locally). These systems may allow the owner to sell PV generated power at the same cost as power is purchased. Other locations have a difference between the price that power is purchased from the utility compared to the price at which it purchases power. In these cases, the latter is less. Some locations may not have a selling price to the utility.

In recent years, more interest in building-integrated photovoltaics (BIPV) has developed. BIPV means that the PV array becomes much more a part of the building design, rather than simply panels that are mounted quite obviously on the roof. This can mean, for example, that the PV units could be of the same size and shape as the roofing tiles. An example of this is shown in the right-hand photograph in Figure 20 of the zero-energy house. In fact, this set of pictures shows the distinction between panels (left hand picture) and BIPV (right hand picture). BIPV could also mean that photovoltaic elements might be used for building siding, skylights, or windows. In some of these applications, the mounted elements may have to be at least partially light transmitting. The upside of these arrangements is that power can be generated by almost any element of the building. However, the downside includes that the sun may not have access to these types of elements for all times of the day; and when it does, the angles might be quite obtuse. Of course, when the element is light transmitting, this means that the power harvest per unit of active area may be quite low compared to conventional arrays located in optimal harvesting orientations. An example of an application is shown in Figure 1.23.

Potential shading is of particular concern in PV applications on buildings. Depending upon the design of the system, it could be the case that partial shading will completely shut down the production in the panel. While there are design features that can minimize this potential problem, shading is undesirable.

The current situation is that some lower cost PV panels (usually based upon cost per peak W generated) have lower efficiency than higher cost panels. Depending upon the limitations of the mounting location (say a smaller roof), higher performance panels may be desirable.

1.6.3.5 Other Solar Applications in Buildings Several other solar applications in buildings are in various stages of development. Only a few will be noted here. Solar lighting is a concept where actual sunlight is brought into the building to provide a natural source. Of course, windows are an example of this; and when the window has a good view from it, this may be irreplaceable. However, windows can be the source of large heat gains or losses.

FIG. 1.23: BUILDING INTEGRATED PHOTOVOLTAICS (BIPV) ARE INCORPORATED IN A NUMBER OF THE FACADES OF THIS BUILDING [Sinopuren 2009].

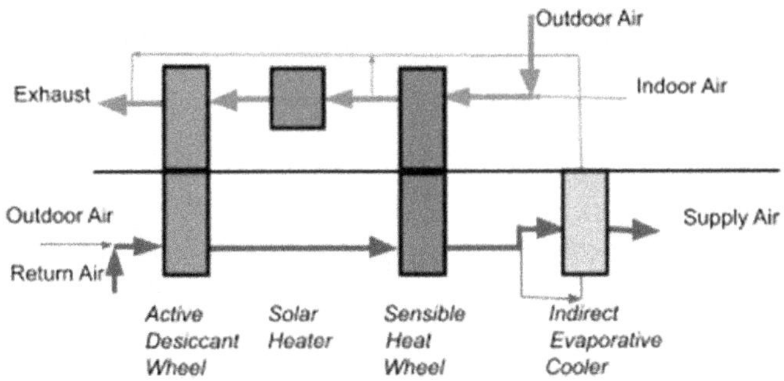

FIG. 1.24: A POSSIBLE SOLAR DRIVEN DESICCANT DEHUMIDIFICATION/COOLING SYSTEM IS SHOWN.

In situations where this is not desirable, some enhancements to the windows may be required. Included is a variety of shading devices (usually preferred on the outside of the window), some type of insulation (like drapes or louver blinds), window tinting, and many other products on the market, as well as design aspects like overhangs and light shelves. If only good quality natural lighting is desired, this might be accomplished with products like light conducting ducts through the roof [Solatube 2010] or solar hybrid lighting [Sunlight Direct 2010]. These devices are designed to minimize the thermal impact of bringing sunlight into buildings. A review of the operational aspects of many of these kinds of systems has been given [Kandilli and Ulgen 2009].

Solar building space cooling is another technology that is moving more toward the market. One of these applications is that of a solar absorption cooling unit that is driven by heat from the sun. Oftentimes these approaches require higher temperatures than are typical from solar water heating systems. Currently the prices of these systems are quite high compared to conventional air conditioning units.

Another approach to cooling is to use a system with desiccants. These materials will soak up water from moisture-laden air when the latter is circulated through the material. Both cooling and removal of water can occur without the use of a vapor compression or absorption cooling system. Desiccant cooling work has been reviewed by Daou et al. [2006]. A possible cooling system using this kind of an approach is shown in Figure 1.24.

1.6.4 Closing Comments

A summary has been given of some of the ways that solar energy may be used in power generation, transportation, and building applications. While solar energy may be more useful in some areas of the country than others, some applications can be adapted in virtually all areas.

The technology has, in general, been expensive because a sizeable market has not developed in the past, which, in turn, kept the prices up. Now, between the encouragement in terms of incentives on the Federal level as well as state and local levels, some technical breakthroughs, and, in some cases, mass production, the markets are developing and some prices have been distinctly decreasing. Better means of storing energy are also coming along, so this periodic resource will be able to deliver useful and economical energy 24 hours a day in the not-too-distant future.

Developments like this can assist the US in becoming much more energy self sufficient in a sustainable way. These are goals that all should desire.

1.6.5 Acknowledgements

I particularly want to credit the excellent assistance by the various students over the years in the Center for Energy Research, without whom the large scope of work we have carried out, some of it touched upon here, would not have been possible. I also appreciate the collaboration of several companies who funded and/or assisted us in a large number of the projects. A great deal of support was received from the US Department of Energy and the National Renewable Energy Laboratory, and that made the whole operation possible.

1.6.6 References

Andujar, J, and F. Segura 2009. "Fuel Cells: History and Updating. A Walk along Two Centuries," *Renewable and Sustainable Energy Reviews*, 13, pp. 2309-2322.

Berger, John J., 1997. *Charging Ahead: The Business of Renewable Energy and What It Means for America,* University of California Press, Berkeley, California, Chapter 4.

Butti, K., and J. Perlin, 1980. *Golden Thread: 2500 Years of Solar Architecture & Technology*, Cheshire Books, Palo Alto, California, p. 29.

Daou, K., R. Z. Wang, Z. Z. Xia, 2006. "Desiccant Cooling Air Conditioning: A Review," *Renewable and Sustainable Energy Reviews*, 10, pp. 55-77.

EIA, 2008. "The Impact of Increased Use of Hydrogen on Petroleum Consumption and Carbon Dioxide Emissions, Appendix C: Existing Hydrogen Production Capacity," EIA Report #: SR-OIAF-CNEAF/2008-04.

ENERGY-10, 2010. Website: http://www.sbicouncil.org/displaycommon.cfm?an=1&subarticlenbr=112

Fraas, L. M, J. Avery, H. Huang, L. Minkin, R. Corio, and J. Fraas, 2007. "Start-Up of First 100 kW System in Shanghai with 3-Sun PV Mirror Modules," *Proceedings of the International Conference on Solar Concentrators (ICSC 4),* El Escorial, Spain, March.

Gordon, J. M., E. A. Katz, D. Feuermann, and M. Huleihil, 2004. "Toward Ultrahigh-Flux Photovoltaic Concentration," *Applied Physics Letters,* 84, May 3.

Jaffe, S., 2010. "Race to the Quarter; Photovoltaic Pricing Trends," paper presented at the Renewable Energy World Conference, Austin, TX.

Kandilli, C., and K. Ulgen, 2009. "Review and Modelling the Systems of Transmission Concentrated Solar Energy via Optical Fibres," *Renewable and Sustainable Energy Reviews*, 13, pp. 67-84.

King, D. L., J. A. Kratochvil, and W. E. Boyson, 1997. "Temperature Coefficients for PV Modules and Arrays: Measurement Methods, Difficulties, and Results," *Proceedings of the 26th IEEE Photovoltaic Specialists Conference,* Anaheim, California, September 29-October 3.

Mahderekal, I., C. K. Halford, and R. F. Boehm, 2006. "Simulation and Optimization of a Concentrated Photovoltaic System," *Journal of Solar Energy Engineering*, 128, p. 139.

Omer, A., 2008. "Renewable Building Energy Systems and Passive Human Comfort Solutions," *Renewable and Sustainable Energy Reviews*, 12, pp. 1562-1587.

Perlin, J., 1999, *From Space to Earth - The Story of Solar Electricity*, Aatec Publications, Ann Arbor, Michigan.

Ricker, Charles, 2008. "Is Solar Thermal Power the Answer?" *Photonics Spectra,* July, pp. 58-59.

Sinopuren, 2009. "Experiencing the Dreams" brochure, Sinopuren Energy Group Ltd., Shenzhen, China.

Slade, A., K. Stone, R. Gordon, and V. Garboushian, 2005. "High Efficiency Solar Cells for Concentrator Systems: Silicon or Multi-Junction," SPIE Optics and Photonics Conference, August, San Diego, California.

Solatube, 2010. Website: http://www.solatube.com/

Sunlight Direct, 2010. Website: http://www.sunlight-direct.com/index.php

Swanson, Richard, 2000. "The Promise of Concentrators," *Progress in Photovoltaics: Research and Applications.* 8, pp. 93-111

Toledo, O. M., F. O. Filio, and A. Diniz, 2010, "Distributed Photovoltaic Generation and Energy Storage Systems: A Review," *Renewable and Sustainable Energy Reviews*, 14, pp. 506-511.

Wang, H., T. Deutsch, and J. Turner, 2008. "Direct Water Splitting under Visible Light with a Nanostructured Photoanode and GaInP2 Photocathode," *ECS Transactions*, 6 (17).

Welford, W., 1983. "Connections and Transitions between Imaging and Nonimaging Optics," Proceedings of the SPIE International Conference on Nonimaging Concentrators, Volume 441, August, San Diego.

Wilson, Max, 2010. "Development of Hydrogen Direct Injection For Conversion of Gasoline Engines," MS Thesis, Mechanical Engineering, UNLV, June.

Zachary, J., 2010. "CSP-Downstream of the Solar Field: Design and Optimization of the Power Plant," presentation at Renewable Energy World Conference, Austin, Texas.

Zhu, L., R. Hurt, D. Correa, and R. Boehm, 2009. "Comprehensive Energy and Economic Analyses on a Zero Energy House Versus a Conventional House," *Energy*, 34, 9, pp. 1043-1053.

Zubi, G., J. Bernal-Agustin, and G. Fracastoro, 2009. "High Concentration Photovoltaic Systems Applying III-V Cells," *Renewable and Sustainable Energy Reviews*, 13, pp. 2645-2652.

CHAPTER

2

SOLAR ENERGY APPLICATIONS AND COMPARISONS: CHINA AND MONGOLIA

Yong X. Tao, Rambod Rayegan, Hongwei Tan and Suraj Talele[1]

2.0 INTRODUCTION

Strictly speaking, all the practical energy sources for applications to human activities on the Planet Earth are from the Sun. Even fossil fuels resulted from millions and millions of years layered deposition which was once living plants (and animals eating plants), obtained their energy intakes and conversion directly and indirectly from solar rays, and reached their maturity before they were buried. From ancient times to today, humans have used various ways to harness direct solar radiation, from using it as a heating source to today's electricity generation systems. Solar energy is also responsible for such renewable energy sources as wind and wave power, hydroelectricity and biomass. Today, although the total energy consumption in the world still predominantly comes from fossil fuels, investing in direct utilization of solar energy has been accelerating on a global scale.

Worldwide, the far most widely used applications for solar energy systems are photovoltaic (PV) systems. 2017 was the watershed year for the world's solar PV installation. 98 GW of solar PV capacity installed increasing the total cumulative solar capacity of the world to 402 GW. 75% of installations came from Asia that led the world in Solar PV capacity addition among which China leads the way. Rising awareness has led to an increase in demand for solar PV installations around the world and especially in Asia. Solar PV is attractive also because of its potential to alleviate pollution, decrease CO_2 emissions and lead the world towards a cleaner and more sustainable environment. Nevertheless, solar power is still dependent on government subsidies, incentives and favorable policies. China added the most solar power (53.1GW) by the end of year 2017 taking its cumulative solar power installation to 131 GW [1]. Distributed systems saw new increases in solar installation in China, where rooftop systems saw increases of three times more. Rising PV equipment prices and political uncertainty caused U.S. solar market to contract 30% more in 2017 compared to 2016. The U.S. is still the second largest installer of solar PV with a total capacity of 10.6 GW. Due to peculiar feature of market trend in Solar PV installations in 2016, market growth was attributed to state mandates in terms of solar installations, but in 2017, maximum installations came through voluntary investments in large-scale solar PV projects.

The weighted average levelised cost of energy (LCOE) on a global scale reaching $100/MWh, which is 73% less compared to that in the year 2010. The U.S. government has announced that the Sun Shot initiative has been successful by achieving 2020 target ahead of time. This is by record-breaking lower bidding prices below $30/MWh, and have been bided for in solar PV auctions in Argentina, Chile, India, Mexico, Saudi Arabia and Emirates of Abu Dhabi and Dubai (UAE). In the year 2017, Solar PV production capacity surpassed 100 GW for cells and 107 GW for modules that are 23% increase and 28% increase respectively, compared to 2016. Manufacturers achieved higher cell and module efficiencies in 2017. A new technology called Passivated Emitter Rear Cell (PERC) has become standard for monocrystalline silicon solar cells because of its higher efficiencies and comparatively less investment. Advanced techniques like multi-bus bars, half-cut cells and Perovskites based solar cells have contributed to increased efficiency and stabilization of the product for commercial deployment.

Although not a dominant market force in the solar energy application, Concentrated Solar Power (CSP) systems also experience a tremendous growth globally. Total global capacity for CSP was at 4.9 GW while the year 2017 saw an addition of 100 MW of CSP. Furthermore, 2 GW of projects are under construction in

[1] This is an update of Chapter 2, "Solar Energy Applications and Comparisons" by Yong X. Tao and Rambod Rayegan that appeared in ASME Press publication, *Energy and Power Generation Handbook: Established and Emerging Technologies*, 2011, New York, NY.
Readers are urged to refer to the 2011 edition of this chapter by Yong and Rayegan which has valuable technical information that would be useful in addition to what is discussed in this update pertaining to "Solar Applications in China." – Editor

China, Middle East and Northern Africa. New additions came from emerging markets but advanced economies did not add fresh capacities in CSP during the year 2017.

In this chapter, energy system applications resulting from the direct solar radiation will be discussed. The particular focus is given to the following applications in China, and an example from Mongolia is also presented.

- Utility-scale solar power systems that generate electricity and feed to the electricity grid. There are photovoltaic (PV) systems and solar thermal power systems; the latter can also produce heat for hot water or air, which is often referred to as the combined solar power and heat systems.
- Building-scale solar power systems, also known as distributed power systems, which generate electricity locally for the building, and may be connected to the grid, or may be a stand-alone systems which require batteries or other electricity storage units. They are primarily photovoltaic systems.
- Solar heating systems for buildings, which are either used as hot water systems, or hot air heating systems.
- Solar high temperature process heating systems for industrial applications.
- Other special solar systems for transportation such as ships and buses.

2.1 OVERVIEW OF CHINA'S SOLAR ENERGY UTILIZATION STRATEGY

Chin is one of the countries that possess significant solar resource with the annual solar hours exceeding 2,200 hr. The regions that have the total annual solar radiation greater than 5,016 MJ/m^2 occupy 2/3 of the country's surface area [2]; therefore, they created favorable conditions for solar energy utilization. From Figure 2.1, it can be seen that the future potential for using solar energy in China is very extensive.

Research and development in solar energy started relatively late in China. It was not until 1970 during the last century that research started to form a certain scale. During earlier stages, China's solar industry exhibited a pattern of two extreme phenomena that relies on the external market support in order to sustain its development. 95% of polysilicon materials used to produce solar PV cells must be imported, and 95% of PV products manufactured in China must be exported to other countries. At that time, China became a manufacture base for the global solar market [3].

Since then, China's solar industry has experienced tremendous progress under the government policy support and investment, and academic research efforts. Both solar research and applications have been catching up to the world standards. As of now, China is the largest producer of solar hot water systems, and largest exporters.

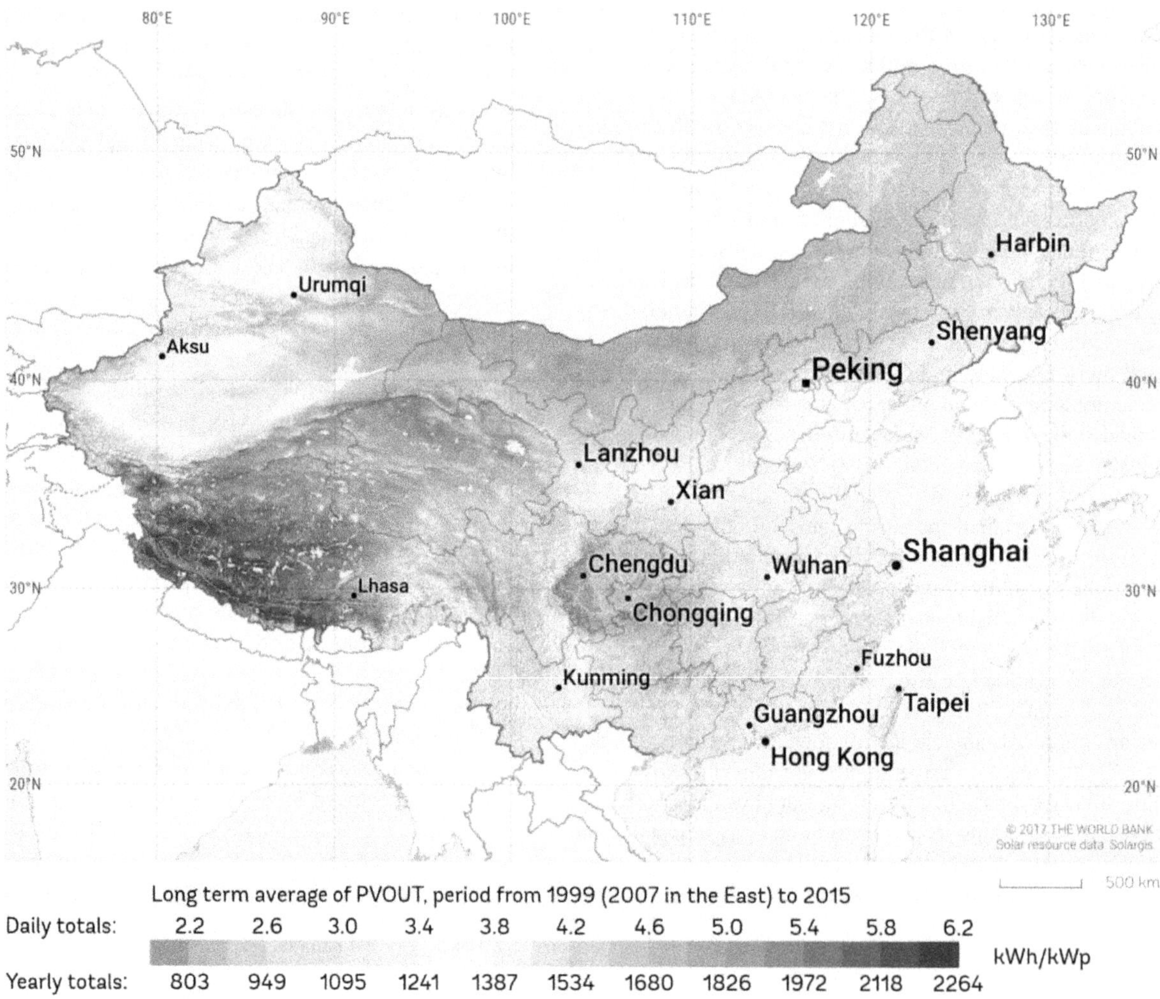

FIG. 2.1: SOLAR RESOURCE DISTRIBUTION IN CHINA

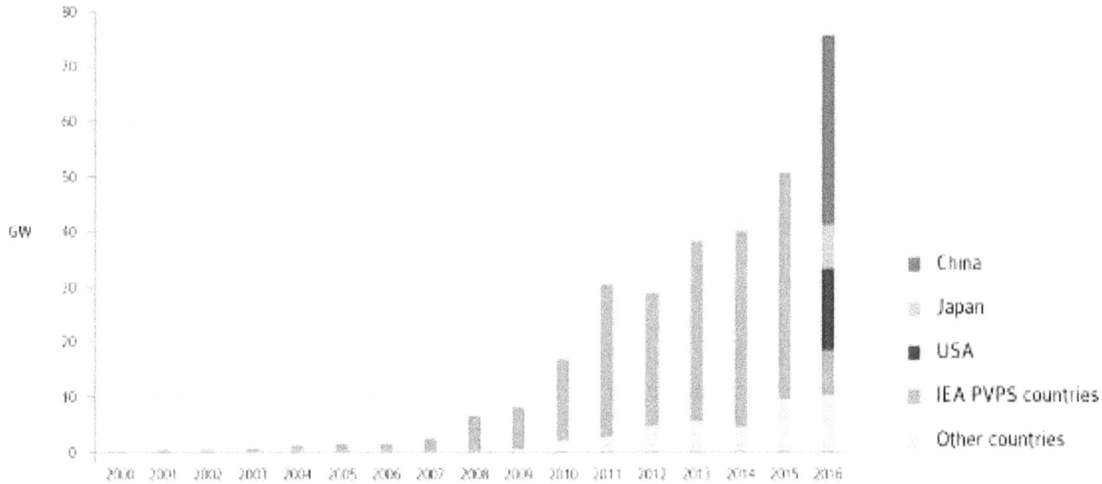

FIG. 2.2: 2000-2016 GLOBAL PV SYSTEM CAPACITY IN GW. (*Source:* International Energy Agency, IEA)

According to the recent report, "2016 Global Renewable Energy Prospects," by Renewable Policy Network for the 21st Century (REN21), the global solar heat collectors increased 6%, in which 77% of the products were from China. In 2012 China became the second largest solar PV market after Germany. Among the new installation of the solar PV systems, China is the number one country in the world for five consecutive years. It is also number one for total accumulative installation of solar PV systems. As seen in Figure. 2.2, 2016's data shows that China once again led the world.

By the end of 2017, China had 130 GW of total system capacity of solar electricity that connected to the grid. By the end of April, 2018, that number exceeded 140 GW [4].

As PV technology and product quality become more and more mature, China evolved from merely a giant assembly-style manufacturing base to become a comprehensive, global force that covered all the elements of innovation, development, manufacturing, complete supply chain, and market penetration. Shown in Figure 2.3, China's solar PV market share is 46% in 2016. In 2017, it exceeded 60%. Eight of the top ten global PV companies are from China, which indicates a strong international competitiveness.

In the meantime, the rapid deployment of solar systems in China has been accompanied by a phenomenon called "Solar Abandoned" (equivalent to Solar Energy Curtailment [5]), which is due to the uneven distribution and mismatch of supply and demand of solar resources in some regions. The Solar Abandoned occurs when the solar PV systems produce more energy than what can be consumed and are unable to transmit the electricity to remote needed areas due to limited transmission lines [6].

Recently the rate of Solar Abandoned became more severe. In 2015, the entire nation has the abandoned rate of 12%. It reduced to 11% in 2016. Further efforts made it to 6% in 2017. However, in some areas, mainly western regions such as Gan Su and Xin Jiang, the worst abandoned rates are 20% and 23%, respectively. Overall, China's rapid PV development and installation have exceeded its grid capability of receiving solar electricity.

Facing the challenges of excess solar electricity in the west and shortage of electricity in the east, China started to transform its deployment strategy from constructing the large-scale ground-based solar electricity plants in the west to strengthening the distributed PV systems.

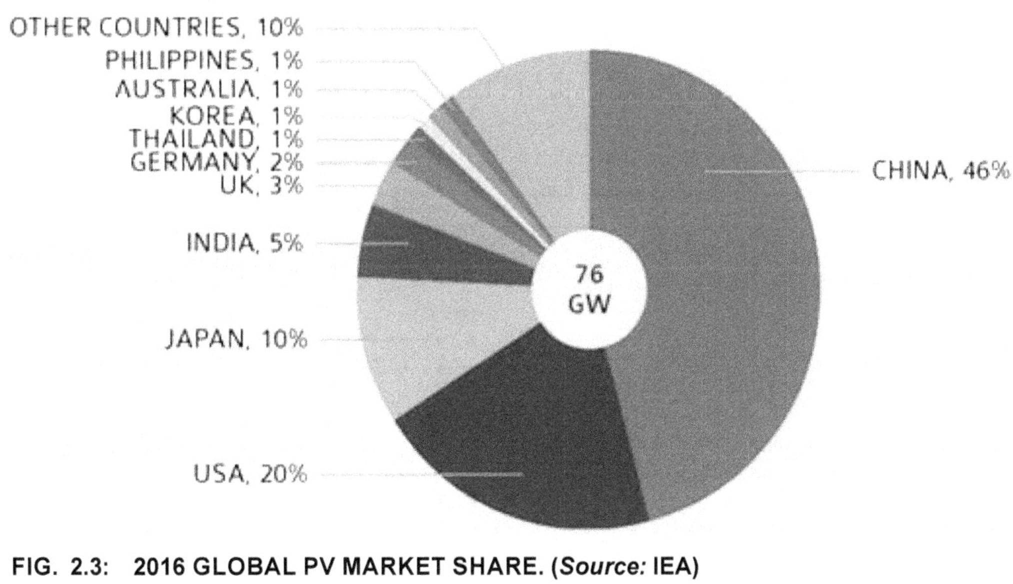

FIG. 2.3: 2016 GLOBAL PV MARKET SHARE. (*Source:* IEA)

Early-stage distributed solar PV systems included the Golden Solar Demonstration project and the Architecture-integrated PV system demonstration projects in 2009. Since 2013, the emphasis has been in development of distributed PV systems. At the end of 2014, the distributed PV systems, including grid-independent systems in China were about 5.0 GW, or 17.62% of total PV systems, as opposed to 82.38% of large-scale ground based systems. However, in 2016 its capacity increased to a total of 10.32 GW with 4.24 GW being the new installations, a 200% increase compared to 2015. In 2017, 500,000 households had PV systems installed [7] with a total of 19 GW capacity. 36% of all the PV systems are now of distributed ones.

The distributed systems brought a new round of significant development for Chinese PV industry. In the meantime, a new round of challenges became unavoidable. According to the standards in the developed countries, a solar energy utilization efficiency between 79% and 83% is considered normal. Anything exceeding 83% is considered to be insufficient solar generation and less than 60% means excessive generation. From 2008 to 2012, approximately 30 GW of total solar electricity was generation in China, which was about 60% of global generation, its solar energy utilization efficiency was only 57%, a severe, excessive solar energy generation and waste. The main reason was the blindly rapid expansion of solar system construction due to lack of planning and questionable government-dominant policies.

Facing such challenges, a major correction was made in 2018. On May 31, a new government policy, named "531" policy was announced that impacted the PV industry tremendously. The 531 policy announced a reduction subsidizing the industry by 0.1 CNY/kW for the ground stations, and 0.05 CNY/kW for the distributed PV systems. The capacity of any solar plant was capped at 10GW, and permits were no longer issued for the ground-based solar stations. A new era of development has started.

The "531" policy not only received strong reactions domestically in China but was also closely watched globally. U.S. Energy and Power Research Institute (EPRI) senior researcher Mary Hutzler, pointed out that due to the adjustment of subsidy of PV power systems, it was predicted that China production of solar power would have reduced by as much as 40%, or about 20 to 35 megawatts. The subsidy reduction policies would not only be a trend for China but also a global trend.

Challenges came with opportunities. On one hand, the large degree of reduction in PV assembly prices provided the possibility of an increase in PV system productions. On the other, local governments and companies were working together to realign their development strategies for new market and new industrial applications. They focused more on increasing production and system efficiencies and improving product qualities. These efforts enabled the industry to push technology innovation, reduced costs and eased the pressure due to subsidy. This trend in streamlining the processes in domestic industries and focusing on quality rather simply on quantity is believed to add a positive, healthy blood to the global PV industry ecosystems.

2.2 LARGE-SCALE SOLAR ENERGY PLANTS FOR POWER GENERATION

Majority of large-scale solar power plants are concentrated in the Northwest regions of China. These areas are generally rich in solar influx yet vast lands and arid desert. There are two types of solar power generation technology deployed in those areas: Solar PV electricity and solar heat to electricity technology.

FIG. 2.4: DUNHUANG PV ELECTRICITY PLANT IN QIUQUAN CITY OF GANSU PROVINCE

2.2.1 Large Scale Solar PV Electricity Plants in China

The total new installation capacity of PV electricity systems was 53.06 GW in 2017, among which 3.362 GW was of PV power plants, an 11% increase compared to the prior year. By the end of 2017, it increased to 100 GW, and 10.059 GW, respectively.

Dunhuang, a township in the city of Jiu Quan, Ganshu Proivnce, located in the most western part of West River Corridor, and on the borders connecting three provinces, Gansu, Qinghai and Xinjiang, the annual solar hours is 3,362 hrs on average. The annual daily radiation is 6,882.1 GW/m^2. Since 2009, when the nation's first 20 MW grid-tie PV electricity power plant was constructed in the Dunhuang PV Industrial Park, the park has developed into a 254 square kilometer establishment (Figure 2.4). In 2017, the Dunhuan reached 780 GW solar power generation capacity. The grid system received more than 546 GWh electricity.

Parallel to the development of PV industry, China emphasized the integration of establishing renewable energy base with the sustainable development and resource utilizations in the arid desert natural ecosystems [8]. The relevant projects achieved good effects on the environment.

The Inner-Mongolia Autonomic Region is located in northern China. Rich in solar energy resources, it is one of the higher solar influx areas in China. The annual solar radiation intensity there is between 4.8 and 6.4 GJ/m^2. The annual solar hours is between 2,600 and 3,300 hours on average. The distribution characteristics

FIG. 2.5: BIRD-EYE VIEW OF EARLY CONSTRUCTION SITE OF SOLAR POWER PLANT IN DENGKOU COUNTY OF INNER-MONGOLIA REGION

FIG. 2.6: TODAY'S VIEW OF SOLAR POWER PLANT IN DENGKOU COUNTY OF INNER-MONGOLIA REGION

FIG. 2.7: 135 MW HIGH EFFICIENCY AGRICULTURAL PV POWER PLANT IN TOKETO COUNTY, INNER MONGOLIA

of solar resource distribution in the Region is that the solar intensities increase from eastern to southwestern areas [9].

One example is the county of Dengkou of Inner-Mongolia Region, which is located near the eastern border of Wulanbuhe desert. The 68.3% of the entire county is desert. The county abandons the traditional anti-desertification efforts, and created new areas that utilize the abundant solar energy. It integrates concentrated PV electricity generation, agriculture facilities and eco-tourism into a systematic economic development with more than 1,667 acres (Figures 2.5 and 2.6). The total investment is 4.9 billion Chinese Yuan with a 500 GW solar plant. Initially, the solar panel covered ground was planted with the grown-cover grass (Sabina Procumbens). The electricity generated by solar energy was used for irrigation. Now 253 agricultural facilities have been constructed to grow tea trees, mushrooms and flowers with additional 30 GW PV electricity system. The success of this park gives hope to the local residents for new economic development ideas and higher requirements for solar power systems and for the better life overall.

Another project in Inner Mongolia, in which solar electricity was complemented by agriculture is in Tokto County, where a 135 MW high efficiency PV electricity power plant was constructed. In addition to the benefit of solar electricity generation, the economic development due to optimal energy infrastructure, improvement in land environment and agriculture development increase the income streams for the local citizens despite heavy saline soil conditions. As a result, the first stage 100 MW system, that was connected to the grid in 2014, alone generated the gross income of more than 120 million CNY in 2015. With 35 MW added in 2015, the entire solar electricity plant runs a smooth operation and achieved the goal of integration of PV industry with agriculture and whole economic ecosystem benefits (Figure 2.7).

From the above examples, it can be seen that the development of solar PV technology in China has reached a more matured stage and commercial grade. The existing problems primarily are how to store excess energy and consequent instability problems of solar energy generation. Therefore, the recent research and development efforts have been shifting towards solar thermal power generation systems which are perceived to have the advantages of capabilities in storing energy, managing peak demands and providing the continuous electricity supplies. These areas have become the hot spots of investment as well.

2.2.2 Large Scale Solar Thermal Power Generation Systems

2.2.2.1 Parabolic Trough Solar Power Technology. Although many solar technologies have been demonstrated, parabolic trough solar thermal electric power plant technology represents one of the major renewable energy success stories of the last two decades. The main reason is that it has been recognized as one of the lowest-cost solar-electric power options available today, and has significant potential for further cost reduction. Nine parabolic trough plants, totaling over 350 megawatts (MW) of electric generation, have been in daily operation in the California Mojave Desert for up to 18 years [6]. These plants provide enough solar electricity to meet the residential needs of a city with 350,000 people. They have demonstrated excellent availabilities (near 100% availability during solar hours) and have reliably delivered power to help California meet its peak electric loads, especially during the California energy crisis of 2000–2001. Growing interest in green power and CO_2 reducing power technologies have helped to increase interest in this technology around the world.

Parabolic trough power plants use concentrated sunlight, in place of fossil fuels, to provide the thermal energy required to drive a conventional power plant. These plants use a large field of parabolic trough collectors, made of non-optical mirrors for low cost, that track the sun during the day and concentrate the solar radiation onto a receiver tube located at the focus of the parabolic shaped mirrors. A heat transfer fluid passes through the receiver and is heated to temperatures required to generate steam and drive a conventional Rankine cycle steam power plant. The largest collection of parabolic systems in the world is the Solar Energy Generating Systems (SEGS) I through IX plants in the Mojave Desert in southern California [6]. The SEGS plants were built in the 1985 to 1991 time frame. All of the SEGS plants are "hybrids," using fossil fuel to supplement the solar output during periods of low solar radiation. Each plant is allowed to generate 25 percent of its energy annually using fossil fuel. With the use of the fossil hybrid capability, the SEGS plants, in Southern California Edison (SCE) on-peak hours, have exceeded 100 percent capacity factor for more than a decade, with greater than 85 percent from solar operation. Since 2011, the following new parabolic trough plants have been built in US: Mojave solar project in Barstow, California, completed in December 2014 with gross capacity of 280 MW corresponds to net capacity of 250 MW, Solana generating station in Gila Bend, Arizona, completed in October 2013, with 6 hours thermal energy storage, and Genesis solar energy project in Blythe, California online on April 24, 2014 [7].

2.2.2.2 Parabolic Trough Solar Power Plants in China China entered the solar thermal power generation market only recently. The number of current systems in operation is relatively small. However, it has gained the dominance in terms of solar thermal power systems that are either under construction or in the planning stages. Up to the end of 2016, the total capacity of solar thermal power plants was 24.2MW, which laid the good foundation for technology development through demonstration projects. In June 2016, the National Energy Bureau announced the first 20 demonstration projects for solar thermal power plants. The total system capacity reached 1.35 GW. In December of the same year, a national strategic plan named "Solar Development Thirteen Five Plan", proposed an aggressive plan to push forward the development of solar thermal power industry. The plan called a total installation capacity of 5 GW by the end of 2020, boasting a new era of solar thermal power technology in China.

The first large scale commercial solar thermal trough project was located in the city of Delingha in Qinghai province. It was put in test run in September 2018. The solar thermal power plant consists of three parcels: Solar Optic Field (SOF), Heat Transfer Fluid and Thermal Energy Storage (HTF+TES) and Balance of Plant (BOP). The first stage capacity was 50 MW with the future expansion to 100 MW. The technology includes parabolic trough collectors, conducting thermal transfer oil, and thermal to electricity conversion. There are 190 trough collector forming a standard loop, collected to a double-tank, binary molten salt (nitride/potassium) thermal storage system of 1,300 MWh (Figure 2.8). It can store the needed energy for 7.5 hours. The generator is a 50 MW system with a mid-temperature, high pressure, single-stage regeneration, and water-cooling steam turbine.

Among four types of solar thermal power generation systems, the most matured system is tough systems, which have a share of 94% among all world solar power systems. The tower systems have only 4.47% share, and Fresnel and Disk systems are less than 1%. The climate and environmental features in China, however, determine that the trend is towards the tower-type, higher concentration ratio, and higher light-thermal conversion ratio technology.

2.2.2.3 Power Tower Systems in China The first commercial solar power tower system in China was connected to power grids in Delingha city of Qinghai province in July, 2013 (Figure 2.9). This 50 MW system was the sixth commercial, grid-connected system in the world. The location of Delingha is on the Qingzang Plateau with 3,554 hours of sun lights. It is the famous sunshine belt with very high solar development value.

By 2017, the investment on this project exceeded 990 million CNY, the permitted land is 3.3 square meters. Using molten salt thermal storage technology, it can stare 7 hour thermal energy with the designed annual electricity generation of 146 GWh. The plant powers 80,000 household and saves the equivalent 46,000 tons of coals. It generated excellent economic and social benefits while reducing about 120,000 tons of CO_2.

In 2016, referred to as the third in the world and the first in Asia, a 10 GW, molten salt, power tower plant started the construction in Dunhuang Solar Electricity Industrial Park in Gansu province. The total investment was 420 million CNY. It has a 140-meter tower absorbing solar radiation energy, carried to

FIG. 2.8: SOLAR THERMAL TROUGH PROJECT IN DELINGHA CITY OF QINGHAI PROVINCE

FIG. 2.9: POWER TOWER PLANT IN CITY OF DELINGHA, QINGHAI PROVINCE

molten salt, from 1,500 plus mirror collectors, beautifully forming a circle of "sunflowers" tracking solar rays (Figure 2.10). By 2018, the second-stage construction has consumed 3.4 billion CNY investment fund. The plant spans over 7.84 square kilometers. This project is not only one of the solar thermal electricity demonstration projects, but also the largest commercial molten salt tow solar thermal electricity plant.

Overall the solar thermal electricity systems in China are still in the stage of demonstration projects even though there exists commercial-scale solar thermal electricity plants. Compared with the solar PV power technology, the scales of investment and development of concentrated solar thermal power systems lag behind significantly. Needless to say, the potentials for development opportunities can never be underestimated.

FIG. 2.10: DUNHUNG SOLAR THERMAL ELECTRICITY PROJECT (*Source:* Dunhung PR)

2.2.3 PV System Application in Republic of Mongolia: Solar Energy Park

In the economic development region of Dornogovi, Zamyn-Üüd, the southeastern part of People's Republic of Mongolia, a 15 MW solar electricity plant has been under construction (Figure 2.11). According to the government publication, the plant installed PV assemblies that can sustain the extreme weather conditions, and is expected to produce annually 32 GWh electricity. It is the second installation of PV power plants, and the larger of two plants.

Mongolia intends to reduce the dependence on the aging coal power plants by utilizing solar power generation. Another project under development is the largest 30 MW project, funded by European Bank of Recovery and Development (EBRD) and other partners. According to EBRD, there are four renewable energy projects being funded with 114 million U.S. dollar investment. In total these four projects produced 180 MW of electricity which is about 16% of the total Mongolian plant capacity. Last year, a FIT plan was implemented to develop further 200 MW solar energy capacity and 450 MW wind energy.

FIG. 2.11: SOLAR ENERGY PARK IN MONGOLIA

2.3 DISTRIBUTED PV SYSTEMS FOR BUILDINGS

2.3.1 Distributed PV Systems for Buildings in China

2.3.1.1 Building Integrated PV Systems. Centralized PV electricity generation plants in China are mostly installed in northwestern arid and desert areas. Due to the low power load demands in those areas, it lacks adequate consumption ability from the produced electricity; therefore exists the "Solar Abandoned" phenomenon. On the contrary, the well-developed eastern regions in China demand high power loads. The lack of large available lands prohibits the deployment of large-scale PV systems. A vital option is to apply the distributed PV systems as an alternative to the buildings and structures that can provide the balance condition of matching the generation capability with the load demands.

The high-degree of urbanization in eastern regions of China provides a favorable development space for building-based distributed PV systems. From "Solar Roof Plan", established in 2009, to 2017, the distributed PV systems have increased to be 23% of all PV systems in China. The rapid progress from simple roof-top systems to building –integrated systems has attracted significant attentions from various stakeholders.

The first adopter of building PV systems is the public building sector. There are two categories of PV applications in buildings for distributed PV systems. One is Building Integrated Photovoltaic (BIPV) system, and the other is Building Attached Photovoltaic (BAPV) system. Figure 12 shows the total of both BIPB and BAPV systems constitutes 22.67% all the PV systems. In addition, there is a small percentage, about 0.27%, of the systems that are grid-independent. It is clear that a need for policy adjustment could lead to further expansion of development for distributed PV systems.

2.3.2. Selected Applications of Distributed PV Systems

2.3.2.1 Educational Buildings. Pingtan High School, affiliated with Fujian Normal University, installed a 1,708-panel solar PV system on the roof of classroom building and laboratory building. The system converts captured solar radiation into DC currents which are in turn processed to AC currents meeting the school's electricity needs. The system is tied to the electric grid and provides excess electricity to the grid. The system produces about 4.8 MWh and contributes to the near zero electricity bill annually with the government subsidy (Figure 2.13).

The "Sunshine Apartment" buildings are located near the bank of Yao River in Zhejiang province. The PV panels and solar heating panels are installed on its roof and the specially designed, 70-degree inclined balconies. According to the users, two apartment buildings can produce 350 kWh electricity per day under normal conditions. Each unit can produce 7 kWh per day on average for daily living, and any excess electricity is contributed to the electrical sub grid in the plant as supplement to industrial production (Figure 2.14). The model of "Sunshine Apartment" is very appropriate for industrial buildings, and is a perfect supplement for peak-load shedding required in many power-thirsty industrial enterprises.

In Tongji University, the Institute of Architecture Design Building was renovated from an original bus parking structure. The original structure has 40,000 square meter occupying space.

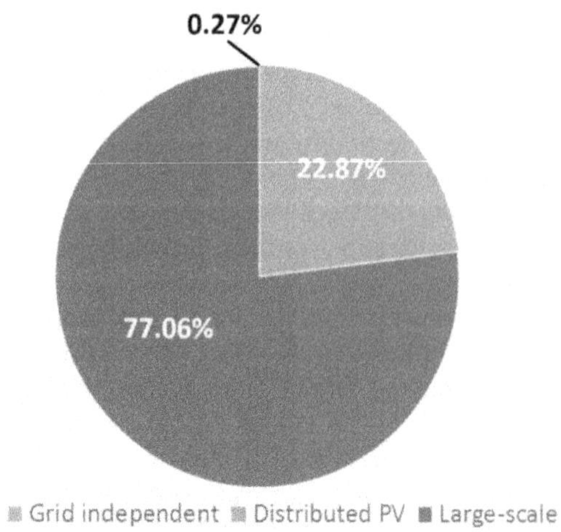

System Type	Annual Capacity MWp	Accumulative Capacity MWp	Market Share %
Grid independent		360	0.27
Distributed PV	19440	29730	22.87
Large-scale	33620	101042	77.06
Total	53060	131140	100

FIG. 2.12: SUMMARY OF SOLAR PV SYSTEM DEVELOPMENT IN CHINA

FIG. 2.13: DISTRIBUTED SOLAR PV SYSTEM IN PINGTAN HIGH SCHOOL OF FU JIAN NORMAL UNIVERSITY

FIG. 2.14: PV AND HEATING PANELS INSTALLED ON THE 70-DEGREE BALCONY SURFACES IN THE "SUNSHINE APARTMENT"

Based on the economic principles of reuse and retrofitting, the renovation of the structure went through a redesign and reutilization engineering process. The renovation adopted the increase in utilization of integrated PV systems consisting of mono silicon, polysilicon and thin-film solar cells on the roof and thin-film façade PV cells. The total PV surface on the roof is 660 square meters (Figure 2.15).

Another one of the large solar system application examples on campus is in Jiangsu College of Urban and Rural Construction. The new campus of this vocational school utilizes roof-integrated PV power systems and is another success story of green campuses.

Jiangsu College of Urban and Rural Construction is located in the city of Changzhou of Jiangsu province. The campus occupies a land of 700 mu (about 115 acres) with the total occupancy area of 281,000 m^2. From the initial design stage, a concepts of green design, green construction, green operation and green life cycle were integrated with the green campus development. The main campus buildings achieving star-level two green architecture design guidelines exceeds more than 48% of the total area. Together dozens of green technologies are applied. The campus 2.2 MWp distributed PV power station received the renewable energy demonstration project of the year in 2014 from the Province of Jiansu. The project won the second place in the SNE 2015 Challenge Cup Solar Energy and Architecture Design and Engineering Competition. The system was completed officially connected to the grid in October 2016.

The new campus buildings are designed as traditional southern China architecture that features white walls, black tiles. The roof structure often has two-slope surfaces. Altogether there are 22 occupied building units and 25,000 m^{2}roof and wall surfaces for PV installation.

Depending on the usage of the buildings such as classrooms, dormitories, dining halls and laboratories, different types of PV components of polysilicon, monosilicon or thin films assemblies

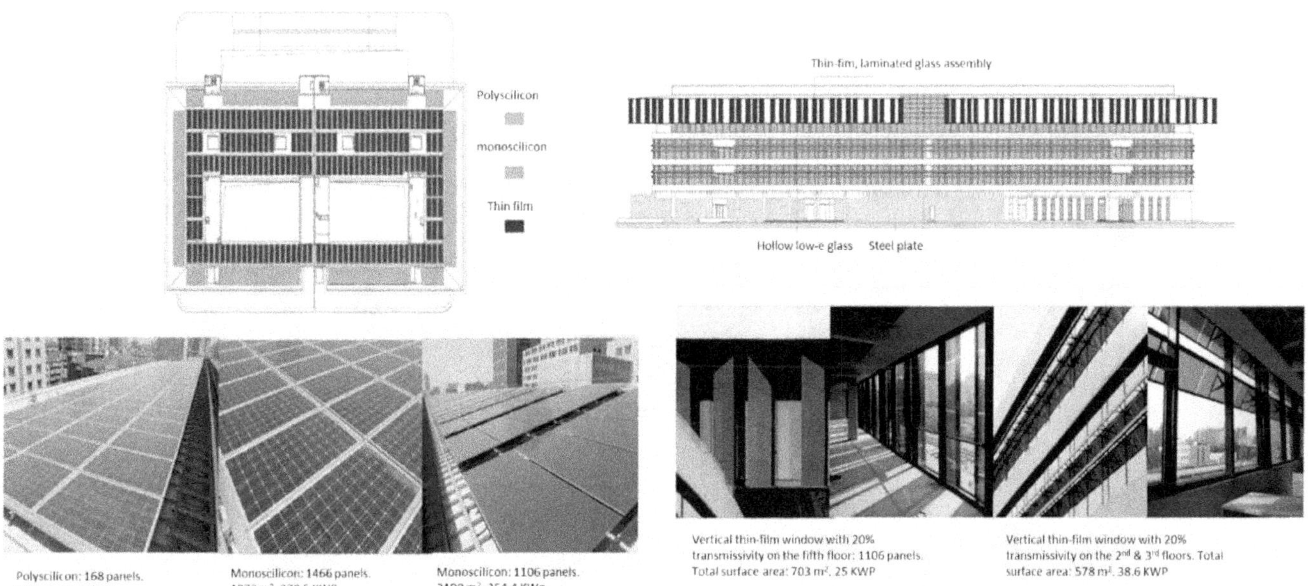

FIG. 2.15: INCORPORATION OF PV SYSTEMS IN THE RENOVATION PROJECT FOR AN EXISTING BUILDING: INSTITUTE OF ARCHITECTURE DESIGN, TONGJI UNIVERSITY

FIG. 2.16: DISTRIBUTED PV SYSTEMS IN JIANGSU COLLEGE OF URBAN AND RURAL CONSTRUCTION

are used for BIPV or BAPV systems. The total capacity of the systems is 2.2 MWp. The operation modes include low voltage grid tie, used as produced and the excess going to the grid (Figure 2.16).

2.3.2.2 PV Applications in Office Buildings. The challenges in PV applications to office buildings are the harmony or clash between aesthetical and PV system requirement and performance. A good example of such application is the R&D building of Xing Ye Solar Energy Science and Technology Company in Zhuhai.

The project useable area is 219,000 m^2 with 17 stories above ground and 67 m in height. The total PV capacity is 155 kWp about 12% of the total power consumption for the building (Figure 2.17). The solar power features include seasonally controlled, ventilated PV curtain walls, vertical elevator ventilated PV skylights, garden-style PV roof shade, laminated PV glass female walls and double layer PV umbrellas. The data center of the building is powered by innovative microgrid or local direct-current microgrid systems.

2.3.2.3 Residential PV Applications. PV applications in residential buildings are primarily in the detached houses, which are not main-street building types in China. Despite this limitation, the development of PV in this sector grows rapidly thanks to the government policy. The main mode of utilization is to generate electricity for house use with limited excess power to the grid.

Chinese roof-top PV market started to grow in 2015 when 30,000 sets of systems were installed. In 2016, it grew by 15,000 systems, and increased to 25,000 systems in 2017. By estimate, at least 400,000 residences could have PV systems installed this year. One of the market players of PV systems is Zhen Tai, which installed more than 30,000 systems in 2017 and hoped to increase by 20,000 systems in 2018. Other manufacturing companies, such as Jingke, Xiexing, Yingli, Tianhe PV and Pingao all planned to enter the residential market directly this year. It has been estimated that about 40,000 new roof-top systems will be installed in China with an accumulated capacity of 2 GW. It has been predicted that China will be ranked number two in the world by 2018 and exceed U.S. to become the largest PV market in the world in 2021. Comparing with commercial sectors' applications due to the continued decrease in FIT (Feed in Tariff), the residential rooftop applications of PV systems will become more

FIG. 2.17: BIPV APPLICATION IN GREEN YES BUILDING OF XING YE SOLAR ENERGY SCIENCE AND TECHNOLOGY COMPANY

(a)

(b)

FIG. 2.18 (A) AND (B): EXAMPLES OF BIPV SYSTEMS IN SHANGHAI'S SINGLE-FAMILY HOUSES

profitable. The policy support from Chinese central government will further boost the related economic development.

Figure 2.18 (a) and (b) show examples of BIPV applications in Shanghai's single-family residential buildings. Shanghai is located in the region III of solar utilization potential. The key factor for BIPV applications is the economic feasibility. The sample projects like those shown in Figure 19 indicated that a 4 kWp capacity per house can reach a 50% total electricity consumption goal with the ROI of 6 years, including the government refund incentives.

2.3.2.4 Large Scale Facility Applications of PV systems The distributed PV systems have demonstrated the success in both small buildings and middle to large size buildings, with the advantages of the rooftop systems occupying much less land than large-scale PV power plants. However, they still face the limitation of available surface areas and scattering. Therefore, large scale facility that have larger surface areas such as in industrial and agricultural buildings (warehouses, factories, fishing facilities, etc.), have proved to be emerging as a vital alternatives for installation of rooftop PV systems. As a result, projects like water-PV, fish-farming and PV, and other related hybrid economic projects start to emerge.

The first rooftop solar power plant in Dalian Port provides a power source of distributed PV to the development of the low-carbon, green cycle port and is connected to the power grid (Figure 2.19). This project utilizes the original roof areas of the port warehouse and does not occupy any additional land resource. Therefore, it significantly enhances the building utilization efficiency.

FIG. 2.19: THE FIRST ROOFTOP SOLAR PV POWER STATION IN DALIAN PORT

The total generation of electricity from the rooftop PV system met the majority of user needs and the remaining power was supplied to the national power grid. In 2017, it produced a total of 2.45 GWh electricity with 2.24 GWh for the self-consumption of buildings. The remaining 200 MWh was supplied to Dalian Utility Company. A 2.576 million RMB of economic benefit was received, equivalent to a reduction of 809.5 tons of standard coal, 2,446 tons of CO_2, 59.7 tons of O_2, 29.8 tons of NO_x, and a significant accomplishment of energy saving.

China's second largest oil field base is in Shandong Dongying City completed its largest rooftop PV power station of 17.7 MW in June 2017 (Figure 2.20). Among the eastern region of China, Dongying receives a sufficient supply of solar energy with an average of 2,700 hours of annual direct solar radiation, which is about 5,224 MJ/m^2 annually, a great potential for PV power generation. The project estimates an overall 580 MWh power generation from solar energy over a 25-year period. Comparing with the traditional fossil-fuel combustion based power plant, it cuts about 490 thousand tons of carbon dioxide and 3,300 tons of carbon sulfoxide.

Further south in Zhejiang province, the Hangzhou International Airport successfully installed its distributed PV power generation system over the roof of freight terminal building (Figure 2.21). The grid-connected system not only produces electricity for the airport, but supplies the access electricity to the national grid. The airport becomes the first "green" airport in Zhejiang province. The total PV surface area is 24,000 M^2 with a capacity of 2.3 GW. It is estimated that the annual electricity generation will be 3.078 MWh with a daily generation of 8.5 kWh. Comparing a traditional power plant, the project saves annually 1,231 tons of coal and cuts the 3.226 tons of CO_2 emission.

FIG. 2.21: DISTRIBUTED PV SYSTEM ON THE ROOF OF FREIGHT TERMINAL OF HANGZHOU INTERNATIONAL AIRPORT IN ZHEJIANG PROVINCE

Academician Chen Yiyu of China Academy of Science presented in a report, entitled "Lake Health and Sustainable Development," that China has more than 80,000 Km^2 of lakes. For example, in Lake Er located near City of Dali in Yunnan province, water surface is about 250 Km^2. In Bohai Bay, the water surface area exceeds 20,000 Km^2, and could have 1 billion kW of solar PV panels installed.

In November 2017, the first ocean fishery project integrated with the grid-connected solar PV project was completed in Guangxi province. The project utilizes the ample daily solar radiation in the bay areas and develops a modern integration of fish farming with PV power generation, a win-win economic development example of power production above water and fish farming under water. A 1,000 mu (about 607 acre) of water surface area is used with a total investment of 0.76 billion RMB (about 119 million USD).

FIG. 2.20: THE LARGEST ROOFTOP PV POWER STATIONS IN DONGYING CITY OF SHANDONG PROVINCE

FIG. 22: FISHERY AND PV POWER GENERATION INTEGRATION PROJECT IN GUANGXI PROVINCE

FIG. 2.23: EXAMPLE OF APPLICATIONS OF A PV SYSTEM IN HOG FARMING ESTABLISHMENT

The total power capacity is 80 MW and 825 mu (136 acres) of fishing area (Figure 2.22).

In summer time when the outdoor temperature becomes 32-33°C, the water temperature underneath PV panels is about 28-30°C, which is very suitable for baby shrimp growth. During winter time, PV panels provide shielding effects to wind and therefore reduce the wind chill effect for fishery products. The PV panel's shielding effects of summer cooling and winter insulating helps fishery farming to reduce the weather induced risks and enhance the yields.

Comparing with the land-based PV power stations, the water surface PV systems are installed on the floating supporting structure (Figure 22). They not only save the land resources but also reduced the panel surface temperature which increases the power conversion efficiency. Because these systems are usually installed adjacent to a city, they tend to provide a good match between power availability and demand usage. Overall, the water surface PV systems have the advantage of large scale construction, easy grid connection and ultimately high power generation capacity.

2.3.2.5. PV-Enabled Poverty-Alleviation Projects Regardless of being installed on residential buildings, industrial buildings or water-surfaces, distributed PV systems demonstrate the advantage of flexibility and adaptivity to local characteristics. By utilizing this feature, Chinese government initiated various projects that combine the distributed PV systems with public economic development. In particular, it focuses on mountainous regions where the regional development and poverty reduction are the priority. By encouraging building wealth through solar PV power system development, the public life can be improved.

The approach is based on the following prediction: As an addition to many established poverty alleviation efforts such as development of enterprises, tourism, electronic commerce, science and technology and housing migration, the PV-enabled projects will solve partially the problems of lacking electricity and boost the enterprise development. Ultimately the goal for the long-term, is that sustainable economic development can be achieved (Figure 2.23).

Ningxia Hui Nationality Autonomous Region is one of the regions in China that has abundant solar radiation resources. It has a yearly solar density of about 4,955 to 6,155 MJ/m^2. Herun Village in Huangzebao Township, which belongs to Guyan City in Ningxia, has a yearly solar density of 5,561.19 MJ/m^2, and its daily solar radiation is also adequate for developing PV electricity stations. Herun Village was therefore chosen for the first PV-Poverty Alleviation project. The government invested to install a 3 kW grid-tie PV electricity system on each household roof. Each system consists of 12 high-efficiency silicon cell assembly and one Sununo Plus 3 kW inverter. After completion of the installation, 30% of the profit from PV electricity generation will belong to each household, providing a guaranteed annual income of no less than 3,000 RMB (Figure 2.24).

Further south, Dawan Village of Jingzhai County in the City of Liu'An of Anhui Province, there is another PV-Poverty Alleviation project, which combines the installation of PV systems with Lingzhi farming. Lingzhi farming itself is a poverty alleviation project, in which local residents are trained to grow a traditional Chinese medicine, called Lingzhi (Ganoderma Lucidum) for the market. Liu'An is one of the poorest regions in Anhui, and Jingzhai, as one of the oldest Chinese revolution regions, was determined

FIG. 2.24: PV-POVERTY ALLEVIATION PROJECT IN HUANGZE VILLAGE

to change the poverty status. The local government combined the two poverty-alleviation projects by building tent-style coverage over the Lingzhi growing areas, on top of which solar panels were installed. The double income from both PV electricity generation and harvesting of Linzhi yields about 3,000 RMB annually for each household from PV electricity and the values of Linzhi per unit mu is between 13,000 to 15,000 RMB.

The area of the PV enterprise is 3,721 mu (about 613 acres) and is the largest PV enterprise in term of area per unit enterprise in China (Figure 2.25). Annual power generation capacity is 220 MWh, which is equivalent to reduction of 100 kilo tons of coal. The project created more than 700 jobs and improved the incomes of more than 200 households.

From two examples illustrated above, it can be found that a great achievement through PV to Poverty alleviation policies has been made. It has been evaluated however that overall the relatively high initial costs are still a main obstacle for wider adoption. It was required that the low-income households needed to share the initial investment fund in addition to government and industry's contributions. Because of the level of poverty, the very poor household cannot afford their share of the initial fund [14]. On the other hand, the central government's subsidies to the projects are the indicators for future expansion and current implementation. It has been recognized that the sustained government subsidies to PV to poverty alleviation projects are the key factor influencing the enthusiasm of low-income people and solar industry.

FIG. 2.25: DAWAN VILLAGE PV-LINZHI PROJECT IN ANHUI PROVINCE

2.4 SOLAR THERMAL SYSTEMS FOR BUILDINGS

2.4.1 Variety of solar heating system applications in China

There are three types of solar heating technology based on the temperature ranges: Low temperature (<100 °C), mid-temperature (100-250 °C) and high temperature (> 250 °C) applications. Based on the utilization categories, solar heating technologies are further divided into variety of systems of solar hot water, building heating, refrigeration, drying, oven, distillation and thermal power plant. The most common applications in building solar heating systems are for low temperature applications such as solar hot water and room heating.

The China's Ten Two Five strategic plan explicitly calls for "wide-applications of solar thermal utilization." This stimulated tremendously the solar thermal industries. Earlier solar heating system development in China was primarily for single-family household hot water supply. In 2014, the national statistical data for solar thermal applications showed that 40% of solar heating systems were used for single family houses; multi-family, tourism business and public buildings had 55% of systems; less than 5% of the systems were installed for district heating and solar refrigeration applications.

The most matured systems for all the solar thermal applications are the solar hot water systems. In the central part of Gurbantunggut Desert in the Junggar Basin of Xinjiang, a solar hot water system was installed in the Xinjiang Oil Field's Carat Beauty Operational District (Figure 2.26). The system consists of 1,500 vacuum thermal absorber tubes, forming thousands of giant mirrors. The total solar absorber area is about 360 m^2. This system supplies daily hot water to the residents and saves about 10,000 kWh electricity annually.

In Beijing, Yanxi International Convention Center installed, on its roof, 138 sets of horizontal heat pipe vacuum tube solar collectors with the total area of 593 m^2 (Figure 2.27). The system provides 55°C domestic hot water for kitchens and rest rooms. The system was designed with the features including central thermal collection, indirect heat exchange for central heating, supplemental boiler and supplemental geothermal heat pump systems. The automated system uses direct solar heat as the priority source and supplies the heat from supplemental sources based on

FIG. 2.26: VACUUM TUBE SOLAR HOT WATER SYSTEM INSTALLED IN CARAT BEAUTY OPERATION DISTRICT OF XINJIAN OIL FIELD

FIG. 2.27: SOLAR HOT WATER SYSTEM IN BEIJING YANXI INTERNATIONAL CONVENTION CENTER

FIG. 2.29: SEASONAL THERMAL STORAGE SYSTEM FOR COMBINED ENVIRONMENTAL HEATING AND HOT WATER SUPPLY IN HEBEI UNIVERSITY OF ECONOMICS AND BUSINESS

the demands. The optimized system makes it possible to maximize the solar energy utilization while maintaining stable heat supplies and achieving energy saving.

One of the largest solar hot water systems is the one installed in Beijing Olympic Village (Figure 2.28). Its features include six control subsystems for solar collector district control, water thermal storage system control, heat exchanger control, thermal sanitization control, freezing and overheating protections. All the subsystems are synchronized to ensure the safe operation with a high efficiency for the entire system. The entire system comprises of a surface area of 6,000 m^2 with 2,187 solar collectors, which were installed over four separate zones on the roofs of 16 buildings. Five heat exchanger stations were established in four car garages corresponding to four zones. During the 2008 Beijing Olympic Games, the entire system supplied shower hot water for 16,800 athletes from all over the world, exceeding the number of athletes served in the past Olympics. After the Olympic Games, the system continued to supply domestic hot water to more than 2,000 households, and saved the annual electricity for more than 5 MWh with the reduction of 3,800 tons of CO_2.

Although China's solar hot water systems have led the rest of world in both production and utilization, the technology development for solar environmental heating and solar registration has been resting at the stage of exploration and project demonstration. Since 2000, China has gradually developed the solar environmental heating technology. During the 125 planning period, this technology is expected to transition from demonstration to important scale-up applications. As of now, very few practical applications for solar environmental heating exist.

The largest demonstration project for seasonal solar thermal storage has been completed in Hebei University of Economics and Business to demonstrate the feasibility of solar thermal

FIG. 2.28: BUILDING-INTEGRATED SOLAR THERMAL SYSTEM IN BEIJING OLYMPIC VILLAGE

storage system for seasonal environmental heating and hot water supplies. This signifies an important transition from solar hot water industry to solar environmental heating industry. The total investment of this project is 40 million RMB. The system consists of 69,000 vacuum tubes with the total thermal concentration area of 11,600 square meters with 228 89-ton water tanks and a total of 20,000 tons of thermal storage capacity (Figure 2.29). This project is also the first commercial project in China that combines both environmental heating and hot water production.

In summary it has been established that China's solar hot water system industry and market is mature. For several years, it keeps the number one rank in technology development speed in the world. In China, the industry has the largest market in the world, and the level of technology is essentially comparable with international markets. However China's solar thermal technology and products are still at a stage of low-level applications, primarily for hot water supplies. The R&D of high tech products lags behind the global development. On per capita basis, China's utilization of solar hot water systems is very low compared with other countries. Within China, there are large discrepancy of applications among domestic regions. This proves that there are still tremendous potentials for China to further broaden its application space.

2.5 SOLAR PROCESS HEAT FOR MANUFACTURING APPLICATIONS

2.5.1 Industrial Application of Solar Thermal Energy

The high temperature application from solar thermal energy is considered to be a high tech area. Because of the capability of reaching very high temperature, the potential for industrial application of solar thermal energy is great.

Today the utilization of solar energy in China is primarily for low-temperature hot water. Very few research has been taken for high temperature thermal energy utilization. In China about 70% of energy consumption is due to industrial sector, in which 50% of industrial energy consumption is due to thermal energy. If only 10% of industrial thermal energy consumption is replaced by solar energy, about 40 million tons of standard coals can be saved and 80 million tons of CO_2 emission can be eliminated. This is significant to achieve the goal of 18% reduction in growth rate per 10,000 RMB of industrial added value.

FIG. 2.30: SHANDONG LINUO SOLAR BOILER PROJECT

The world first industrial thermal application of Composite Parabolic-light Collector (CPC) mid-to-high temperature system was initiated in Shandong Jinan. Shangong Linuo solar boiler project uses concentrated solar energy to raise the water temperature from 15 °C to 95 °C and then heats the 95 °C water to 150 °C steam (Figure 2.30). This system breaks the limit of standard hot water solar collectors and solves a difficult technology problem in mid-to high temperature solar thermal applications. As a result, the project marks the new era for industrial solar thermal energy industry.

The CPC mid-to-high solar boiler system has a total concentrated thermal collector area of 5,200 M^2. It saves 1,156 tons of standard coals annually, and eliminates 3,005 tons of CO_2. As the largest province in China that both generate and consume heat, Shandong province has 50,000 of variety of coal-fired boilers. If all of them connected to CPC mid-to high temperature industrial solar thermal systems, about 88.4 million $M^2$of solar collectors can be installed. Therefore, about 19.6 million tons of standard coals can be saved and 44.2 million tons of CO_2 can be eliminated.

2.6 OTHER SOLAR ENERGY APPLICATIONS

2.6.1 Solar Hybrid Ships

One of China's high tech projects in ships is the application of solar power in a large-scale Roll-on/roll-off ship. Zhong Yuan Tengfei is the world largest solar powered hybrid ship, which has a 143 kW solar PV system on board and sails to all over the world. The solar power system installed is a grid-independent and grid-tie capability (Figure 2.31). The net effect of this solar PV system is to reduce the Diesel power consumption by about 120 to 143 kW. According to the measurement data, Zhong Yuan Tengfei ship can store energy in batteries in 5 hours of direct solar radiation conditions. It can supply lighting energy needs for 24 hours. In a normal sun shine day, the system can provide 16 hour electricity, equivalent to the daily saving of 0.46 ton of Diesel fuel. The economic benefit is obvious.

2.6.2 Thin-Film Solar Electric Curtain Walls

In a recent renovation project for the National Electricity Investment Company building, which is located in Beijing Finance Street South, thin-film B IPV panels were installed on the curtain walls and roofs where sufficient sunrays can be received (Figure 2.32). The resulted surfaces possess multiple functions of day lighting, insulation and electricity generation. Both energy consumption reduction and solar electricity generation were achieved, and represent an effective, complementary scenario. After the renovation, the total solar curtain wall system has a capacity of 170.66 kW. Within the projected 25 year usage, it will generate 126.9 MWh annually, equivalent to a saving of 275.8 tons of standard coals, removal of 99.84 tons of ashes, 12.53 tons of NOx, or 825.9 tons of CO_2. The net effect can also be equivalent to plant 45,000 trees on Finance Street, and remove emission from 150 cars. The economic benefit is also very promising. A 9.37% of investment return is projected within 25 years, an accumulative income of 20 million RMB.

FIG. 2.31: ZHONG YUAN TENGFEI SOLAR HYBRID ROLL-ON/ROLL-OFF SHIP

(a) (b)

FIG. 2.32: NATIONAL ELECTRICITY INVESTMENT BUILDING: (A) BEFORE AND (B) AFTER THE RENOVATION, WHICH INSTALLED THIN-FILM BIPV PANELS

FIG. 2.33: SOLAR BUS IN LASA CITY, TIBETTHERMAL SYSTEM IN BEIJING OLYMPIC VILLAGE

2.6.3 Solar Bus

In Tibet Lasa City, the first solar powered bus was placed in service. From the outside the solar bus looks almost the same as the regular electrically powered bus. However, it has 32 solar PV panels covering its roof (Figure 2.33). Aided by solar power, the bus can charge its batteries while running. The fully charged batteries allow the bus to cruise for more than 200 km.

2.7 SCALING- UP RENEWABLE ENERGY PLAN (SREP) IN MONGOLIA:

Mongolia has tremendous solar energy resources. Particularly the South Gobi region is very attractive from the solar energy-harnessing point of view. With average sunny days of 270 to 300 days per year, Mongolia enjoys about 2250 to 3300 sunshine hours annually. Solar intensity over the region is estimated to be 4.3-4.7 KH/hr while, the solar radiations received are in the range of 1200-1600/m^2. Majority of Mongolian region received incoming solar radiation within the range of 5.5-6.0 KWh/m^2. According to U.S. National Renewable Energy Laboratory (NREL), solar energy potential from Mongolia is about 1500 GW.

Small-scale solar PV systems have been very successful in Mongolia as they were supplied to nomadic herders. Nevertheless, large-scale grid connected solar PV projects are still not present in the country. Although, various utility-scale solar PV projects by private sector are under development. District heating is also in pilot-testing stage in some locations. The solar insolation, although, not at par with the requirement of Concentrated Solar Power (CSP) operation, its integration with solar thermal systems using concentrated solar panels for district heating is a possibility.

Electricity supplied by solar power is priced at U.S.$ 0.15-0.18/KWh. Nevertheless, as per national Mongolian law, in order to compensate for the price difference of grid-connected electricity produced by renewable sources, selling prices of other sources that are connected to transmission network need to be adjusted. Mongolian energy sector faces electric supply deficit on account of limited investment in capacity augmentation and relies on electric import from Russia.

Long distances and intermittent population separate Mongolian rural areas. This makes it ideal for distributed renewable energy systems. The mismanagement in direct subsidy transfer to the renewable energy generators caused negative impact on balance sheets of the generators. Mongolian government has targeted to increase the share of renewable energy up to 20% by 2020 and 30% by 2030 in the total energy mix of the country. Currently, due to generous Feed in Tariffs (FiTs), the total solar power capacity has been increased to 200 MW.

2.8 SOLAR MEASUREMENTS, MODELING AND FORECASTING

Solar energy has become a vital alternative energy source due to its availability and its improved affordance from economic perspective. Therefore, the significant increase in solar energy applications and planning of new solar power plant installation calls for advance in solar measurement techniques and mathematical modeling to develop reliable forecasting methodologies for solar data analysis. In this section, we present the recent developments in solar energy measurement techniques, mathematical models, forecasting methodologies and data analysis procedures.

2.8.1 Solar Radiation Measurement and Models

Heliostat power plants are a type of solar power plants, which utilize concentrated solar thermal power to produce electricity by focusing solar rays from solar troughs to a centrally located mirror located at a certain height. The major problem currently faced by plant operators is that the solar rays, as Direct Normal Irradiance (DNI), reflected from the concentrated solar trough panels reduce the concentration flux as it should have reached to the mirrors or the heliostat. This loss in the concentration flux is attributed to the atmospheric extinction. To quantify this loss due to atmospheric extinction, SW2011 transmittance model has been refined to include different perceptible water vapor (PWR) amounts and it is adjusted to the height profile of the investigated site at Plataforma Solar de Almeria (PSA), a research center in Europe. The model performs better than the previous SW2011 model with mean difference of 0.01, 0.05 and 0.03 with respect to reference data. The model is versatile and can be synchronized with existing ray tracing and transmittance tools used at various locations [15]. Solar measurements were performed to quantify the global horizontal irradiances (GHI) and direct normal irradiances (DNI). These measurements were done in clear sky conditions, partly cloudy and mostly cloudy conditions for one year. 51 radiometers were used which included pyranometers, pyrheliometers, rotating shadow band radiometers (RSR). The measurement differences between GHI and DNI values to reference values are very similar for solar zenith angles below 60°. When the solar zenith angle increased above 60° GHI values showed difference of +/- 17% while DNI values changed +/- 40% with respect to reference values [16]. Such measurements are important for understanding the resemblances between GHI and DNI radiometers. At the same time we can also understand contrasts and dissimilarities existing between them. Accurate measurements are of extreme importance for calculating solar outputs. A Fast All-sky Radiation Model for Solar applications (FARMS) has been developed which takes utilizes existing Radiative Transmittance (RT) models by calculating the cloud transmittance and reflectance in advance. This allows the model not to rely on the lookup tables (LUTs) which are more computationally heavy and time consuming. The FARMS model utilizes the 16-strem discrete ordinate radiative transfer model, which is more efficient than the current two-stream discrete ordinate model. The FARMS model is able to predict solar irradiance on the terra firma surface with less than 5% difference with respect to observed data [17]. A procedure has been developed to calculate longwave broadband irradiance in a direct solar beam during sunrise to sunset. This is significant because consensus exists for night time longwave irradiance but for daytime it does not exist. Two pyrgeometers were spectrally matched and they were installed on the solar tracker model. Out of these two pyrgeometers was shaded. Only daytime and clear sky conditions were used for collecting data. Data was collected from June 17 to July 24, 2014. The measured data was verified by performing simulations using radiative transmittance models. The irradiances varied from 1 W-m^{-2} to 16 W-m^{-2} [18]. In multi-junction solar cells, thermalization losses are minimized to obtain high efficiencies during their operation. However their operation is highly sensitive to solar spectrum variations. These variations are affected mainly by air mass (AM) and aerosol optical depth (AOD) during the course of entire day. To optimize the behavior of multi-junction solar cells their behavior should be characterized for spectra in the environment in which they operate instead of a single reference spectrum. Along with this issue current practices ignore the Luminescence Coupling (LC) in the multi-junction solar cell operation. To address this a new characterization technique has been developed which utilizes computation of the solar cell of multiple solar spectral compositions and intensities. Performance maps for solar cells are then developed to predict the annual energy output [19].

Various mathematical models has been described and compared such as Atwater and Ball Model, Davies and Hay Model, Watt Model, Hoyt Model, Lacis and Hansen Model. A new model has been proposed which utilizes three codes namely Soltran, Brite

Monte Carlo Code and Dave Spherical Harmonics Code. The new model is called Bird Simplified Clear Sky Model. Each of three codes captures various atmospheric layers in order to describe the direct normal irradiance and diffuse irradiance. In order to solve radiative transfer issue, each code implements its own technique [20-21]. An improved diffuse fraction model has been presented which is advanced version of the previously published same model. The advanced model, known as Direct Insolation Simulation Code (DISC) model, includes effects of hour-to-hour variability index and regional surface albedo in order to predict accurately the diffuse fraction. The model is correlated corresponding to 32 year data from Bergen, Norway. The model was also tested corresponding to datasets from four other European weather stations. The model shows improvement at high clearness index with Root Mean Square Difference (RMSD) decreased by 25%-30% [22]. The solar grid can be victim of intermittent supply of solar radiations due to cloudy climate or occurrences such as solar eclipses. Solar eclipses may partially or completely shade the solar surface due to moon position. A new algorithm is developed to understand Moon's position during eclipses. The algorithm is called Solar and Moon Position Algorithm (SAMPA). The algorithm calculates Moon's zenith angle and azimuth angle with their uncertainties of +/- 0.001° and +/- 0.003° respectively [23]. A Solar Position Algorithm (SPA) has been developed to calculated solar position with predictability range +/- 0.0003°. Major changes include considering direction from north and eastward for solar zenith angles and azimuth angle calculations instead of South and eastward. The observer's geographical location in terms of longitude is considered positive instead of negative with respect to Greenwich meridian [24].

2.8.2 Recent Advances in Solar Forecasting.

Forecasting is extremely important for independent system operators (ISO), utilities and distributed system operators. Due to intermittent nature of the wind and solar resources, the electricity grids follow a reserve-based strategy. This strategy allows to allocate different types of reserves like spinning contingency reserve, non-spinning contingency reserve and regulating reserve. Solar forecasting is performed at different timescales and as per requirements of geographies and customers. Therefore, solar forecasting is very important and needs accurate models which fit to different time scales and identify reserves which need to be allocated. A new methodology has been developed by to address this issue by identifying solar forecasting metrics like excess kurtosis, skewness, Renyi entropy, swinging door algorithm among others. Along with climatology, model and persistence approach has been used for solar forecasting using North American Mesoscale Forecast System (NAM) model, streamer radiative Transfer Model (RTM) and PV-Lib toolbox for irradiance to power modeling. The methodology forecasts global horizontal irradiance (GHI) and solar PV power [25]. The production cost model of an independent system operator at New England (ISO-NE) was implemented in PLEXUS (2015) in order to simulate its one-year operation on sub hourly basis to demonstrate value of Day Ahead (DA) improvements due to solar forecasting. Total installed capacity was 34.2 GW. Different solar penetration levels were considered as 4.5%, 9%, 13.5% and 18% based on NREL data for total installed solar capacity of 4,874 MW. Weather Research Forecasting Model (WRM) was used for solar forecasting. In order to establish importance of solar forecasting in grid operations, steady solar forecast improvements were achieved by hindcasting approach which resulted in reducing the generation costs of the grid and demonstrated financial importance of solar forecasting [26]. Traditional solar forecast model employ artificial neural network (ANN) and support vector regression (SVR). But these technique computationally very heavy and complicated. In order to circumvent this difficulties and extreme learning machine (ELM) based forecaster is proposed. This forecaster utilizes optimal synchrophasor sensor placement (OSSP) based state estimator for high speed and accurate estimation of solar forecast for distributed systems [27]. We are currently in smart grid era and due to increasing trend towards distributed power plant systems, we will transition to smart grid structure of future. This future structure will involve smart grid to be divided into multiple micro grids and ultimately this micro grids divided into nanogrids. Key factors affecting the operating behaviors are weather, temporal factors, volatility in generation and demand profiles, electricity price and demand response, renewable energy resources, energy storage, electric vehicles, smart and efficient appliances, virtual power plants/ micro grids/ nano grids. Key challenges facing solar forecasting are large volume of data from smart meters, including small time duration of demand forecast, identifying impacts of climate change, renewable energy integration, government policies and technology developments and identifying customer behavior in terms of technology adoption, behavioral change towards sustainable habits and energy related customer behaviors [28].

2.9 SUMMARY

In this chapter, recent applications of solar energy systems in China and Mongolia are presented in comparison with the rest of world. By far, large-scale, photovoltaic-driven applications are dominant in China's present markets in terms of annual productions of cells, panels and systems and the number of installations in utility scale systems, buildings and other applications. This is primarily due to the government initiatives and subsidy policies. The impacts of distributed PV and thermal systems are less than large-scale applications in terms of total power generation capacity but are still significant. While solar hot water systems are relatively less expensive, the challenge to significantly bring down the cost of PV systems to compete against the conventional fossil fuel-based electricity is still there and requires both technological breakthroughs and non-technical measures.

China has established a clear target for the renewable energy utilization. By 2030 China will have its 20% of total energy generation from non-fossil energy sources. Under this ambitious plan, solar energy as a renewable energy source has a tremendous development potential. The key infrastructure investment in ultra-high voltage transmission line, research and development in energy storage technology and boost in economic development will alleviate the "Solar Abandoned" (Solar Energy Curtailment) phenomenon in China's western regions and result in optimal energy portfolio establishment. Republic of Mongolia faces the similar challenges and opportunities to harvest its rich resource of solar energy.

2.10 REFERENCES

1. O'Shaughnessy, Eric, Ardani, Kristen, Cutler, Dylan and Margolis, Robert (2017). Solar Plus: A Holistic Approach to Distributed Solar PV. Technical Report No. TP-6A20-68371. National Renewable Energy Laboratory.
2. Zhang, Jie, Hodge, Bri-Mathias, Lu, Siyuan, Hamann, Hendrik, Lehman, Brad, Simmons, Joseph, Campos, Edwin, Banunaraynan, Venkat (2015). Baseline and Target Values for PV Forecasts: Toward Improved Solar Power Forecasting. Solar Energy Vol. 122, pp. 804-819.
3. Leisch, Jennifer and Cochran, Jaquelin (2016). Forecasting Wind and Solar Generation: Improving System Operations". U.S. Agency for International Development (USAID).
4. Wang, Gang, Quan, Zhenhua, and Zhao, Yaohua (2014). Solar-air composite heat source heat pump hot water system. CIESC Journal, 65 (3), pp.1033-1039.
5. Lori Bird, Jaquelin Cochran, and Xi Wang (2014). Wind and Solar Energy Curtailment: Experience and Practices in the United States. Technical Report No. NREL/TP-6A20-60983. National Renewable Energy Laboratory.
6. Xie, Zhijia, Li, Jianlin, Cheng, Wei, and Liu, Zhaoliang. Demand Analysis of Energy Storage Station Reducing Photovoltaic Power Plant Abandoned Electricity. Electrical & Energy Management Technology. 2018(1): 18-24.
7. National Energy Administration. http://www.nea.gov.cn/2018-06/01/c_137223463.htm
8. Lv, Fang, Xu, Honghua, and Wang, Sicheng. National Survey Report of Photovoltaic Applications in CHINA 2017. http://www.iea-pvps.org/
9. Yang, Zhong (2008). Development State and Trends of Photovoltaic Power. Journal of Jinling Institute of Technology. 24 (1), pp. 9-13.
10. Yan, Lu-Guang, Xia, Xun-Cheng, and Zhou, Feng-Qi. A development study for Chinese large scale renewable energy source base and technology. Advanced Technology of Electrical Engineering and Energy. 2007, 26 (2):13-24.
11. DOE NREL (2006). Parabolic Trough Solar Thermal Electric Power Plants. Publication: FS-550-40211; DOE/GO-102006-2339.
12. DOE NREL (2019). Parabolic Trough Projects, https://solarpaces.nrel.gov/by-technology/parabolic-trough. Accessed April 24, 2019.
13. Li, Changqing, Su, Meiling, and Yang, Xinjiletu. Carbon sink industry development potential in Inner Mongolia. Journal of Arid Land Resources and Environment. 2012, 26 (5):162-168.
14. Wang, Wenjing, and Wang, Sicheng. Status and Prospect of Chinese Distributed Photovoltaic Power Generation System. Bulletin of Chinese Academy of Sciences. 2016 (2):165-172.
15. Jiang, An-Yin, and Liu, Bo. Poverty Alleviation in the Context of Precision Poverty Alleviation. Journal of Agro – Forestry Economics and Management. 2017, 16 (6):789-794.
16. Hanrieder, N, Sengupta, M, Xie, Y, Wilbert, S, and Pitz-Paal, R. Modelling Beam Attenuation in Solar Tower Plants using Common DNI Measurements. Solar Energy. 2016, 129: 244-255.
17. Habte, A, Sengupta, M, Andreas, A, Wilcox, S, and Stoffel T. Intercomparison of 51 Radiometers for Determining Global Horizontal Irradiance and Direct Normal Irradiance Measurements. Solar Energy. 2016, 133: 372-393.
18. Xie, Y, Sengupta, M, and Dudhia, J. A Fast All-sky Radiation Model for Solar Applications (FARMS): Algorithm and Performance Evaluation. Solar Energy. 2016, 135: 435-445.
19. Reda, I, Konings, J, and Xie, Y. A Method to Measure the Broadband Longwave Irradiance in the Terrestrial Direct Solar Beam. Journal of Atmospheric and Solar- Terrestrial Physics. 2015, 129: 23-29.
20. Garcia, I, Macmahon, W, Steiner, M, Geisz, J, Habte, A, and Friedman, D. Optimization of Miltijunction Solar Cells Through Indoor Energy Yield Measurements. IEEE Journal of Photovoltaics. 2015, 5 (1): 438-445.
21. Bird, R, and Hulstorm, R. A Simplified Clear Sky Model for Direct and Diffuse Insolation on Horizontal Surfaces. Solar Energy Research Institute. 1981
22. Bird, R, and Riordan, C. Simple Solar Spectral Model for Direct and Diffuse Irradiance on Horizontal and Tilted Planes at the Earth's Surface for Cloudless Atmospheres. Journal of Climate and Applied Meteorology. 1986, 25: 87-97.
23. Skartveit A, Olseth J, Tuft M. An Hourly Diffuse Fraction Model with Correction for Variability and Surface Albedo. Solar Energy. 1998, 63 (3): 173-183.
24. Reda I. Solar Eclipse Monitoring for Solar Energy Applications Using the Solar and Moon Position Algorithms. National Renewable Energy Laboratory (NREL). 2010.
25. Reda I, Andreas A. Solar Position Algorithm for Solar Radiation Applications. National Renewable Energy Laboratory. 2008.
26. Zhang J, Hodge B, Lu S, Hamann H, Lehman B, Simmons J, Campos E, Banunarayanan V, Black J, Tedesco J. Baseline and Target Values for Regional and Point PV power forecasts: Toward Improved Solar Forecasting. Solar Energy. 2015, 122: 804-819.
27. Martinez-Anido C, Botor B, Florita A, Draxl C, Lu S, Hamann H, Hodge B. The Value of Day-Ahead Solar Forecasting Improvement. Solar Energy. 2016, 129: 192-203.
28. Jiang H, Zhang Y. Short-Term Distribution System State Forecast Based on Optimal Synchrophasor Sensor Placement and Extreme Learning Machine. 2016 IEEE Power and Energy Society General Meeting (PESGM), 17-21 July 2016.
29. Khodayar, M, and Wu, H. Demand Forecasting in the Smart Grid Paradigm: Features and Challenges. The Electricity Journal. 2015, 28 (6): 51-62.

CHAPTER

3

Solar Thermal Power Plants: from Endangered Species to Bulk Power Production in Sun Belt Regions

Manuel Romero José González-Aguilar and K. R. Rao[1]

3.1 INTRODUCTION

Solar thermal power plants, due to their capacity for large-scale generation of electricity and the possible integration of thermal storage devices and hybridization with backup fossil fuels, are meant to supply a significant part of the demand in the countries of the solar belt [1]. Nowadays, the high-temperature thermal conversion of concentrated solar energy is rapidly increasing with many commercial projects taken up in Spain, USA, and other countries such as India, China, Israel, Australia, Algeria, and Italy. This is the most promising technology to follow the pathway of wind energy in order to reach the goals for renewable energy implementation in 2020 and 2050.

Spain with 2400 MW connected to the grid in 2013 is taking the lead on current commercial developments, together with USA where a target of 4500 MW for the same year has been fixed and other relevant programs like the "Solar Mission" in India recently approved and going for 22 GW-solar, with a large fraction of thermal [2].

Solar Thermal Electricity or STE (also known as CSP or Concentrating Solar Power) is expected to impact enormously on the world's bulk power supply by the middle of the century. Only in Southern Europe, the technical potential of STE is estimated at 2,000 TWh (annual electricity production) and in Northern Africa it is immense [3]. Worldwide, the exploitation of less than 1% of the total solar thermal power plant potential would be enough to meet the recommendations of the United Nations' Intergovernmental Panel on Climate Change for long-term climate stabilization [4]. One MW of installed concentrating solar thermal power avoids 688 tons of CO_2 compared to a Combined Cycle conventional plant and 1360 tons of CO_2 compared to a conventional coal/steam plant. A 1-m^2 mirror in the primary solar field produces 400 kWh of electricity per year, avoids 12 tons of CO_2 and contributes to a 2.5-ton savings of fossil fuels during its 25-year operation lifetime. The energy payback time of concentrating solar power systems is less than 1 year, and most solar-field materials and structures can be recycled and used again for further plants.

But in terms of electric grid and quality of bulk power supply, it is the ability to provide dispatch on demand that makes STE stand out from other renewable energy technologies like PV or wind. Thermal energy storage systems store excess thermal heat collected by the solar field. Storage systems, alone or in combination with some fossil fuel backup, keep the plant running under full-load conditions. This capability of storing high-temperature thermal energy leads to economically competitive design options, since only the solar part has to be oversized. This STE plant feature is tremendously relevant, since penetration of solar energy into the bulk electricity market is possible only when substitution of intermediate-load power plants of about 4000–5000 hours/year is achieved.

The combination of energy on demand, grid stability, and high share of local content that lead to creation of local jobs provide a clear niche for STE within the renewable portfolio of technologies. Because of that, the European Commission is including STE within its Strategic Energy Technology Plan for 2020, and the US DOE is launching new R&D projects on STE. A clear indicator of the globalization of such policies is that the International Energy Agency (IEA) is sensitive to STE within low-carbon future scenarios for the year 2050. At the IEA's Energy Technology Perspectives 2010 [5], STE is considered to play a significant role among the necessary mix of energy technologies needed to halving global energy-related CO_2 emissions by 2050, and this scenario would require capacity additions of about 14 GW/year (55 new solar thermal power plants of 250 MW each).

1 The editor has provided in Section 3.8 relevant write-ups from online sources in the public domain for solar energy applications pertaining to Spain, rest of Europe, China, India, Australia, South Africa and Chile, mostly from 2009 to date.

STE systems consist of a large reflective surface collecting the incoming solar radiation and concentrating it onto a solar receiver with a small aperture area. The solar receiver is a high-absorptance radiative/convective heat exchanger that emulates as closely as possible the performance of a radiative black body. An ideal solar receiver would thus have negligible convection and conduction losses. In the case of a solar thermal power plant, the solar energy is transferred to a thermal fluid at an outlet temperature high enough to feed a heat engine or a turbine that produces electricity. The solar thermal element can be a parabolic trough field, a linear Fresnel reflector field, a central receiver system or a field of parabolic dishes, normally designed for a normal incident radiation of 800–900 W/m^2. Annual normal incident radiation varies from 1600 to 2800 kWh/m^2, allowing from 2000 to 3500 annual full-load operating hours with the solar element, depending on the available radiation at the particular site [6].

The first generation of commercial STE projects is mainly based on technological developments and concepts that matured after more than two decades of research. Nevertheless, the current solar thermal power plants are still based on conservative schemes and technological devices which do not exploit the enormous potential of concentrated solar energy. Commercial projects use technologies of parabolic troughs with low concentration in two dimensions and linear focus, or systems of central tower and heliostat fields, operating with thermal fluids at relatively modest temperatures, below 400°C [7]. The most immediate consequence of these conservative designs is the use of systems with efficiencies below 20% nominal in the conversion of direct solar radiation to electricity; the tight limitation in the use of efficient energy storage systems; the high water consumption and land extension due to the inefficiency of the integration with the power block; the lack of rational schemes for their integration in distributed generation architectures and the limitation to reach the temperatures needed for the thermochemical routes used to produce solar fuels like hydrogen.

In the first commercial projects involving parabolic trough technology, some improvements are being introduced, such as the use of large molten salt heat storage systems able to provide high degrees of dispatch for the operation of the plant, such as the plants Andasol 1 and 2 in Guadix, Spain, with 7.5 hours of nominal storage, or the use of direct steam generation loops to replace thermal oil at the solar field. Central towers are opening the field to new thermal fluids like molten salts (Gemasolar tower plant in Seville, Spain) and air, and new solar receivers like volumetric absorbers.

In parallel, a new generation of concentrating solar thermal power systems is starting deployment. This new generation is characterized by its modularity and higher conversion efficiencies. The design strategy is based on the use of highly compact heliostat fields, using mirrors and towers of small size, and looking for integration into high-temperature thermodynamic cycles. There are currently some initiatives with prototypes at an experimental stage and announcing large commercial projects like the one proposed by BrightSource with a prototype of 6 MWth at the Neguev desert in Israel, the 100 kWe prototype promoted by the AORA company and researchers of the Weizmann Institute in Israel, and the 5 MWe prototype built by the company eSolar in California.

3.2 SOLAR THERMAL POWER PLANTS: SCHEMES AND TECHNOLOGIES

3.2.1 Solar Concentration: Fundamentals for STE Plants

Solar energy has a high exergy value since it originates from processes occurring at the sun's surface at a black-body equivalent temperature of approximately 5777 K. Because of this high exergetic value, more than 93% of the energy may be theoretically converted to mechanical work by using thermodynamic cycles [8], or to Gibbs free energy of chemicals by solarized chemical reactions [9]. According to thermodynamics and Planck's equation, the conversion of solar heat to mechanical work or Gibbs free energy is limited by the Carnot efficiency, and, therefore, to achieve maximum conversion rates, the energy should be transferred to a thermal fluid or reactants at temperatures close to that of the sun.

Even though solar radiation is a source of high temperature and exergy at origin, with a high radiosity of 63 MW/m^2, sun-to-earth geometrical constraints lead to a dramatic dilution of flux and to irradiance available for terrestrial use; only slightly higher than 1 kW/m^2 with a consequent supply of low temperatures to the thermal fluid. It is, therefore, an essential requisite for solar thermal power plants and high-temperature solar chemistry applications to make use of optical concentration devices that enable the thermal conversion to be carried out at high solar fluxes and with relatively low heat losses. A simplified model of a STE plant is depicted in Fig. 3.1.

Solar reflective concentrators follow the basic principles of Snell's Law of reflection [10]. In a specular surface, like the mirrors used in solar thermal power plants, the reflection angle

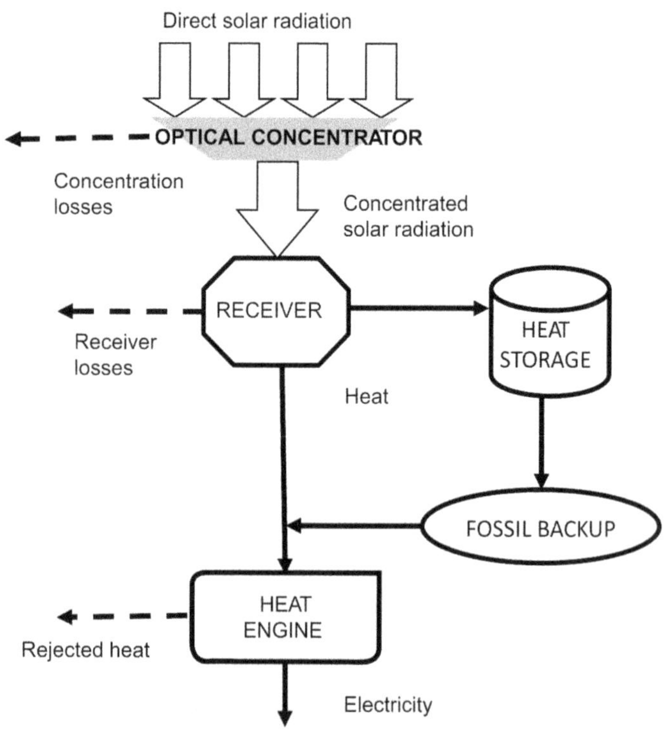

FIG. 3.1: FLOW DIAGRAM FOR A TYPICAL SOLAR THERMAL POWER PLANT

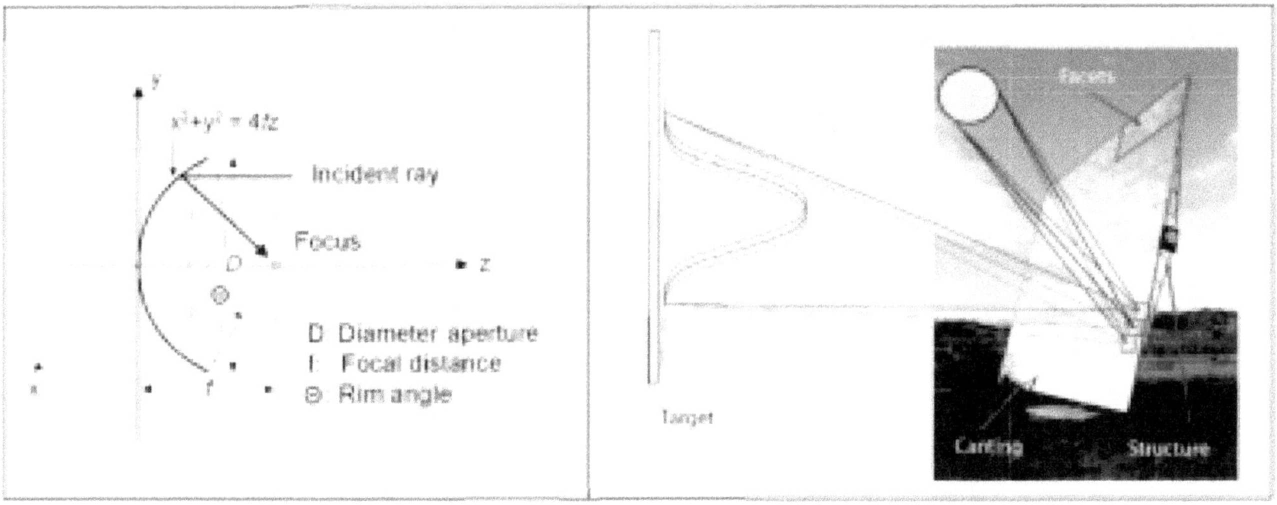

FIG. 3.2: CONFIGURATION OF IDEAL PARABOLIC CONCENTRATOR (LEFT) AND FLUX PROFILE REFLECTED BY A TYPICAL CONCENTRATOR INCLUDING SUNSHAPE (RIGHT)

equals the angle of incidence. The most practical and simplest primary geometrical concentrator typically used in STE systems is the parabola. Even though there are other concentrating devices like lenses or Compound Parabolic Concentrators [11], the reflective parabolic concentrators and their analogues are the systems with the greatest potential for scaling up at a reasonable cost. Parabolas are imaging concentrators able to focus all incident paraxial rays onto a focal point located on the optical axis (see Fig. 3.2).

The paraboloid is a surface generated by rotating a parabola around its axis. The parabolic dish is a truncated portion of a paraboloid. For optimum sizing of the parabolic dish and absorber geometries, the geometrical ratio between the focal distance, f, the aperture diameter of the concentrator, d, and the rim angle, Θ, must be taken into account. The ratio can be deducted from the equation describing the geometry of a truncated paraboloid, $x^2 + y^2 = 4fz$, where x and y are the coordinates on the aperture plane and z is the distance from the plane to the vertex. For small rim angles, the paraboloid equals to a sphere and in many cases spherical facets are used, therefore, in most solar concentrators, the following correlation is valid:

$$f/d = \frac{1}{4\tan(\Theta/2)} \tag{3.1}$$

For example, a paraboloid with a rim angle of 45 deg has f/d of 0.6. The ratio f/d increases as the rim angle decreases. A parabolic concentrator with a very small rim angle has very little curvature and the focal point far from the reflecting surface. Because of that, STE systems making use of cavity receivers with small apertures should use small rim angles. Conversely, those STE systems using external or tubular receivers will make use of large rim angles and short focal lengths.

The thermodynamic limit or maximum concentration ratio for an ideal solar concentrator would be set by the size of the sun. By applying the geometrical conservation of energy in a solar concentrator, the following expressions are obtained for 3D and 2D systems (for a refraction index $n = 1$, and assuming a flat aperture area):

$$C\max,\ 3D = \frac{1}{\sin^2\theta_S} \leq 46{,}200 \tag{3.2}$$

$$C\max,\ 2D = \frac{1}{\sin\theta_S} \leq 215 \tag{3.3}$$

Then the semi-angle subtended by the sun is $\theta_S = 4.653 \times 10^{-3}$ rad (16'), and the maximum concentration values 46,200 for 3D and 215 for 2D. For real concentrators, the maximum ratios of concentration are much lower because of microscopic and macroscopic, tracking and mechanical, sunshape, and other errors.

In a real mirror with intrinsic and constructional errors, the reflected ray distribution can be described with "cone optics" and, as depicted in Fig. 3.2, the flux profile obtained onto the target can be described by a Gaussian shape. The reflected ray direction has an associated error that can be described with a Normal Distribution Function. The errors of a typical reflecting solar concentrator may be either microscopic (specularity) or macroscopic (waviness of the mirror and error of curvature). All the errors together end up modifying the direction of the normal compared to the reference reflecting element. However, it is necessary to discriminate between microscopic and macroscopic errors. Microscopic errors are intrinsic to the material and depend on the fabrication process, and can be measured at the lab with mirror samples. Macroscopic errors are characteristic of the concentrator and the erection process, therefore, they should be measured and quantified with the final system in operation [12]. It is relatively easy to quantify all that random errors onto the target by using the standard deviation of the reflected rays ($\sigma^2 = \Sigma\, \sigma_i^2$) or by fitting the solar concentrated flux to a normal distribution function with a standard deviation σ, commonly known as beam quality. Typical beam qualities of solar reflective concentrators used in STE plants are of a few milliradians and their values determine the size of the absorber or cavity. Beam quality convolves with the sunshape leading to an increment of the acceptance angle of

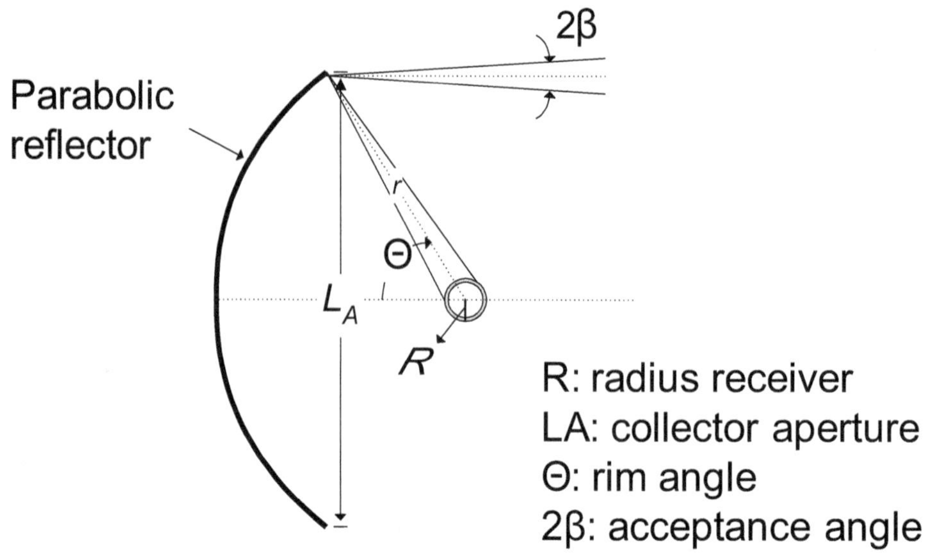

FIG. 3.3: RIM ANGLE AND ACCEPTANCE ANGLE IN A PARABOLIC TROUGH CONCENTRATOR

the absorber, receiver aperture, or cavity from $2.\theta_S$ (ideal value corresponding to the size and errors of the sun) to $2.\beta$ (real value of the acceptance angle including all sources of errors).

The concentrator with more extended use is the parabolic trough it is a 2D parabolic reflector with a linear focus and single axis tracking. The concentration ratio C for this system is determined by the ratio between the aperture of the parabolic surface and the total area of the tubular receiver (Eq. 3.4).

$$C = \frac{A_C}{A_R} = \frac{L_A \cdot L}{2\pi R \cdot L} = \frac{L_A}{2\pi R} \tag{3.4}$$

Where A_C is the aperture area of the concentrator, A_R is the area of the receiver or absorber, L_A is the width of the aperture of the concentrator, L is the length of the concentrator and R is the radius of the receiver tube (Fig. 3.3). According to this, the following relations are obtained:

$$\frac{R}{r} = \sin\beta \tag{3.5}$$

$$\frac{L_A}{2r} = \sin\Theta \tag{3.6}$$

Where r is the distance from the edge of the parabola to the center of the receiver, 2β is the acceptance angle and Θ is the rim angle of the concentrator.

Then the Concentration Ratio C can be expressed also as:

$$C = \frac{\sin\Theta}{\pi \sin\beta} = \frac{\sin\Theta}{\pi} C_{2D\max} \tag{3.7}$$

The concentration ratio for a parabolic trough can reach a maximum value of 67.7 (for a refractive index n = 1), considering the best case when the rim angle $\Theta = \pi/2$ rad and the acceptance angle is $\beta = \theta s = 4.7 \times 10^{-3}$ rad. In practice and because of imperfections of the reflective surface, errors on tracking and absorber tube positioning for large collector assemblies of 5.8 m of aperture and 100 m length, the typical values of C are between 20 and 30.

If we repeat the same analysis of C for a 3D parabolic reflector, the expression leading to the maximum concentration is:

$$C = \frac{\sin^2\Theta \cdot \cos^2(\Theta + \beta)}{\sin^2\beta} \tag{3.8}$$

Figure 3.4 represents C versus the rim angle Θ obtained from Eq. 8. As it can be observed the maximum concentration ratio is reached for a rim angle of 45 deg. and, according to equation 1, for a focal ratio of $f/D = 0.6$. This optimum concentration is independent of the acceptance angle of the system. The acceptance half-angle β is the sum of the half-angle subtended by the Sun, θ_s, and the optical error associated to the concentrator system, ψ, as $\beta = \theta_s + \psi$. Typical values of C are 1000 to 3000 for parabolic dishes between 5 and 10 m diameter.

3.2.2 Solar Thermal Power Plant Technologies

Four concentrating solar power technologies are today represented at pilot and commercial-scale [13]: parabolic trough collectors (PTC), linear Fresnel reflector systems (LFR), power towers or central receiver systems (CRS), and dish/engine systems (DE). All the existing pilot plants mimic parabolic geometries with large mirror areas and work under real operating conditions (Fig. 3.5).

PTC and LFR are 2-D concentrating systems in which the incoming solar radiation is concentrated onto a focal line by one-axis tracking mirrors. They are able to concentrate the solar radiation flux 30 to 80 times, heating the thermal fluid up to 450°C, with power conversion unit sizes of 30 to 80MW, and therefore, they are well suited for centralized power generation at dispatchable markets with a Rankine steam turbine/generator cycle. CRS optics is more complex, since the solar receiver is mounted on top of a tower and sunlight is concentrated by means of a large paraboloid that is discretized into a field of heliostats. This 3-D concentrator is, therefore, off-axis and heliostats require two-axis tracking. Concentration factors are between 200 and 1000 and unit sizes are between 10 and 200 MW, and they are, therefore, well suited for dispatchable markets and integration into

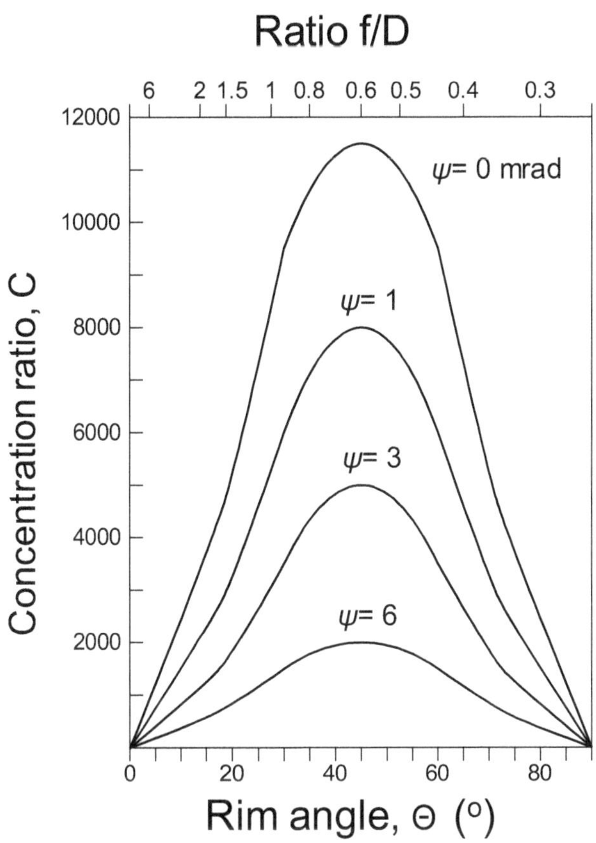

FIG. 3.4: CONCENTRATION RATIO VERSUS RIM ANGLE AT DIFFERENT BEAM QUALITIES OF THE CONCENTRATOR (Ψ = 0 CORRESPONDS TO AN IDEAL CONCENTRATOR OF $\beta = \theta$ S)

advanced thermodynamic cycles. A wide variety of thermal fluids, like saturated steam, superheated steam, molten salts, atmospheric air or pressurized air, can be used, and temperatures vary between 300°C and 1000°C. Finally, DE systems are small modular units with autonomous generation of electricity by Stirling engines or Brayton mini-turbines located at the focal point. Dishes are parabolic 3D concentrators with high concentration ratios (1000–3000) and unit sizes of 5–25 kW. Their current market niche is in both distributed on-grid and remote/off-grid power applications.

Typical solar-to-electric conversion efficiencies and annual capacity factors are listed in Table 3.1 [6]. The values for parabolic troughs, by far the most mature technology, have been demonstrated commercially. Those for linear Fresnel, dish and tower systems are, in general, projections based on component and early commercial projects and the assumption of mature development of current technology. With current investment costs, all STE technologies are generally thought to require a public financial support strategy for market deployment. At present direct capital costs of STE and power generation costs are estimated to be 2–3 times those of fossil-fueled power plants, however, industry roadmaps advance 60% cost reduction before 2025 [7]. In fact governments at some countries like Spain are already accelerating the process of drastic tariff reduction with the goal of STE, PV, and wind energy becoming tariff-equivalent in less than one decade.

Every square meter of STE field can produce up to 1200 kWh thermal energy per year or up to 500 kWh of electricity per year. That means a cumulative savings of up to 12 tons of carbon dioxide and 2.5 tons of fossil fuel per square meter of CSP system over its 25-year lifetime [14]. After two decades of frozen or failed

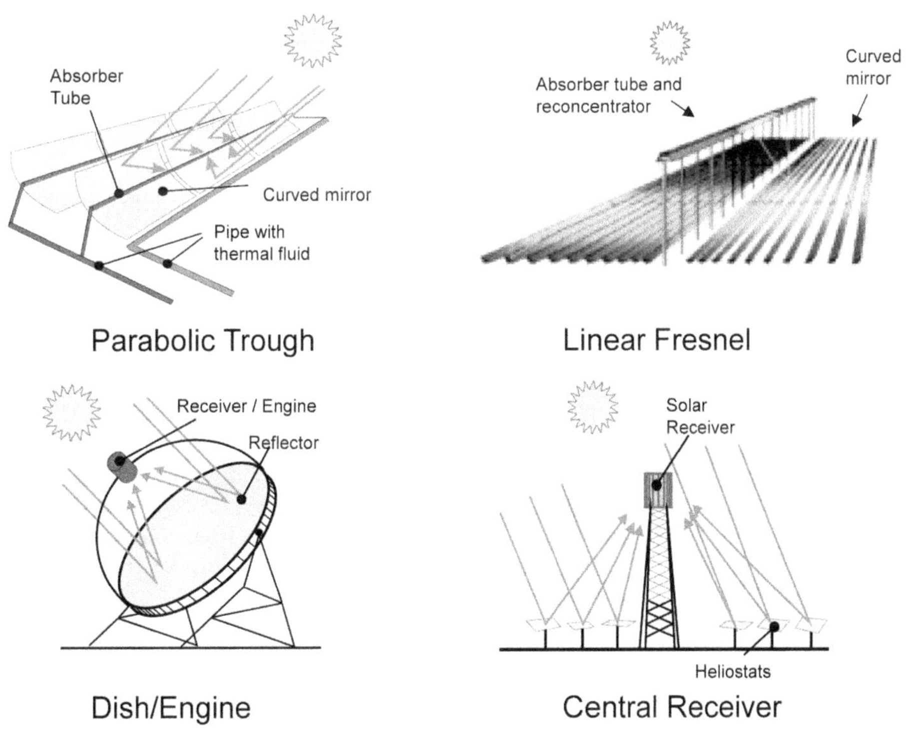

FIG. 3.5: SCHEMATIC DIAGRAMS OF THE FOUR STE SYSTEMS CURRENTLY SCALED UP TO PILOT AND DEMONSTRATION SIZES

TABLE 3.1

CHARACTERISTICS OF SOLAR THERMAL ELECTRICITY SYSTEM (ADAPTED FROM REF. [15]). FRESNEL SYSTEMS ARE NOT INCLUDED SINCE PERFORMANCE DATA AVAILABLE ARE NOT CONCLUSIVE FOR A COMPARATIVE ASSESSMENT

	Parabolic troughs	Central receiver	Dish–Stirling
Power Unit	30–80 MW*	10–200 MW*	5–25 kW
Temperature operation	390°C	565°C	750°C
Annual capacity factor	23–50%*	20–77%*	25%
Peak efficiency	20%	23%	29.4%
Net annual efficiency	11–16%*	7–20%*	12–25%
Commercial status	Mature	Early projects	Prototypes-demonstration
Technology risk	Low	Medium	High
Thermal storage	Limited	Yes	Batteries
Hybrid schemes	Yes	Yes	Yes
Cost W installed			
$/W	3.49–2.34*	3.83–2.16*	11.00–1.14*
$/Wpeak**	3.49–1.13*	2.09–0.78*	11.00–0.96*

*Data interval for the period 2010–2025.

** Without thermal storage.

projects, approval in the past few years for specific financial incentives in Europe, the US, India, Australia and elsewhere, is now paving the way for launching of the first commercial ventures. Spain with 2400 MW connected to the grid in 2013 is taking the lead on current commercial developments, together with USA where a target of 4500 MW for the same year has been fixed and other relevant programs such as the "Solar Mission" in India has been recently approved for 22 GW-solar, with a large fraction being thermal [2].

3.3 PARABOLIC-TROUGHS

3.3.1 The parabolic-trough collector (PTC)

Parabolic-trough collectors are linear-focus concentrating solar devices suitable for working in the 150°C–400°C temperature range [16]; current research with new thermal fluids intends to increase the operating temperature up to 500°C [17]. The concentrated radiation heats the fluid that circulates through the receiver tube, thus transforming the solar radiation into thermal energy in the form of the sensible heat of the fluid. Figure 3.6 shows a typical PTC and its components.

Collector rotation around its axis requires a drive unit. One drive unit is usually sufficient for several parabolic-trough modules connected in series and driven together as a single collector. The type of drive unit assembly depends on the size and dimensions of the collector. Drive units composed of an electric motor and a gearbox combination are used for small collectors (aperture area < 100 m^2), while powerful hydraulic drive units are required to rotate large collectors. A drive unit placed on the central pylon is commanded by a local control unit in order to track the sun. At present, all commercial PTC designs use a single-axis sun-tracking system [18].

Thermal oils are commonly used as the working fluid in these collectors for temperatures above 200°C, because at these operating temperatures, normal water would produce high pressures inside the receiver tubes and piping. This high pressure would require stronger joints and piping, and thus raise the price of the collectors and the entire solar field. However, the use of demineralized water for high temperatures/pressures is currently under investigation in Spain at the Plataforma Solar de Almería (PSA) and the feasibility of direct steam generation at 100 bar/400°C in the receiver tubes of parabolic trough collectors has already been proven in an experimental stage [19]. For temperatures below 200°C, either a mixture of water/ethylene glycol or pressurized liquid water can be used as the working fluids because the pressure required in the liquid phase is moderate.

The oil most widely used in parabolic-trough collectors for temperatures up to 395°C is VP-1, which is a eutectic mixture of 73.5% diphenyl oxide/26.5% diphenyl. The main problem with this oil is its high-solidification temperature (12°C), which requires an auxiliary heating system when oil lines run the risk of cooling below this temperature. Since the boiling temperature at 1013mbar is 257°C, the oil circuit must be pressurized with nitrogen, argon, or some other inert gas when oil is heated above this temperature. Blanketing of the entire oil circuit with an oxygen-free gas is a must when working at high temperatures because high-pressure mists can form an explosive mixture with air. Though there are other suitable thermal

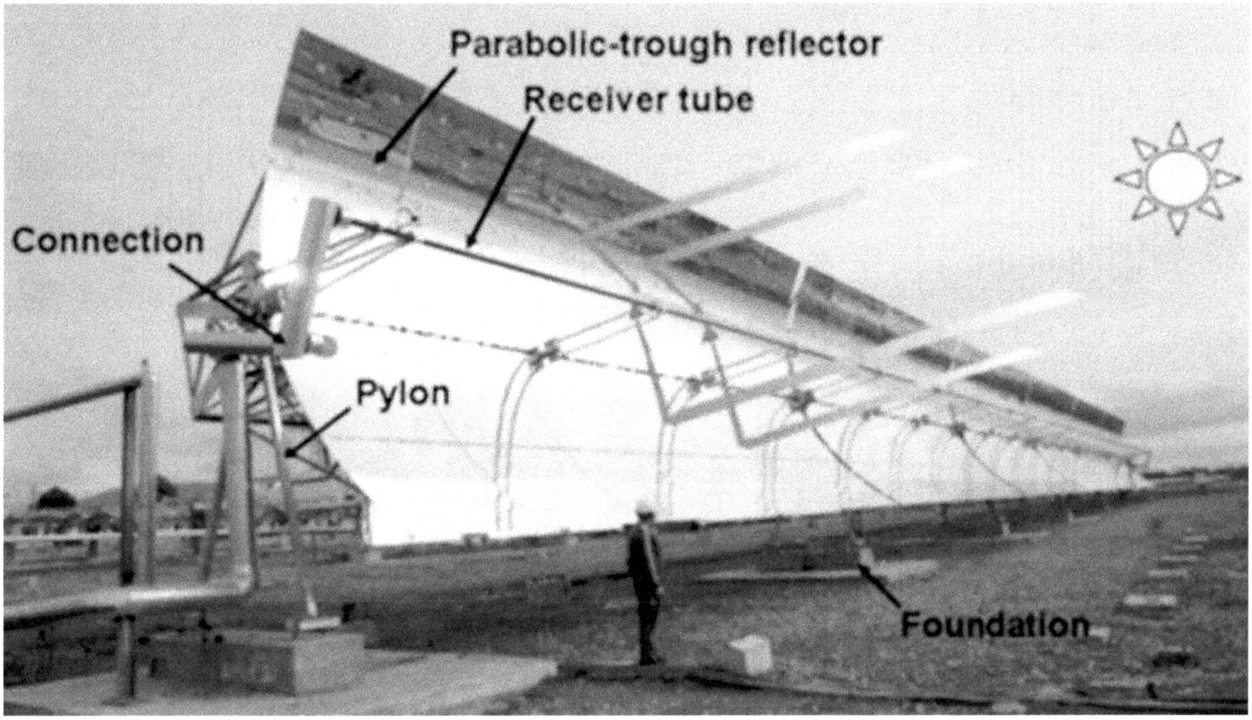

FIG. 3.6: TYPICAL PARABOLIC-TROUGH COLLECTOR (PTC TEST FACILITY AT THE PLATAFORMA SOLAR DE ALMERÍA, SPAIN) (*Source:* Eduardo Zarza, Plataforma Solar De Almería)

oils for slightly higher working temperatures with lower solidification temperatures, they are too expensive for large solar plants.

The typical PTC receiver tube is composed of an inner steel pipe surrounded by a glass tube to reduce convective heat losses from the hot steel pipe [6]. The steel pipe which has a selective high-absorption (>90%), low-emission (<30% in the infrared) coating that reduces radiation thermal losses. Receiver tubes with glass vacuum tubes and glass pipes with an anti-reflective coating achieve higher PTC thermal efficiency and better annual performance, especially at higher operating temperatures. Receiver tubes with no vacuum are usually for working temperatures below 250°C because thermal losses are not so critical at these temperatures. Due to manufacturing constraints, the maximum length of a single receiver pipe is less than 6 m, so that the complete receiver tube of a PTC is composed of a number of single receiver pipes welded in series up to the total length of the PTC. The total length of a PTC is usually within 25–150 m.

Figure 3.7 shows a typical PTC vacuum receiver pipe. The outer glass tube is attached to the steel pipe by means of flexible metal differential expansion joints which compensate for the different thermal expansion of glass and steel when the receiver tube is working at nominal temperature. At present there are only two manufacturers of PTC vacuum absorber tubes: the German company Schott and the Israeli company SOLEL. The flexible expansion joint used by these two manufacturers is shown in Fig. 3.7. The glass-to-metal-welding used to connect the glass tube and the expansion joint is a weak point in the receiver tube and has to be protected from the concentrated solar radiation to avoid high thermal and mechanical stress that could damage the welding. An aluminum shield is usually placed over the joint to protect the welds.

As seen in Fig. 3.7, several chemical getters are placed in the gap between the steel receiver pipe and the glass cover to absorb gas molecules from the fluid that get through the steel pipe wall to the annulus.

Parabolic-trough collector reflectors have a high specular reflectance (>88%) to reflect as much solar radiation as possible. Solar reflectors commonly used in PTC are made of back-silvered glass mirrors, since their durability and solar spectral reflectance are better than the polished aluminum and metallized acrylic mirrors also available in the market. Solar spectral reflectance is typically 0.93 for silvered glass mirrors and 0.87 for polished aluminum. Low-iron glass is used for the silvered glass reflectors and the glass to improve solar transmission.

Two PTC designs specially conceived for large solar thermal power plants are the LS-3 (owned by the Israeli company SOLEL Solar Systems) and EuroTrough (owned by the EuroTrough Consortium), both of which have a total length of 100 m and a width of 5.76 m, with back-silvered thick-glass mirrors and vacuum absorber pipes. American Solargenix design has an aluminum structure. However, other collector designs are recently becoming commercially available in the short-to-medium term like the ones developed by the companies Solargenix, Albiasa, or Sener [17]. The main constraint when developing the mechanical design of a PTC is the maximum torsion at the collector ends, because high torsion would lead to a smaller intercept factor and lower optical efficiency.

In a typical parabolic-trough collector field, several collectors connected in series make a row, and a number of rows are connected in parallel to achieve the required nominal thermal power output at design point. The number of collectors connected in series in every row depends on the temperature increase to be achieved between the row inlet and outlet. In every row of collectors, the receiver tubes in adjacent parabolic-trough collectors

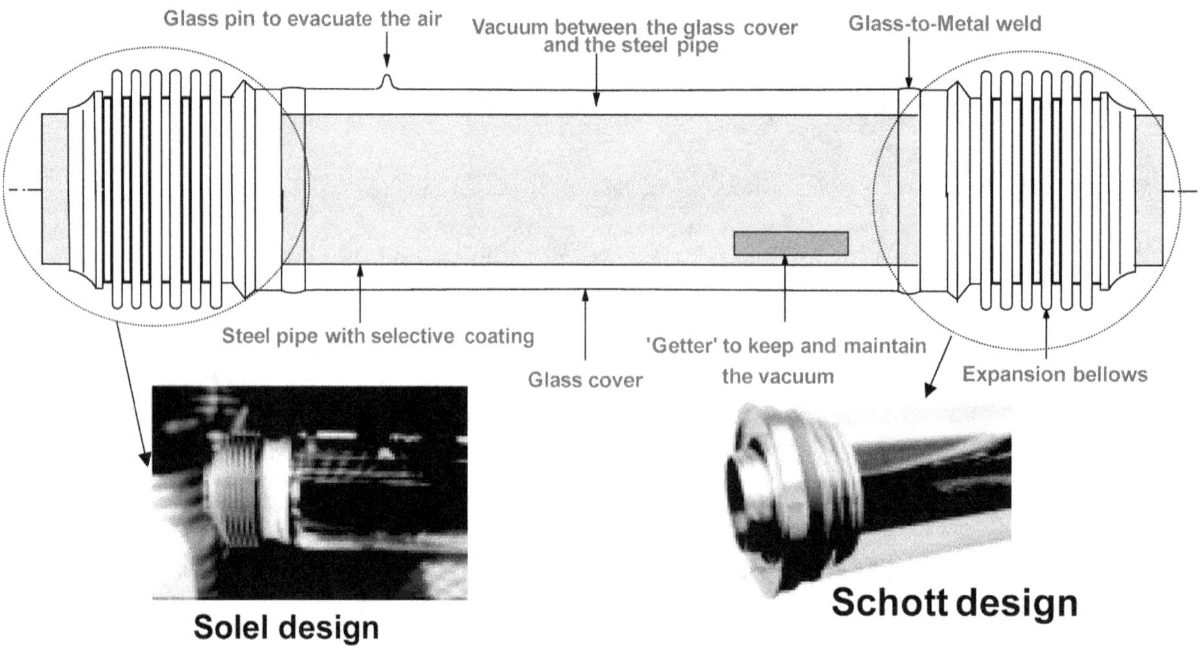

FIG. 3.7: EVACUATED RECEIVER TUBE DESIGN WITH GLASS-TO-METAL WELDS FOR PARABOLIC TROUGH COLLECTORS, AND DETAILS OF COMMERCIAL TUBES SUPPLIED BY THE COMPANIES SOLEL AND SCHOTT

have to be connected by flexible joints to allow independent rotation of both collectors as they track the sun during the day. These flexible connections are also necessary to allow the linear thermal expansion of the receiver tubes when their temperature increases from ambient to nominal temperature during system start-up. Two main types of flexible connections are available: flexible hoses and ball joints.

Flexible connections are also needed to connect the row to the main field pipe header inlet and outlet.

3.3.2 Electricity Generation with Parabolic-Trough Collectors

The suitable parabolic-trough collector temperature range and their good solar-to-thermal efficiency up to 400°C make it possible to integrate a parabolic-trough solar field in a Rankine water/steam power cycle to produce electricity. The simplified scheme of a typical solar thermal power plant using parabolic-troughs integrated in a Rankine cycle is shown in Fig. 3.8. The technology commercially available at present for parabolic-trough power plants is the HTF (Heat Transfer Fluid) technology, which uses oil as the heat carrier between the solar field and the power block. The solar field collects the solar energy available in the form of direct solar radiation and converts it into thermal energy as the temperature of the oil circulating through the receiver tubes of the collector's increases. Once heated in the solar field, the oil goes to the steam generator, which is an oil-water heat exchanger where the oil transfers its thermal energy to the water that is used to generate the superheated steam required by the turbine. The steam generator is, therefore, the interface between the solar system (solar field + oil circuit) and the power conversion system (PCS) itself. Normally, the steam generator used in these solar power plants consists of three stages: preheater, evaporator, and superheater.

Though parabolic-trough power plants usually have an auxiliary gas-fired heater to produce electricity when direct solar radiation is not available, the amount of electricity produced with natural gas is always limited to a reasonable level. This limit changes from one country to another: 25% in California (USA), 15% in Spain, and no limit in Algeria. Typical solar-to-electric efficiencies of a large solar thermal power plant (>30 MWe) with parabolic-trough collectors is between 15% and 22%, with an average value of about 17%. The yearly average efficiency of the solar field is about 50%.

The maturity of PTC systems is confirmed by the SEGS plants. The plants SEGS (Solar Electricity Generating Systems) II to IX, which use thermal oil as the working fluid (HTF technology), were designed and implemented by the LUZ International Limited company from 1985 to 1990. All the SEGS plants are located in the Mojave Desert, Northwest of Los Angeles (California, USA). With their daily operation and over 2.2 million m^2 of parabolic-trough collectors, SEGS plants are this technology's best example of commercial maturity and reliability. Their plant availability is over 98% and their solar-to-electric annual efficiency is in the range of 14–18%, with a peak efficiency of 22% [16].

The electricity produced by the SEGS plants is sold to the local utility, Southern California Edison, under individual 30-year contracts for every plant. To optimize the profitability of these plants, it is essential to produce the maximum possible energy during peak-demand hours, when the electricity price is the highest. The gas boilers can be operated for this, either to supplement the solar field or alone. Nevertheless, the total yearly electricity production using natural gas is limited by the Federal Commission for Energy Regulation in the United States to 25% of the overall yearly production.

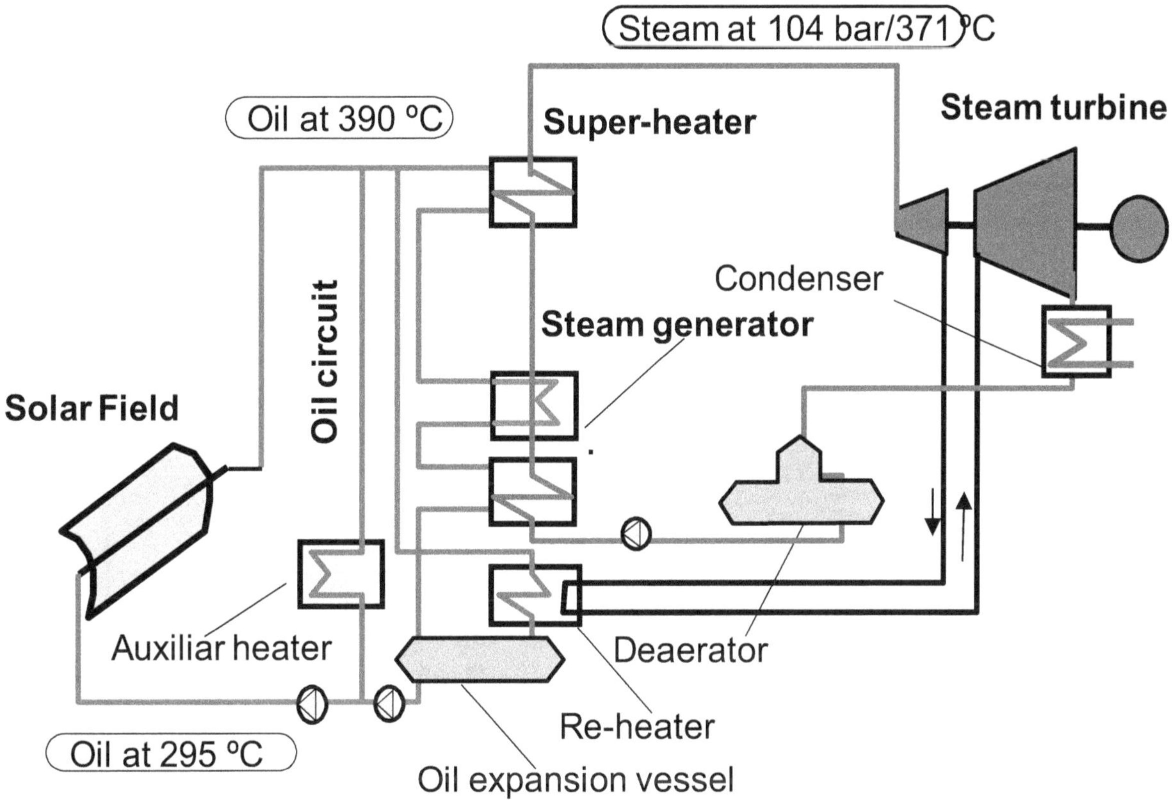

FIG. 3.8: TYPICAL FLOW DIAGRAM OF A SEGS PLANT WITH OIL AS HEAT TRANSFER FLUID IN THE SOLAR FIELD. (*Source:* Eduardo Zarza, Plataforma Solar De Almería)

Peak-demand hours are when electricity consumption is more and, therefore, the tariff is the highest. Off-peak and super off-peak hours are when electricity consumption is low, and the electricity price is, therefore, also lower. At present, 16% of the SEGS plants' annual net production is generated during summer peak-demand hours, and the revenues from this are in the order of 55% of the annual total. These figures show how important electricity generated during peak-demand hours is for the profitability of these plants. Thanks to the continuous improvements in the SEGS plants, the total SEGS I cost of \$0.22/kWh_e for electricity produced was reduced to \$0.16/kWh_e in the SEGS II and down to \$0.09/kWh_e in SEGS IX [20].

With the revival of commercial STE projects since 2006 in US and Spain, a new generation of SEGS-type plants has come to the arena. This is the case of the Nevada Solar One project of 75 MWe in US, the Ibersol project in Puertollano, Spain or the Shams One 100 MW plant in Abu Dhabi [21]. Nevada Solar One started grid-connected operation in June 2007 and it is considered a milestone in the opening of the second market deployment of PTC technology in the world after SEGS experience. Since then, more than 40 PTC plants (about 50 MWe each) are constructed and started operation between the period 2007 and 2013 and more than 2 GW on track in the US [2]. Main characteristics of NSO Plant are shown in Table 3.2.

In spite of their environmental benefits, there are still some obstacles to the commercial use of this technology. The main barriers at present are the high investment cost (2500–4000 \$/kW, depending on plant size and thermal storage capacity) and the minimum size of the power block required for high thermodynamic efficiency. However, these barriers are shared by all the solar thermal power technologies currently available [22].

One of the strategies to mitigate risk perception and increment the capacity factor of the power block is to integrate a parabolic-trough solar field in the bottoming cycle of a combined-cycle gas-fired power plant. This configuration is called the Integrated Solar Combined Cycle System (ISCCS). Though the contribution of the solar system to the overall plant power output is small (10%–15% approximately) in the ISCCS configuration, it seems to be a good approach to market penetration in some developing countries, which is why the Government of Algeria has promoted an ISCCS plant and the World Bank, through its Global Environment Facility (GEF), is promoting ISCCS plants in Morocco and Egypt [23]. Figure 3.9 shows the schematic diagram of a typical ISCCS plant.

3.3.3 Thermal Energy Storage and New Thermal Fluids

In many countries, the market penetration of STE systems is based on feed-in-tariffs or green certificates linked to significant restrictions or regulations regarding the use of hybrid concepts like ISCCS schemes. Because of that the use of thermal energy storage systems with an oversized solar field is pursued to optimize economics and dispatchability of PTC plants.

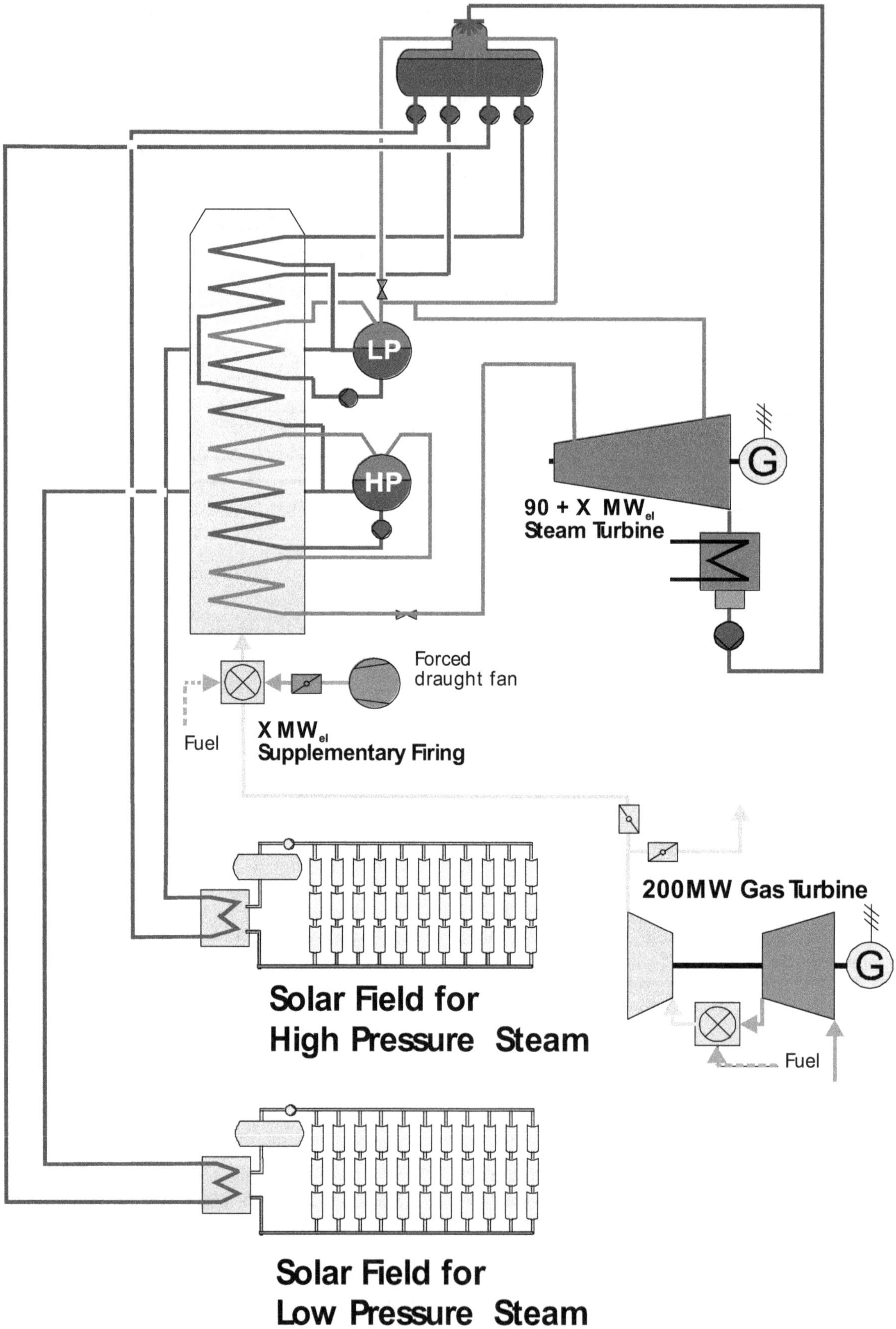

FIG. 3.9: HYBRID STE PLANT MAKING USE OF THE ISCCS SCHEME (INTEGRATED SOLAR COMBINED CYCLE SYSTEM). THE SOLAR FIELD IS INTEGRATED IN THE BOTTOMING CYCLE OF A COMBINED-CYCLE GAS-FIRED POWER PLANT. (*Source:* Solarpaces)

TABLE 3.2

MAIN CHARACTERISTICS OF NEVADA SOLAR ONE PLANT

Solar field		Power block	
Solar collector assemblies	760	Turbine generator gross output	75 MWe
Aperture area (m)	5	Net output to utility	72 MWe
Aperture area (m^2)	470	Solar steam inlet pressure	86.1 bar
Length (m)	100	Solar steam reheat pressure	19.5 bar
Concentration ratio	71	Solar steam inlet temperature	371°C
Optical efficiency	0.77		
Number of mirror segments	182,400		
Number of receiver tubes	18240		
Field aperture (m^2)	357,200		
Site area (km^2)	1.62		
Field inlet temperature (°C)	300		
Field outlet temperature (°C)	390		

In this case, the solar field has to be oversized so that it can simultaneously feed the power conversion system and charge the storage system during sunlight hours. Thermal energy from the storage system is then used to keep the steam turbine running and producing electricity after sunset or during cloudy periods. Yearly hours of operation can be significantly increased and plant amortization is thus enhanced when a storage system is implemented. However, the required total investment cost is also higher.

For temperatures of up to 300°C, thermal mineral oil can be stored at ambient pressure, and is the most economical and practical solution. Synthetic and silicone oils, available for up to 410°C, have to be pressurized and are expensive. Then, molten salts can be used between 220 and 560°C at ambient pressure, but require parasitic energy to keep them liquid.

Two pioneer projects introducing thermal storage are the plants Andasol-I and Andasol-II. This PTC plants installed in Guadix, Spain, have a nominal power of 50 MWe each and an oversized solar field (510,120 m^2 mirrors surface area) with an integrated 1010 MWth molten-salt thermal storage system to extend the plant's full-load operation 7.7 hours beyond daylight hours leading to a capacity factor of 41%. Basically, the system is composed of [24]:

- 2 tanks (cold and hot) containing 28,500 tons of nitrate molten salts at 2 different temperature levels (292 and 386°C). Molten salt composition: 60% $NaNO_3$ + 40% KNO_3.
- 6 heat exchangers (series connected) located between the tanks, which allow storage charging and discharging from and to the Heat Transfer Fluid (HTF) of the plant, in this case, Dowtherm A.
- 3 cantilever pumps in each tank to transfer the molten salt from one tank to the other through the 6 heat exchangers (counterflow). There is a spare pump unit in the cold tank.
- 6 electrical heating devices in the bottom of each tank to maintain salt temperature above 292°C during long maintenance periods (salts crystallize at 238°C and solidify at 221°C).
- Electrical heat tracing in piping and heat exchangers parts to keep temperatures at adequate levels.
- A natural convection system through tanks foundations in order to decrease concrete temperature.
- A drainage tank to ease draining of piping and heat exchangers.

Figure 3.10 shows Andasol plant flow diagram and Fig. 3.11 an aerial view of the two tanks, pumps and salt circuit. As it can be observed, the introduction of a third circuit with molten salts adds more complexity with a number of new heat exchangers and heat tracing to avoid the salts thawing. It is still early to know the final result in terms of energy management and operational robustness. Early operational data demonstrate the capacity of the plant to supply electricity several hours after sunset.

The company ACS-Cobra is constructing three more 50-MWe plants similar to the Andasol projects, Extresol 1 (in collaboration with the company Sener) and Extresol 2 in Torre de Miguel Sesmero (Badajoz, Spain), and Manchasol 1 in Alcázar de San Juan (Toledo, Spain). This company is also planning 2 more plants, Extresol 3 in Badajoz, and Manchasol 2 in Ciudad Real [17]. Other countries such as Italy are planning further steps with the use of molten salts both for thermal storage and solar field, like the 5-MW demonstration project Archimede [25].

Even though molten salts are nowadays the preferred option for demonstration and first commercial projects with storage in the US and Spain, there are other options under development and assessment like the use of concrete or other solid bed materials [26], and the use of Phase change media (PCM) [27]. PCM provide a number of desirable features, e.g., high volumetric storage capacities and heat availability at constant temperatures. Energy storage systems using the latent heat released on melting eutectic salts or metals have often been proposed, but never carried

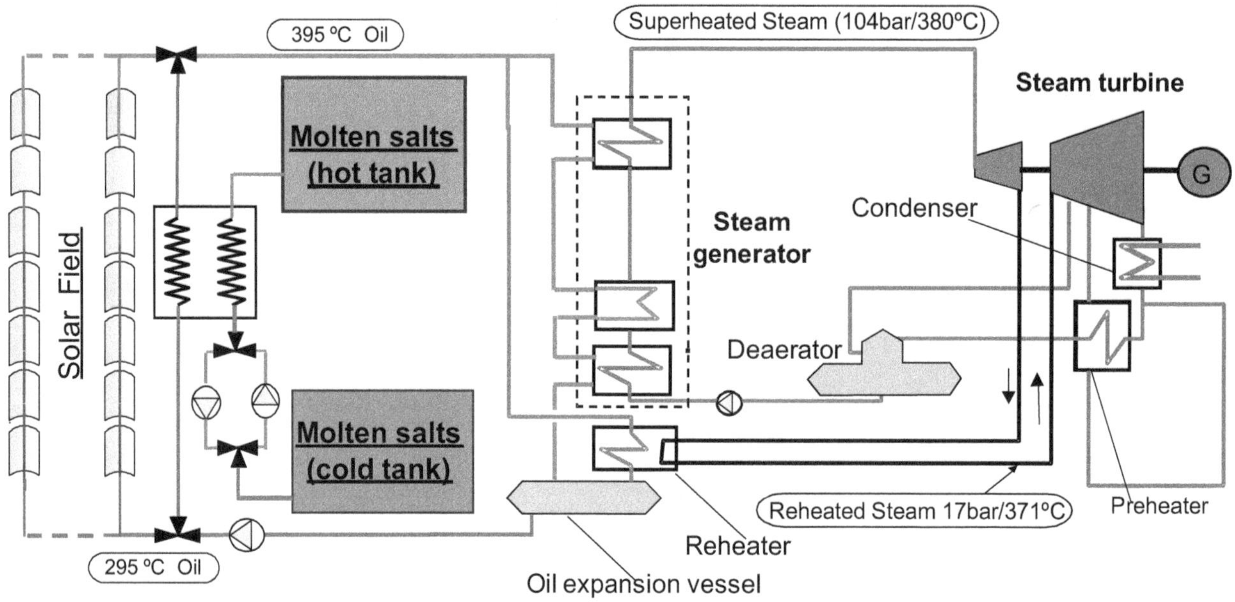

FIG. 3.10: ANDASOL PLANT SCHEME INTRODUCING A MOLTEN-SALT CIRCUIT WITH TWO STORAGE TANKS TO INCREMENT CAPACITY FACTOR. (*Source*: Eduardo Zarza, Plataforma Solar de Almería)

FIG. 3.11: AERIAL VIEW OF POWER ISLAND AND DETAIL OF MOLTEN-SALT STORAGE TANKS AND HEAT EXCHANGERS OF ANDASOL PLANT. THIS 50 MW PLANT STORES THERMAL ENERGY EXCESS IN 28,500 TONNES OF NITRATE MOLTEN SALTS ABLE TO PROVIDE UP TO 7.7 EQUIVALENT HOURS OF OPERATION AT NOMINAL CAPACITY. (ANDASOL-I PLANT, GUADIX, SPAIN) (*Source*: ACS/Cobra Energía (Spain))

out on a large scale due to difficult and expensive internal heat exchange and cycling problems. Heat exchange between the heat transfer fluid (HTF) and the storage medium is seriously affected when the storage medium solidifies. Encapsulation of PCMs has been proposed to improve this. Combining the advantages of direct-contact heat exchange and latent heat, hybrid salt-ceramic phase change storage media have recently been proposed. The salt is retained within the submicron pores of a solid ceramic matrix such as magnesium oxide by surface tension and capillary force. Heat storage is then accomplished in two modes: by the latent heat of the salt and also the sensible heat of the salt and the ceramic matrix.

PCM are extremely suitable for direct steam generation PTC plants. If the superheated steam required to feed the steam turbine in the power block were produced directly in the receiver tubes of the parabolic-trough collectors, the oil would be no longer necessary, and temperature limitation and environmental risks associated with the oil would be avoided. Simplification and cost reduction of overall plant configuration is then evident since only one fluid is used. This effect combined with the increment of efficiency after removing intermediate heat exchanger might lead to a reduction of 15% of the cost of the electricity produced. The disadvantages of this concept originate from the thermo-hydraulic problems associated with the two-phase flow existing in the evaporating section of the solar field. Nevertheless, experiments performed at in a 2-MWth loop at the Plataforma Solar de Almería (PSA) in Spain have proven the technical feasibility of direct steam generation with horizontal parabolic-trough collectors at 100 bar/400°C within the framework of two European projects, Direct Solar Steam (DISS) and Integration of DSG Technology for Electricity Production (INDITEP) [28]. Two pre-commercial projects, Puertollano GDV and Real-DISS, are under development in Spain to demonstrate the technical feasibility of direct steam generation combined with a power block [17].

3.4 LINEAR-FRESNEL REFLECTORS

3.4.1 Historical Evolution of Linear-Fresnel Reflector Systems

Conceptually, Linear-Fresnel Reflectors (LFR) are optical analogues of parabolic troughs. They are 2-D concentrating reflectors with linear focus, where the parabolic reflective surface is obtained by an array of linear mirror strips which independently move and collectively focus on absorber lines suspended from elevated towers [13, 29]. They are fix focus reflectors where the absorber is static. Reflective segments are close to the ground and can be assembled in a compact way up to 1 ha/MW. The objective is to reproduce the performance of parabolic troughs though with lower costs. However, optical quality and thermal efficiency is lower because of a higher influence of the incidence angle and the cosine factor, and therefore, the temperature obtained at the working fluid is also lower (150–350°C). Because of that, LFR are mainly oriented to produce saturated steam via direct in-tube steam generation, and application into ISCCS or in regenerative Rankine cycles, though current R&D is aiming at higher temperatures above 400°C.

In contrast, LFRs typically use

- lower-cost non-vacuum thermal absorbers where the stagnant air cavity provides significant thermal insulation
- light reflector support structures close to the ground
- low-cost flat float glass reflector
- low-cost manual cleaning, because the reflectors are at human height.

The LFRs also have much better ground utilization, typically using 60–70% of the ground area compared to about 33% for a trough system, and lower O&M costs due to more accessible reflectors.

After some pioneering experiences [29], the first serious development began on the CLFR Australian design at the University of Sydney in 1993, in which a single field of reflectors used multiple linear receivers and reflectors which change their focal point from one receiver to another during the day in order to minimize shading in the dense reflector field. These are called the Compact Linear Fresnel Reflector systems, or CLFR systems and cover about 71% of the ground compared with 33% for trough systems [30]. In 1994, a company called Solel Europe entered into a commercial in-confidence agreement with Sydney University regarding CLFR technology. This company was later reconstituted as Solarmundo and it built a 2400 m^2 LFR prototype collector field at Liege in Belgium 2000, but test results were not reported. Later the company moved to Germany and was renamed the Solar Power Group. SPG signed an exclusivity cooperation agreement with DSD Industrieanlagen GmbH a company of the MAN/Ferrostaal group. A 800kW Linear Fresnel pilot operating at 450°C has already been tested in Plataforma Solar de Almería with components developed by Solar Power Group, and the technology is expected to be commercially available by 2012 [7, 31].

Back in Australia, in early 2002, a new company, Solar Heat and Power Pty Ltd (SHP), was formed after acquiring the patents covering the Australian CLFR work. SHP immediately made extensive changes to the engineering design of the reflectors to lower cost and has become the first to commercialize LFR technology. SHP initiated in 2003 for Macquarie Generation, Australia's largest generator, a demonstration project of 103 MWth (approximately 39 MWe) plant with the aim of supplying preheat to the coal fired Liddell power station. Phase 1 of the project, completed in 2004, resulted in a 1350 m^2 segment not connected to the coal fired plant, and was used to trial initial performance, and it first produced steam at 290°C in July 2004. The expansion to 9 MWth was completed by 2008. Activities of the company moved to the US and were continued by Ausra, Palo Alto. Ausra established a factory of components, tubular absorbers and mirrors in Las Vegas and built the Kimberlina 5-MW demonstration plant in Bakersfield at the end of 2008. In 2010, Aura has been purchased by AREVA that is committed to the commercial deployment of this technology.

The third technology player after SPG/MAN and AREVA is the German company NOVATEC Biosol [7]. The technology of NOVATEC is based upon its collector Nova-1 aimed to produce saturated steam at 270°C. They have developed a serial production factory for pre-fabricated components and a 1.4 MW small commercial plant in Puerto Errado, Murcia, Spain. This plant is grid connected since March 2009 [32]. NOVATEC is promoting 50 MW plants mixing PTC and LFR fields, where LFR provides pre-heating and evaporation and PTC field takes over superheating. The company claims that this hybridization results in 22% less land use and higher profitability.

3.4.2 Future Technology Development and Performance Trends

Even though some solid commercial programmes are underway on LFR, still it is early to have consolidated performance data with respect to electricity production. The final optimization would integrate components development to increment temperature of operation and possible hybridization with other STE systems like parabolic troughs.

There are many possible types of receiver, including evacuated tube and PV modules, but the most cost-effective system seems to be an inverted cavity receiver. In the case of SHP technology, the absorber is a simple parallel array of steam pipes at the top of a linear cavity, with no additional redirection of the incoming light from the heliostats to minimise optical losses and the use of hot reflectors. In the case of SPG technology, the absorber is a single tube surmounted by a hot non-imaging reflector made of glass, which must be carefully manufactured to avoid thermal stress under heating, and exhibit some optical loss. Both systems can produce saturated steam or pressurized water. At present, AREVA and NOVATEC are looking for new absorbers able to work at temperatures above 450°C. By 2015, according to developers, Linear Fresnel can be expected to be operating with superheated steam at 500°C yielding an efficiency improvement of up to 18.1% relative to current saturated steam operation at 270°C [7].

For reflectors, automation is a key issue that has been demonstrated by NOVATEC. Additional effort should be given to the optimized demonstration of multi-tower arrays to maximize ground coverage ratios (Fig. 3.12).

However, it is the lack of reliable information regarding annual performance and daily evolution of steam production that should be targeted as a first priority. Still, some concerns remain regarding the ability to control steam production, because of the pronounced effect of cosine factor in this kind of plants. This dynamic performance would also affect the potential integration with other STE systems such as PTC or CRS. Until now most comparative assessments vis-à-vis parabolic troughs are not economically conclusive, revealing the need to use much larger fields to compensate lower efficiencies [33]. In order to achieve break even costs for electricity with current LFR technology, the cost target for the Fresnel solar power plants need to be about 55% of the specific costs of parabolic trough systems [33].

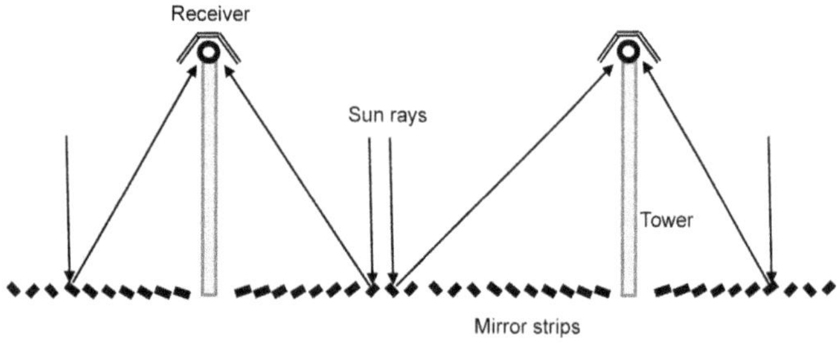

FIG. 3.12: SCHEME OF COMPACT LFR SYSTEM WITH A MULTITOWER ARRAY AND DYNAMIC AIMING STRATEGY OF MIRROR STRIPS

3.5 CENTRAL RECEIVER SYSTEMS (CRS)

In power towers or central receiver systems, incident sunrays are tracked by large mirrored collectors (heliostats), which concentrate the energy flux onto radiative/convective heat exchangers called solar receivers, where energy is transferred to a thermal fluid, mounted on top of a tower (Fig. 3.13). Plant sizes of 10 to 200 MW are chosen because of economy of scale, even though advanced integration schemes are claiming the economics of smaller units as well [34]. The high solar flux incident on the receiver (averaging between 300 and 1000 k W/m^2) enables operation at relatively high temperatures of up to 1000°C and integration of thermal energy into more efficient cycles in a step-by-step approach. CRS can be easily integrated in fossil plants for hybrid operation in a wide variety of options and has the potential for generating electricity with high annual capacity factors through the use of thermal storage. With storage, CRS plants are able to operate over 4500 hours per year at nominal power [35].

The heliostat field and solar receiver systems distinguish solar thermal tower power plants from other STE plants, and are, therefore, given more attention below. In particular, the heliostat field is the single factor with the most impact on plant investment. Collector field and power block together represent about 72% of the typical solar-only plant (without fossil backup) investment, of which heliostats represent 60% of the solar share. Even though the solar receiver impacts the capital investment much less (about 14%), it can be considered the most critical subsystem in terms of performance, since it centralizes the entire energy flux exchange. The largest heliostat investment is the drive mechanism and reflecting surface, which alone are almost 70% of the total.

3.5.1 Technology of Heliostats and Solar Receivers

The collector field consists of a large number of mirrors, called heliostats, with two-axis tracking and a local control system to continuously focus direct solar radiation onto the receiver aperture area. During cloud passages and transients the control system must defocus the field and react to prevent damage to the receiver and tower structure. Heliostats fields are characterized by their off-axis optics. Since the solar receiver is located in a fixed position, the entire collector field must track the sun in such a way that each and every heliostat individually places its surface normal to the bisection of the angle subtended by sun and the solar receiver. Each and every heliostat has its own elevation and azimuthal angle updated about every 4 sec.

Heliostat field optical efficiency includes the cosine effect, shadowing, blocking, mirror reflectivity, atmospheric attenuation and receiver spillage [36]. Because of the large area of land required, complex optimization algorithms are used to optimize the annual energy produced by unit of land, and heliostats must be packed as close as possible so the receiver can be small and concentration high. Since the reflective surface of the heliostat is not normal to the incident rays, its effective area is reduced by the cosine of the angle of incidence; the annual average cosine varies from about 0.9 at two tower heights north of the tower to about 0.7 at two tower heights south of the tower. Of course, annual average cosine is highly dependent on site latitude. Consequently, in places close to the Equator a surround field would be the best option to make best use of the land and reduce the tower height. North fields improve

FIG. 3.13: AERIAL VIEW OF THE FIRST COMMERCIAL SOLAR TOWERS IN THE WORLD. PS20 WITH 20 MW AT THE FRONT AND THE PS10 WITH 11 MW AT UPPER LEFT. HELIOSTAT FIELD LAYOUT IS FORMED BY HUNDREDS OF TRACKING MIRRORS FOCUSING CONCENTRATED LIGHT ONTO THE RECEIVER APERTURE. PS20 AND PS10 HAVE BEEN DEVELOPED AND THE TECHNOLOGY IS OWNED BY ABENGOA SOLAR S.A. AND ARE LOCATED IN SANLUCAR LA MAYOR, SEVILLE, SPAIN. (*Source*: Abengoa Solar, S.A.)

performance as latitude increases (South fields in the Southern hemisphere), in which case, all the heliostats are arranged on the North side of the tower.

The combination of all the above-mentioned factors influencing the performance of the heliostat field should be optimized to determine an efficient layout. There are many optimization approaches to establish the radial and azimuthal spacing of heliostats and rows [37]. One of the most classic, effective and widespread procedures is the "radial staggered" pattern, originally proposed by the University of Houston in the 1970s [38]. Integral optimization of the heliostat field is decided by a tradeoff between cost and performance parameters. Heliostats, land and cabling network must be correlated with costs. Cost and performance also often have reverse trends, so that when heliostats are packed closer together, blocking and shadowing penalties increase, but related costs for land and wiring decrease. A classical code in use since the 1980s for optimization of central receiver subsystems is DELSOL3 [39].

Mature low-cost heliostats consist of a reflecting surface, a support structure, a two-axis tracking mechanism, pedestal, foundation and control system (Fig. 3.14). The development of heliostats shows a clear trend from the early first generation prototypes, with a heavy, rigid structure, second-surface mirrors and reflecting surfaces of around 40 m^2 [40] , to designs with large 100–120 m^2 reflecting surfaces, lighter structures and lower-cost materials [41]. Since the first-generation units, heliostats have demonstrated beam qualities below 2.5 mrad (standard deviation of reflected rays including all heliostat errors but not including intrinsic to the solar disk) that are good enough for practical applications in solar towers, so the main focus of development is directed at cost reduction.

In Spain, some developments worthy of mention are the 105-m^2 GM-100 [42] and more recently, the 90-m^2 and 120-m^2 Sanlúcar heliostats finally adopted for the first commercial tower power plant PS10 promoted by the company Abengoa Solar [43]. Recently, the company SENER has developed a similar 115-m^2 heliostat for its Gemasolar plant. Estimated production costs of large area glass/metal heliostats for sustainable market scenarios are around \$130–200/m^2. Large-area glass/metal units make use of glass mirrors supported by metallic frame facets.

Recently, some developers are introducing substantial changes in the conception of heliostat design. A number of projects based upon the paradigm of maximum modularity and mass production of components are claiming small-size heliostats as a competitive low-cost option. Companies like Brightsource, eSolar, Aora,

FIG. 3.14: LATERAL VIEW OF A TYPICAL GLASS-METAL HELIOSTAT WITH T-SHAPE SUPPORTING STRUCTURE FORMED BY PEDESTAL, TORQUE TUBE AND TRUSSES. REFLECTIVE SURFACE IS MADE OF FACETS WITH A METALLIC FRAME AND GLASS MIRROR. LOCAL CONTROL BOX IS LOCATED AT THE BOTTOM OF THE PEDESTAL. (*Source*: Abengoa Solar, S.A.) (All rights reserved on Figure 3.14 to Abengoa Solar, S.A. © 2010)

or Cloncurry are introducing heliostat units of only a few square meters. Brightsource with an ambitious program of large projects is making use of single-facet 7.3-m^2 heliostats [44] and the company eSolar with a multi-tower plant configuration presents a highly innovative field with ganged heliostats of extremely small size (1.14 m^2 each) that implies the large number of 12,180 units for a single 2.5 MW tower [45]. If such small heliostats may reach installed costs below 200 \$/m^2, it can only be understood under aggressive mass production plans and pre-assembly during manufacturing process by reducing on-site mounting works. Annual performance and availability of those highly-populated fields are still under testing.

In a solar power tower plant, the receiver is the heat exchanger where the concentrated sunlight is intercepted and transformed into thermal energy useful in thermodynamic cycles. Radiant flux and temperature are substantially higher than in parabolic troughs, and therefore, high technology is involved in the design, and high-performance materials should be chosen. The solar receiver should mimic a black body by minimizing radiation losses. To do so, cavities, black-painted tube panels, or porous absorbers able to trap incident photons are used. In most designs, the solar receiver is a single unit that centralizes all the energy collected by the large mirror field, and therefore, high availabilities and durability are a must. Just as cost reduction is the priority for further development in the collector field, in solar receivers, the priorities are thermal efficiency and durability. Typical receiver absorber operating temperatures are between 500°C and 1200°C and incident flux covers a wide range between 300 and over 1000 kW/m^2 [6, 46].

There are different solar receiver classifications criteria depending on the construction solution, the use of intermediate absorber materials, the kind of thermal fluid used, or heat transfer mechanisms. According to the geometrical configuration, there are basically two design options, external and cavity-type receivers. In a cavity receiver, the radiation reflected from the heliostats passes through an aperture into a box-like structure before impinging on the heat transfer surface. Cavities are constrained angularly and subsequently used in North-field (or South-field) layouts. External receivers can be designed with a flat plate tubular panel or are cylindrically shaped. Cylindrical external receivers are the typical solution adopted for surround heliostat fields. Figure 3.15 shows examples of cylindrical external, billboard external and cavity receivers.

Receivers can be directly or indirectly irradiated depending on the absorber materials used to transfer the energy to the working fluid [47]. Directly irradiated receivers make use of fluids or particle streams able to efficiently absorb the concentrated flux. Particle receiver designs make use of falling curtains or fluidized beds. In many applications, and to avoid leaks to the atmosphere, direct receivers should have a transparent window. Windowed receivers are excellent solutions for chemical applications as well, but they are strongly limited by the size of a single window, and therefore, clusters of receivers are necessary.

The key design element of indirectly heated receivers is the radiative/convective heat exchange surface or mechanism. Basically, two heat transfer options are used, tubular panels and volumetric surfaces. In tubular panels, the cooling thermal fluid flows inside the tube and removes the heat collected by the external black panel surface by convection. It is, therefore, operating as a recuperative heat exchanger. Depending on the heat transfer fluid properties and incident solar flux, the tube might undergo thermomechanical stress. Since heat transfer is through the tube surface, it is difficult to operate at an incident flux above 600 kW/m^2 (peak). Table 3.3 shows how only with high thermal conductivity liquids like sodium it is possible to reach operating fluxes above 1 MW/m^2. Air-cooled receivers have difficulties working with tubular receivers because of the lower heat transfer coefficients. To improve the contact surface, a different approach based on wire, foam or appropriately shaped materials within a volume are used. In volumetric receivers, highly porous structures operating as convective heat exchangers absorb the concentrated solar radiation. The solar radiation is not absorbed on an outer surface, but inside the structure "volume." The heat transfer medium (mostly air) is forced through the porous structure and is heated by convective heat transfer [58]. Volumetric absorbers are usually made of thin heat-resistant wires

FIG. 3.15: DIFFERENT CONFIGURATIONS OF SOLAR RECEIVERS. FROM LEFT TO RIGHT AND TOP TO BOTTOM: A) EXTERNAL TUBULAR CYLINDRICAL, B) CAVITY TUBULAR, C) BILLBOARD TUBULAR AND D) VOLUMETRIC

(in knitted or layered grids) or either metal or ceramic (reticulated foams, etc.) open-cell matrix structures. Good volumetric absorbers are very porous, allowing the radiation to penetrate deeply into the structure. Thin substructures (wires, walls or struts) ensure good convective heat transfer. A good volumetric absorber produces the so-called volumetric effect, which means that the irradiated side of the absorber is at a lower temperature than the medium leaving the absorber. Under specific operating conditions, volumetric absorbers tend to have an unstable mass flow distribution [49]. Receiver arrangements with mass flow adaptation elements (e.g., perforated plates) located behind the absorber can reduce this tendency, as well as appropriate selection of the operating conditions and the absorber material [53]. A number of initiatives have formulated air-cooled volumetric schemes for STE plants, both for atmospheric pressure ([51], [52]) and for pressurized systems [46], though still commercial plants with these technologies are missing. Though air-cooled open volumetric receivers are a promising way of producing superheated steam, the modest thermal efficiency at the receiver (74% nominal and 61.4% annual average) must still be improved [53]. At present, all the benefits from using higher outlet temperatures are sacrificed by radiation losses at the receiver, leading to low annual electricity production, so it is clear that volumetric receiver improvements must reduce losses.

Selection of a particular receiver technology is a complex task, since operating temperature, heat storage system and thermodynamic cycle influence the design. In general, tubular technologies allow either high temperatures (up to 1000°C) or high pressures (up to 120bar), but not both [54]. Directly irradiated or volumetric receivers allow even higher temperatures but limit pressures to below 15 bar.

TABLE 3.3

OPERATING TEMPERATURE AND FLUX RANGES OF SOLAR TOWER RECEIVERS

Fluid	Water/ steam	Liquid sodium	Molten salt (nitrates)	Volumetric air
Flux (MW/m^2)				
- Average	0.1–0.3	0.4–0.5	0.4–0.5	0.5–0.6
- Peak	0.4–0.6	1.4–2.5	0.7–0.8	0.8–1.0
Fluid outlet temperature (°C)	490–525	540	540–565	(700–1000)

3.5.2 Experience in central receiver systems

Although there have been a large number of STE tower projects, only a few have culminated in the construction of an entire experimental system. Table 3.4 lists systems that have been tested all over the world along with new early commercial plants. In general terms, as observed, they are characterized as being small demonstration systems between 0.5 and 10 MW, and most of them were operated in the 1980s ([6], [36], [55]). The thermal fluids used in the receiver are liquid sodium, saturated or superheated steam, nitrate-based molten salts and air. All of them can easily be represented by flow charts, where the main variables are determined by working fluids, with the interface between power block and the solar share.

The set of experiences referred to has served to demonstrate the technical feasibility of the CRS power plants, whose technology is sufficiently mature. The most extensive experience has been collected by several European projects located in Spain at the premises of the Plataforma Solar de Almería [55] and the 10 MW Solar One [56] and Solar Two plants [57] in the US. At present, water/steam and molten salts are the heat transfer fluids being selected for the first generation of commercial plants.

3.5.3 Water/Steam Plants from PS 10 Project to Superheated Steam

Production of superheated steam in the solar receiver has been demonstrated in several plants, such as Solar One, Eurelios, and CESA-1, but operating experience showed critical problems related to the control of zones with dissimilar heat transfer coefficients like boilers and superheaters [55]. Better results regarding absorber panel lifetime and controllability have been reported for saturated steam receivers. The good performance of saturated steam receivers was qualified at the 2-MW Weizmann receiver that produced steam at 15bar for 500 hours in 1989 [58]. Even though technical risks are reduced by saturated steam receivers, the outlet temperatures are significantly lower than those of superheated steam, making applications where heat storage is replaced by fossil fuel backup necessary.

PS10, the first commercial CRS plant in the world, adopted the conservative scheme of producing saturated steam to limit risk perception and avoid technology uncertainties. The 11 MW plant, located near Seville in South Spain, was designed to achieve an annual electricity production of 23 GWh at an investment cost of less than 3500€/kW. The project made use of available, well-proven technologies like the glass-metal heliostats developed by the Spanish INABENSA company and the saturated steam cavity receiver developed by the TECNICAL company to produce steam at 40 bar and 250°C [43]. The plant is a solar-only system with saturated steam heat storage able to supply 50 minutes of plant operation at 50% load [6]. The system makes use of 624 heliostats of 121 m^2 each, distributed in a North-field configuration, a 100-m high tower, a 15 MWh heat storage system and a cavity receiver with four 4.8 × 12 m tubular panels. The basic flow diagram selected for PS10 is shown in Fig. 3.16. Though the system makes use of a saturated steam turbine working at extremely low temperature, the nominal efficiency of the power block (30.7%) is relatively good. This efficiency is the result of optimized management of waste heat in the thermodynamic cycle. As summarized in Table 3.5, the combination of optical, receiver and power

TABLE 3.4

POWER TOWERS IN THE WORLD (ONLY PS10, PS20 AND GEMASOLAR ARE COMMERCIAL GRID-CONNECTED PROJECTS)

Project	Country	Power (MWe)	Heat transfer fluid	Storage media	Beginning operation
SSPS	Spain	0.5	Liquid Sodium	Sodium	1981
EURELIOS	Italy	1	Steam	Nitrate salt/water	1981
SUNSHINE	Japan	1	Steam	Nitrate salt/water	1981
Solar One	U.S.A.	10	Steam	Oil/rock	1982
CESA-1	Spain	1	Steam	Nitrate salt	1982
MSEE/Cat B	U.S.A.	1	Nitrate salt	Nitrate salt	1983
THEMIS	France	2.5	Hitec salt	Hitech salt	1984
SPP-5	Russia	5	Steam	Water/steam	1986
TSA	Spain	1	Air	Ceramic	1993
Solar Two	U.S.A.	10	Nitrate salt	Nitrate salt	1996
Consolar	Israel	0.5*	Pressurized air	Fossil hybrid	2001
Solgate	Spain	0.3	Pressurized air	Fossil hybrid	2002
PS10	Spain	11	Water/steam	Saturated steam	2006
PS20	Spain	20	Water/steam	Saturated steam	2009
Gemasolar**	Spain	17	Nitrate salt	Nitrate salt	2011

* Thermal

** Project under construction.

block efficiencies lead to a total nominal efficiency at design point of 21.7%. Total annual efficiency decreases to 15.4%, including operational losses and outages. PS10 is a milestone in the CRS deployment process, since it is the first solar power tower plant developed for commercial exploitation. Commercial operation started in June 21st 2007. Since then the plant is performing as designed. The construction of PS20, a 20 MWe plant with the same technology as PS10, followed. PS20 started operation in May 2009. With 1,255 heliostats (120 m^2 each) spread over 90 ha and with a tower of 165 m high, the plant is expected to produce 48.9 GWh/year.

Saturated steam plants are considered a temporary step to the more efficient superheated steam systems. Considering the problems found in the 1980s with superheated steam receivers, the current trend is to develop dual receivers with independent absorbers, one of them for the preheating and evaporation and another one for the superheating step. The experience accumulated with heuristic algorithms in central control systems applied to aiming point strategies at heliostat fields allows achieving a flexible operation with multi-aperture receivers. The company Abengoa Solar, developer of PS10 and PS20, is at present designing a superheated steam receiver for a new generation of water/steam plants [59].

China also has selected superheated steam for the first tower construction (Dahan) of 1 MW located in Yanqing, 65 km North of Beijing. It is a small project promoted by the Chinese Ministry of Science and Technology and the Chinese Academy of Science. The water/steam receiver is a joint development with Korea.

But the most advanced strategy is the program announced by the BrightSource Industries (Israel) Ltd (BSII), is that it has already built a demonstration plant of 6 MWth located at the Neguev dessert in June 2008 [44]. The final objective of BSII is to promote plants producing superheated steam at 160 bar and temperature of 565°C (named DPT550). With those characteristics, they expect up to 40% conversion efficiency at the power block for unit sizes between 100 and 200 MW. The receiver is cylindrical, dual and with a drum. The first commercial Project under development in California is Ivanpah Solar in California of 400 MW and expected to commission in 2012, followed by a 1.8 GW unit more in California and 400 MW unit in Nevada [2]. Planned conversion efficiency from solar to electricity is 20%.

The combination of recent initiatives on small heliostats, compact modular multi-tower fields and production of superheated steam may be clearly visualized in the development program of the company eSolar. This company proposes a high degree of modularity with power units of 46 MW covering 64 ha, consisting of 16 towers and their corresponding heliostat fields sharing a single central power block. With replication, modularity sizes up to 500 MW and upward may be obtained [60]. Within the development program of STE in the US, eSolar has four plants totalizing 334 MW [2]. Figure 3.17 depicts the first two modules of 2.5 MW each installed in 2009 by eSolar in Lancaster, California, and already in operation. It can be observed that each receiver has two independent cavities and the heliostat layout consists of identical arrays to maximize replication and modularity that clearly penalizes optical efficiency and cosine factor. Each tower is associated with 12,180 flat heliostats of 1.14 m^2 each and a single 46 MW commercial plant would contain 194,880 heliostats [45]. The receivers are dual-cavity, natural-circulation boilers. Inside the cavity, the feedwater is preheated with economizer panels before entering the steam drum. A downcomer supplies water to evaporator panels where it is boiled. The saturated water/vapor mixture returns to the drum where the steam is separated, enters superheater panels, and reaches 440°C at 6.0 MPa. Each receiver absorbs a full-load power of 8.8 MWth. Overall plant efficiency expected by eSolar would be 23% solar to electricity [60].

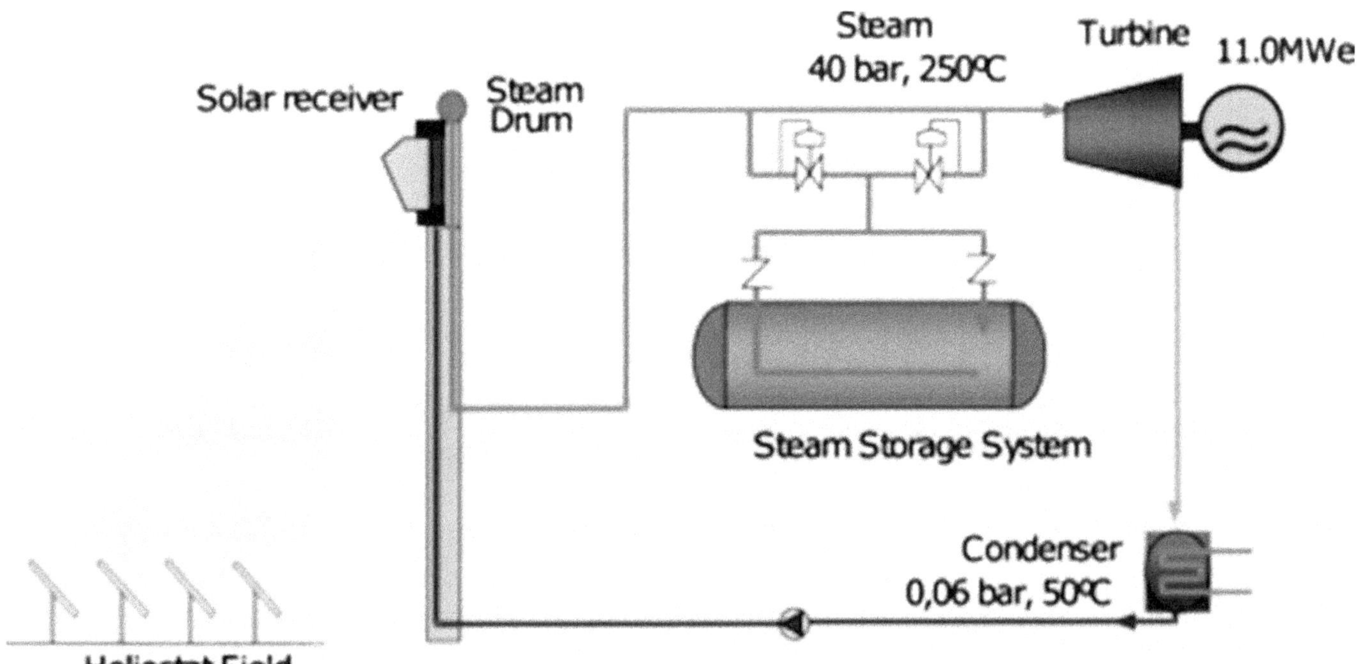

FIG. 3.16: BASIC SCHEME OF PS10 SOLAR THERMAL POWER PLANT WITH SATURATED STEAM AS THERMAL FLUID. (*Source*: Abengoa Solar, S.A.)

TABLE 3.5

ANNUAL ENERGY BALANCE FOR THE PS10 PLANT AT NOMINAL CONDITIONS

	Nominal rate operation	
Optical efficiency	77.0%	67.5 MW → 51.9 MW
Receiver and heat handling efficiency	92.0%	51.9 MW → 47.7 MW
Thermal power to storage		11.9 MW
Thermal power to turbine		35.8 MW
Thermal power/electric power efficiency	30.7%	35.8 MW → 11.0 MW
Total efficiency at nominal rate	**21.7%**	
	Energetical balance in annual basis	
Mean annual optical efficiency	64.0%	148.63 GWh → 95.12 GWh
Mean annual receiver & heat handling efficiency	90.2%	95.12 GWh → 85.80 GWh
Operational efficiency (start-up/stop)	92.0%	85.80 GWh → 78.94 GWh
Operational efficiency (breakages, O&M)	95.0%	78.94 GWh → 75.00 GWh
Mean annual thermal energy /electricity efficiency	30.6%	75.00 GWh → 23.0 GWh
Total annual efficiency	**15.4%**	

3.5.4 Molten Salt Systems: Solar Two and Gemasolar

For high annual capacity factors, solar-only power plants must have an integrated cost-effective thermal storage system. One such thermal storage system employs molten nitrate salt as the receiver heat transfer fluid and thermal storage media. To be usable, the operating range of the molten nitrate salt, a mixture of 60% sodium nitrate and 40% potassium nitrate, must match the operating temperatures of modern Rankine cycle turbines. In a molten-salt power tower plant, cold salt at 290°C is pumped from a tank at ground level to the receiver mounted atop a tower where it is heated by concentrated sunlight to 565°C. The salt flows back to ground level into another tank. To generate electricity, hot salt is pumped from the hot tank through a steam generator to make superheated steam. The superheated steam powers a Rankine-cycle turbine. The collector field can be sized to collect more power than is demanded by the steam generator system, and the excess salt is accumulated in the hot storage tank. With this type of storage system, solar power tower plants can be built with annual capacity factors up to 70%. As molten salt has a high-energy storage capacity per volume (up to 500–700 kWh/m^3), they are excellent candidates for solar thermal power plants with large capacity factors. Even though nitrate salt has a lower specific heat capacity per volume than carbonates, they still store 250 kWh/m^3. The average heat conductivity of nitrates is 0.52 W/mK and their heat capacity is about 1.6 kJ/kgK. Nitrates are a cheap solution for large storage systems.

Several molten salt development and demonstration experiments have been conducted over the past two decades in the US and Europe to test entire systems and develop components. The largest demonstration of a molten-salt power tower was the Solar Two project, a 10-MW power tower located near Barstow, CA. The purpose of the Solar Two project was to validate the technical characteristics of the molten-salt receiver, thermal storage, and steam generator technologies, improve the accuracy of economic projections for commercial projects by increasing the capital, operating, and maintenance cost database, and distribute information to utilities and the solar industry to foster a wider interest in the first commercial plants. The Solar Two plant was built at the same site as the Solar One pilot plant and reused much of the hardware including the heliostat collector field, tower structure, 10 MW turbine and balance of plant. A new, 110 MWh$_t$ two-tank molten-salt thermal storage system was installed, as well as a new 42 MWth receiver, a 35 MWth steam generator system (535°C, 100 bar), and master control system [61].

The plant began operating in June 1996. The project successfully demonstrated the potential of nitrate-salt technology. Some of the key results were: receiver efficiency was measured at 88%, the thermal storage system had a measured round-trip efficiency of over 97%, and the gross Rankine-turbine cycle efficiency was 34%, all of which matched performance projections. On one occasion,

FIG. 3.17: FIRST EXPERIMENTAL EXAMPLE IN THE WORLD OF A MULTITOWER SOLAR ARRAY CORRESPONDING TO ESOLAR'S SIERRA SOLAR GENERATING STATION WITH TWO FULL-SIZED HELIOSTAT FIELDS IN LANCASTER, CALIFORNIA. (*Source*: eSolar)

TABLE 3.6

TECHNICAL SPECIFICATIONS AND DESIGN PERFORMANCE OF THE GEMASOLAR PROJECT

Technical specifications	
Heliostat field reflectant surface	304,750 m^2
Number heliostats	2650
Land area of solar field	142 ha
Receiver thermal power	120 MWth
Tower height	145 m
Heat storage capacity	15 hours
Power at turbine	17 MWe
Power NG burner	16 MWth
Operation	
Annual electricity production	112 MWhe
Production from fossil (annual)	15%
Capacity factor	74%

the plant operated around-the-clock for 154 hours straight [62]. On April 8, 1999, testing and evaluation of this demonstration project was completed and subsequently was shut down.

To reduce the risks associated with scaling up hardware, the first commercial molten-salt power tower should be approximately three times the size of Solar Two. One attempt to prove scaled-up molten-salt technology is the Gemasolar project being promoted by Torresol Energy, a joint venture between the Spanish SENER and MASDAR initiative from Abbu Dhabi ([63], [64]). Table 3.6 summarizes the main technical specifications of Gemasolar project.

With only 17 MWe, the plant is predicted to produce 112 GWh_e/year. A large heliostat field of 304,750 m^2 (115 m^2 each heliostat) is oversized to supply 15-hour equivalent heat storage capacity.

The plant is designed to operate around-the-clock in summertime, leading to an annual capacity factor of 74%. Fossil backup corresponding to 15% of annual production will be added. The levelized energy costs are estimated to be approximately $0.16/kWh. Gemasolar might represent a breakthrough for STE technology. Construction works started in November 2008 and commercial operation is scheduled for May 2011.

3.6 DISH/STIRLING SYSTEMS

3.6.1 Dish–Stirling Technology: Subsystems and Components

Dish–Stirling systems track the sun and focus solar energy onto a cavity receiver, where it is absorbed and transferred to a heat engine/generator. An electrical generator, directly connected to the crankshaft of the engine, converts the mechanical energy into electricity (AC). To constantly keep the reflected radiation at the focal point during the day, a sun-tracking system continuously rotates the solar concentrator on two axes following the daily path of the sun. With current technologies, a 5 kW_e Dish–Stirling system would require 5.5 m diameter concentrator, and for 25 kW_e, the diameter would have to increase up to 10 m. Stirling engines are preferred for these systems because of their high efficiencies (40% thermal-to-mechanical), high-power density (40–70 kW/liter), and potential for long-term, low-maintenance operation. Dish–Stirling systems are modular, i.e., each system is a self-contained power generator, allowing their assembly in plants ranging in size from a few kilowatts to tens of megawatts [65].

The concentrator is a key element of any Dish–Stirling system. The curved reflective surface can be manufactured by attached segments, by individual facets or by stretched membranes shaped by a continuous plenum. In all cases, the curved surface should be coated or covered by aluminum or silver reflectors. Second-surface glass mirrors, front surface thin-glass mirrors or polymer films have been used in various different prototypes.

First-generation parabolic dishes developed in the 1980s were shaped with multiple, spherical mirrors supported by a trussed structure [66]. Though extremely efficient, this structure concept was costly and heavy. The introduction of automotive industry concepts and manufacturing processes has led to optimized commercial versions like the 25-kW SunCatcher system developed by the company Stirling Energy Systems (Fig. 3.18) that has earned the world's highest recorded peak efficiency rating, 31.25%, for converting solar energy-to-grid quality electricity. Another good example of the latest developments in Dish–Stirling systems is the 10-kW Eurodish prototype. The Eurodish concentrator consists of 12 single segments made of fiber glass resin. When mounted, the segments form an 8.5-m diameter parabolic shell. The shell rim is stiffened by a ring truss to which bearings and the Stirling support structure are later attached. Thin 0.8-mm-thick glass mirrors, are glued onto the front of the segments for durable and high reflectivity around 94% [67].

As in central receivers and parabolic trough absorbers, the receiver absorbs the light and transfers the energy as heat to the engine's working gas, usually helium or hydrogen. Thermal fluid working temperatures are between 650°C and 750°C. This temperature strongly influences the efficiency of the engine. Because of the high operating temperatures, radiation losses strongly penalize the efficiency of the receiver; therefore, a cavity design is the optimum solution for this kind of system.

Two different heat transfer methods are commonly used in parabolic dish receivers [68]. In directly illuminated receivers, the same fluid used inside the engine is externally heated in the receiver through a pipe bundle. In indirect receivers, an intermediate fluid and a heat pipe is used to decouple solar flux and working temperature from the engine fluid. The phase change guarantees good temperature control, providing uniform heating of the Stirling engine [69].

Stirling engines solarized for parabolic dishes are externally heated gas-phase engines in which the working gas is alternately heated and cooled in constant-temperature, constant-volume processes (Fig. 3.19). This possibility of integrating additional external heat what makes it an ideal candidate for solar applications. Since the Stirling cycle is very similar to the Carnot cycle, the theoretical efficiency is high. High reversibility is achieved since work is supplied to and extracted from the engine at isothermal conditions. The clever use of a regenerator that collects the heat during constant-volume cooling and heating substantially

FIG. 3.18: ROW OF DISH-STIRLING SYSTEMS (STIRLING ENERGY SYSTEMS SUNCATCHER) OF 25 KW EACH AT THE MARICOPA SOLAR PLANT IN ARIZONA. (*Source*: Stirling Energy Systems)

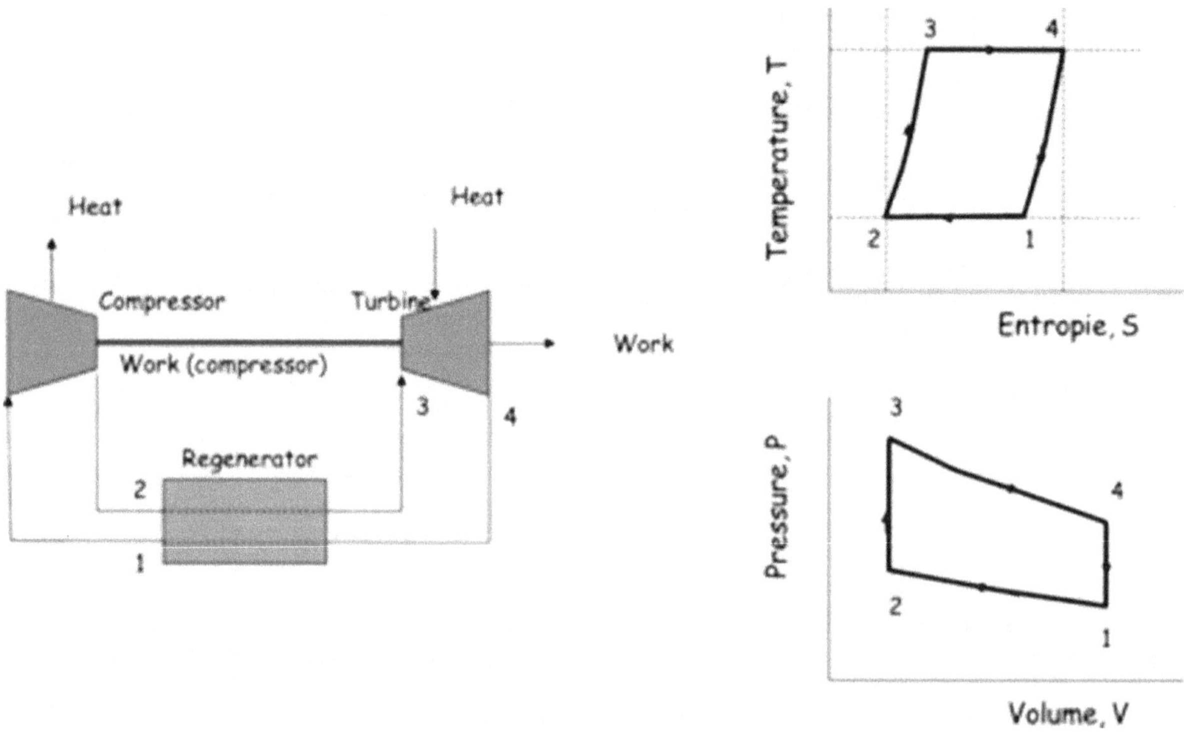

FIG. 3.19: IDEAL STIRLING CYCLE PRESENTS ISOTHERMAL COMPRESSION AND EXPANSION PLUS A REGENERATOR THAT LEADS TO A HIGH THERMODYNAMIC EFFICIENCY

enhances the final system efficiency. For most engine designs, power is extracted by kinematics of the rotating crankshaft connected to the pistons by a connecting rod. An example of a kinematic Stirling engine is shown in Fig. 3.20. Though theoretically, Stirling engines may have a high life-cycle projection, the actual fact is that today their availability is still not satisfactory, since an important percentage of operating failures and outages are caused by pistons and moving mechanical components. Availability is, therefore, one of the key issues, since it must operate for more than 40,000 hours in 20-year lifetime, or ten times more than an automobile engine. One option to improve availability is the use of free-piston designs. Free-piston engines make use of gas or a mechanical spring so that mechanical connections are not required to move reciprocating pistons. Apparently, they are better than kinematic engines in terms of availability and reliability. The most relevant program in developing dishes with free-piston technology is promoted by the company Infinia.

3.6.2 Dish–Stirling Developments

Like the other CSP technologies, practical Dish–Stirling development started in the early 1980s. Most development has been concentrated in the US and Germany, though developed for commercial markets they have been tested in a small number of units [70].

The first generation of dishes was a facet-type concentrator with second-surface mirrors that already established concentration records ($C = 3000$), and had excellent performances, though their estimated costs for mass production were above $300/m^2. Their robust structures were extremely heavy, weighing at 100 kg/m^2 [55]. The 25 kW Vanguard-1 prototype built by Advanco was operated at Rancho Mirage, California, in the Mojave desert in production mode for 18 months (February 1984 to June 1985). This system was 10.7 m in diameter with a reflecting surface of 86.7 m^2 and a 25 kW PCU made by United Stirling AB (USAB) model 4-95 Mark II. This engine had four cylinders with a 95 cm^3 cylinder displacement. The working gas was hydrogen at a maximum pressure of 20 MPa and temperature of 720°C. Engine power was controlled by varying the working-gas pressure. McDonnell Douglas later developed another somewhat improved dish system making use of the same technology and the same engine. The dish was 10.5 m and 25 kW. The 88 m^2 parabolic dish consisted of 82 spherically curved glass facets. Reported performances and efficiencies were similar to those of Advanco/Vanguard [66]. The project was frozen for several years until in 1996, Stirling Energy Systems (SES) acquired the intellectual and technology rights to the concentrator and the US manufacturing rights to what is now called the Kockums, 4-95 Stirling engine-based PCU [65]. The re-designed 25-kWe system named SunCatcher is being qualified in a commercial basis at the Maricopa Solar Plant in Arizona (Fig. 3.20). The plant totalizes 1.5 MW with 60 dishes and is in operation since January 2010. Commercialization is expected in California with two large plants of about 800 MW each.

In Europe the most important development is the EuroDish system. The EuroDish project is a joint venture undertaken by the European Community, German/Spanish Industry (SBP, MERO, Klein+Stekl, Inabensa), and research institutions DLR and CIEMAT. The engine used in the EuroDish is SOLO Kleinmotoren 161. Two new 10-kW EuroDish units were installed at the Plataforma Solar de Almeria, Spain, early in 2001 for testing and demonstration. In a follow-up project called EnviroDish, additional units were deployed in France, India, Italy and Spain to accumulate operating experience at different sites. The peak system efficiency was first measured at 20%. The estimated annual production of a EuroDish system operating in Albuquerque, New Mexico, is 20,252 kWh of electricity with 90% availability and an annual efficiency of 15.7% [65]. SBP and the associated EuroDish industry have performed cost estimates for a yearly production rate of 500 units per year (5 MW/year) and 5000 units per year, which corresponds to 50 MW/year. The actual cost of the 10-kW unit without transportation and installation cost and excluding foundations is approximately US$10,000/kW. The cost projections at production rates of 500 and 5000 units per year are US$2,500/kW and US$1,500/kW, respectively based upon replication of automotive mass-production processes.

3.7 TECHNOLOGY DEVELOPMENT NEEDS AND MARKET OPPORTUNITIES FOR SOLAR THERMAL ELECTRICITY (STE)

From the 1970s to the 1990s the development of solar thermal electricity technologies remained restricted to a few countries and only a few, though important, research institutions and industries were involved. The situation has dramatically changed since 2006 with the approval of specific feed-in-tariffs or power purchase agreements in Spain and the US. Both countries with more than 6 GW of projects under development and more than 1 GW in operation at the end of 2010 are undoubtedly leading the commercialization of STE [2]. Other countries such as India, China, Australia and Italy adopted the STE process. Subsequently a number and variety of engineering and construction companies, consultants, technologists and developers committed to STE are rapidly growing.

A clear indicator of the globalization of STE commercial deployment for the future energy scenario has been elaborated by the International Energy Agency (IEA). This considers STE

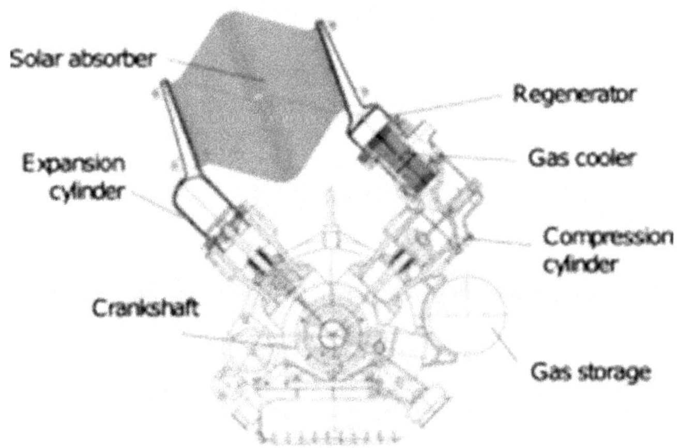

FIG. 3.20: KINEMATIC STIRLING ENGINE V-161 OF 10 KWE MANUFACTURED BY SOLO KLEINMOTOREN WITH PISTONS SITUATED IN V AND CONNECTED TO A TUBULAR ARRAY HEAT EXCHANGER. (*Source*: Solo Kleinmotoren)

to play a significant role among the necessary mix of energy technologies for halving global energy-related CO_2 emissions by 2050 [5]. This scenario would require capacity addition of about 14 GW/year (55 new solar thermal power plants of 250 MW each). However, this new opportunity is introducing an important stress to the developers of STE. In a period of less than 5 years, in different parts of the world, these developers of STE are forced to move from strategies oriented to early commercialization markets based upon special tariffs, to strategies oriented to a massive production of components and the development of large amounts of projects with less profitable tariffs. This situation is speeding up the implementation of second generation technologies even though in some cases still some innovations are under assessment in early commercialization plants or demonstration projects. The projected evolution of Levelized Electricity Costs of different STE technologies is depicted in Fig. 3.21. LEC value may be reduced an additional 30% when moving to future sites with very high Direct Normal Irradiance.

The reduction in electricity production costs should be a consequence, not only of mass production but also of scaling-up and R&D. A technology roadmap promoted by the European Industry Association ESTELA [7] states that by 2015, when most of the improvements currently under development are expected to be implemented in new plants, energy production boosts greater than 10% and cost decreases up to 20% are expected to be achieved. Furthermore, economies of scale resulting from plant size increase will also contribute to reduce plants' CAPEX per MW installed up to 30%. STE deployment in locations with very high solar radiation further contributes to the achievement of cost competitiveness of this technology by reducing costs of electricity up to 25%. All these factors can lead to electricity generation cost savings up to 30% by 2015 and up to 50% by 2025, reaching competitive levels with conventional sources (e.g., coal/gas with stabilized Electricity Costs <10€c/kWh). Similar projections are published in another recent roadmap issued by the IEA [71]. Other roadmaps coordinated by R&D centers expect larger influence of innovations (up to 25%) in cost reduction [72]. Some of the key general topics on medium to long-term R&D proposed by the STE community are [22]:

Build confidence in the technology through:

- pilot applications based on proven technologies;
- high reliability of unattended operation;
- increased system efficiency through higher design temperatures;
- hybrid (solar/fossil fuel) plants with small solar share.

Reduce costs through:

- improved designs, materials, components, subsystems and processes;
- exploitation of economies of scale.

Increase solar share through:

- suitable process design;
- integration of storage.

In all cases, R&D is multi-disciplinary, involving optics, materials science, heat transfer, control, instrumentation and measurement techniques, energy engineering and thermal storage.

3.7.1 Trough and Linear Fresnel Power Plants

To further reduce costs and increase reliability in next generation PTC and LFR technology, the following are expected:

- Lighter and lower-cost structural designs including front surface mirrors with high solar-weighted reflectivity of about 95%.
- Development of high-absorptance coatings for tube receivers (96% and higher) able to work efficiently at over 500°C.
- Development of medium temperature thermal energy storage systems (Phase Change Materials, molten salts, concrete) suitable for solar-only systems;
- Continued improvement in overall system O&M, including mirror cleaning, integral automation and largely unattended control;
- System cost reductions and efficiency improvements from substituting water for synthetic oil as the heat-transfer fluid (Direct Steam Generation technology)

3.7.2 Power Tower Plants

Power tower R&D in the United States, Europe, and Israel is concentrated in the two most relevant subsystems with regard to costs: heliostat field and solar receiver. The following improvements are expected:

Improvements in the heliostat field as a result of better optical properties, lower cost structures, and better control. Improvements in materials should be analogous to those for trough collectors. In general terms, optical performance and durability of existing heliostats is acceptable (95% availability and beam quality below 2.5 mrad), therefore, R&D resources should focus basically on cost reduction.

Development of water/superheated steam and advanced air-cooled volumetric receivers using both wire-mesh absorbers and ceramic monoliths is the subject of various projects. Dual-aperture receivers for water/steam and volumetric receivers for air still need further development for scale-up, materials durability and thermal efficiency.

Heat storage is another key issue for CRS development. The new developments in air-cooled receivers have led to the development of advanced thermocline storage systems making use of packed-bed ceramic materials. This system has shown excellent performance for small units of a few MWh but pressure losses and design restrictions appear when size is increased.

Finally, more distributed control architectures, system integration and hybridization in high-efficiency electricity production schemes should be developed as already mentioned for trough systems.

3.7.3 Dish/Engine Power Plants

Several dish/engine prototypes have operated successfully during the last 20 years in the US and Europe, but there is no large-scale deployment yet. This situation may change soon with some existing commercialization initiatives. The use of dishes for standalone or grid-support installations will reach near-term markets as costs drop to less than 12¢/kWh. This lower cost can be achieved through:

- improvements in mirrors and support structures, improvements in hybrid heat-pipe and volumetric receivers coupled to Stirling and Brayton engines, and development of control systems for fully automatic operation; and
- improvements in system integration by reduction of parasitic loads, optimization of startup procedures, better control strategies, and hybrid Stirling-Brayton operation.

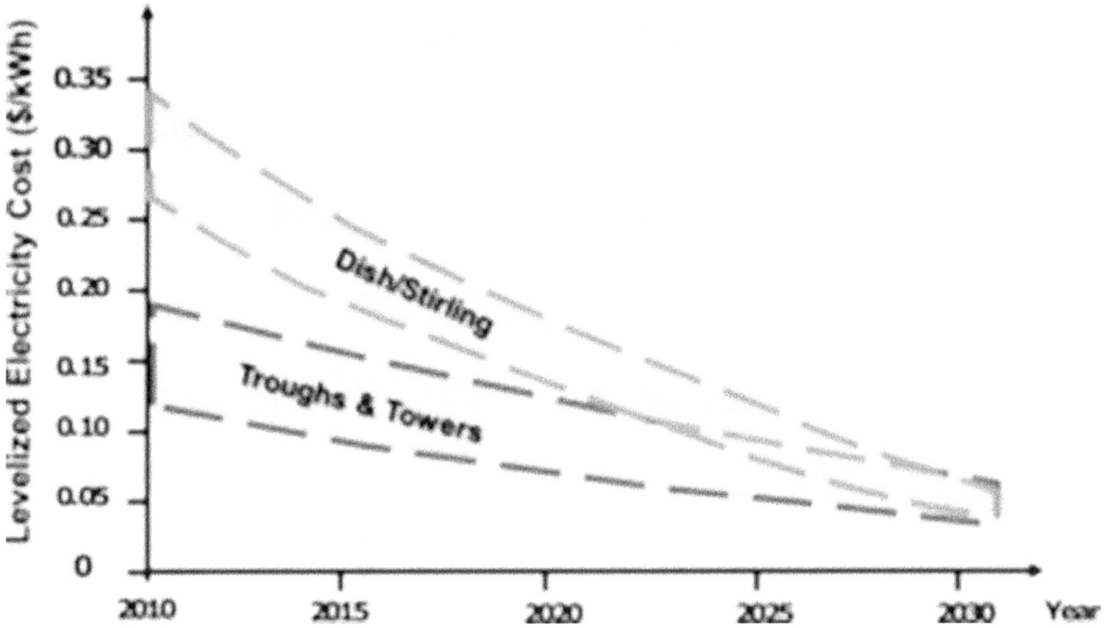

FIG. 3.21: EVOLUTION OF LEVELIZED ELECTRICITY COST FOR STE TECHNOLOGIES BASED UPON TECHNOLOGY ROAD-MAPS AND INDUSTRY

3.8 RECENT SOLAR APPLICATIONS IN SPAIN, REST OF EUROPE, ASIA, AFRICA, MIDDLE EAST, SOUTH AND CENTRAL AMERICA AND AUSTRALASIA[2]

In chapter 3 titled "Solar Thermal Power Plants: From Endangered Species to Bulk Power Production in Sun Belt Regions" authored by Manuel Romero and José González-Aguilar in the reference [73] the authors dealt with developments in Spain and Europe. In this section of this chapter update is provided to reflect on considerable changes in Solar applications in Spain and Europe since the initial edition of this publication in 2011. The advances in the US have been covered elsewhere in this publication with significant changes in Europe captured in this section.

3.8.1. Objectives of CSP/STE Research Community in the European Union (EU) [74]

Concentrated Solar Power (CSP), also called Solar Thermal Electricity (STE) has been developed in European Union (EU) which plays a relevant role in any future energy scenario of EU because of its capability of base load dispatchability for integrating thermal energy storage. The most important objective for the CSP/STE community, within the European Energy Research Alliance (EERA) Joint Program is to foster and consolidate European industrial players with worldwide competitiveness in the field. So far, Europe is still a technological leader in this sector, but this situation could quickly change due to the ambitious initiatives recently launched in other world regions (China being the most remarkable one). Thus most important objective is to foster and consolidate European industrial players with worldwide competitiveness in the field while developing in Europe advanced and novel solutions for STE plants, matching costs and efficiency targets [74].

3.8.2 Recent Solar Applications in Spain [75]

Initiative for Global Leadership in Concentrated Solar Power (CSP) with an 'Implementation Plan' has been developed to help meet the energy needs of several parts of the world, for creating an export sector for the European renewable industry in supporting the decarburization agenda of the Paris Agreement. Integrated SET Plan with ten key priority actions, Concentrated Solar Power (CSP) technology, SET Plan strategic targets for CSP have been documented in this reference [75]. Figure 3.22 shows SET Plan Key Priority Actions [75].

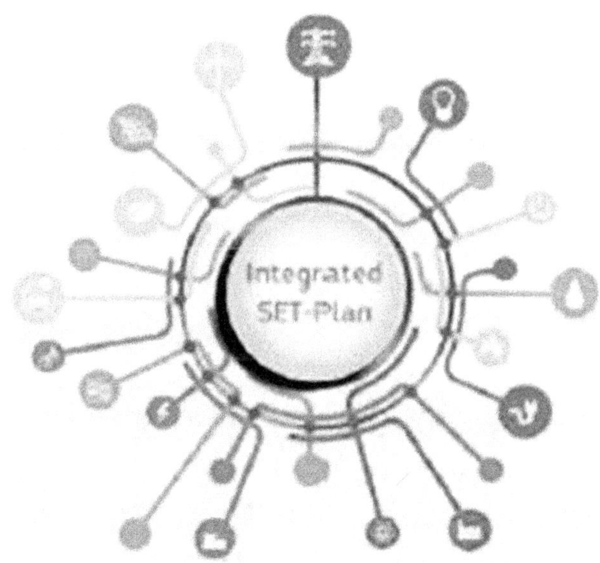

FIG. 3.22: FIGURE OF SET PLAN KEY PRIORITY ACTIONS. (*Source*: [75])

2 Editor has provided the entire update provided in this Section 3.8 for this publication.

The Communication for an Integrated Strategic Energy Technology (SET) Plan for 'Accelerating the European Energy System Transformation' was adopted by the European Commission (EC) on 15 September 2015 for European Community, SET Plan countries and stakeholders to co-operate to implement the SET Plan [75].

Targets have been set for several renewable energy technologies, including Solar Energy with significant potential for cost reduction and large-scale deployment worldwide. This calls for higher level of alignment between EU and individual national efforts for public investments and leverage of private investments. Implementation Plans (IP) are developed by Temporary Working Groups (TWG) led by the SET Plan country. TWG on Concentrated Solar Power was the first one to be launched in April-May 2016 [75].

Concentrated Solar Power (CSP) technology involves thermal energy storage and Solar Thermal Electricity (STE) which can combinedly make a significant contribution to the transformation of the European energy system by providing an important share of dispatchable renewable electricity. By providing flexibility for grid services, CSP can facilitate the integration of variable output renewables such as photovoltaic (PV) or wind energy, thereby contributing to the reliability of the transmission grid. The best solar resources for CSP are to be found in Southern Europe, which makes this technology complementary to those renewable energy technologies that find their best resources in other regions of Europe. According to the forecasts of the International Energy Agency (IEA), CSP could account for up to 11% of the electricity generated worldwide and up to 4% of the electricity generated in Europe by 2050 [75].

Figure 3.23 shows "IEA 2050 Roadmap – Generation mix in %" and "Main elements of a CSP plant that hold potential for cost reduction" are depicted in reference [75]. CSP/STE composition of governments and funding agencies are from: Spain, Portugal, France, Germany, Italy, Cyprus and Turkey and Belgium), with leadership from Ministry of Economy (Spain), [75]. The Ministry of Economy and Business (MINECO) is the department of the Government of Spain responsible for the proposing and carrying out the government policy on economic affairs and reforms to improve competitiveness, telecommunications and the information society, development of the Digital Agenda as well as the policy on business support [75].

A table based on the discussions in the TWG, specific twelve R&I Activities have been defined for the Implementation Plan and ranked according to their potential contribution to achieve the targets is shown in Table 3.7.

Representatives from Member States, France, Greece, Italy, Germany, Lithuania, Malta, Portugal, Cyprus and Turkey were involved in an Initiative for Global Leadership in CSP/STE in May 2015 with Switzerland and UK active in CSP. Other countries such as Norway, Denmark or the Netherlands were also interested in joining the Commission and the main stakeholders were ESTELA, EUTurbines and EERA-CSPas of December 2015. The Strategic targets ('the Targets') were (1) 40% cost reduction by 2020 (from 2013) translating into a supply price* < 10 c€/kWh for a radiation of 2050 kWh/m2/year (average in Southern Europe) and (2) New cycles (including supercritical ones) with a first demonstrator by 2020 [76].

Solar Thermal Technology is a relatively new renewable energy technology, unlike the more mainstream solar PV technology, CSP (Concentrated Solar Power) hasn't gone main stream yet. This is partly due to cost, as Solar Thermal Technology is still relatively expensive when looking at kWp implementation cost. Also, CSP plants are generally utility sized, although eight CSP companies to "Look Out For in 2018" [77].

There are different variations of CSP plants however all follow the same process of using solar energy captured by a receiver to produce steam that is then used to drive a turbine. The real advantage of solar thermal plants is that they can also include thermal storage systems to store energy and continue producing power

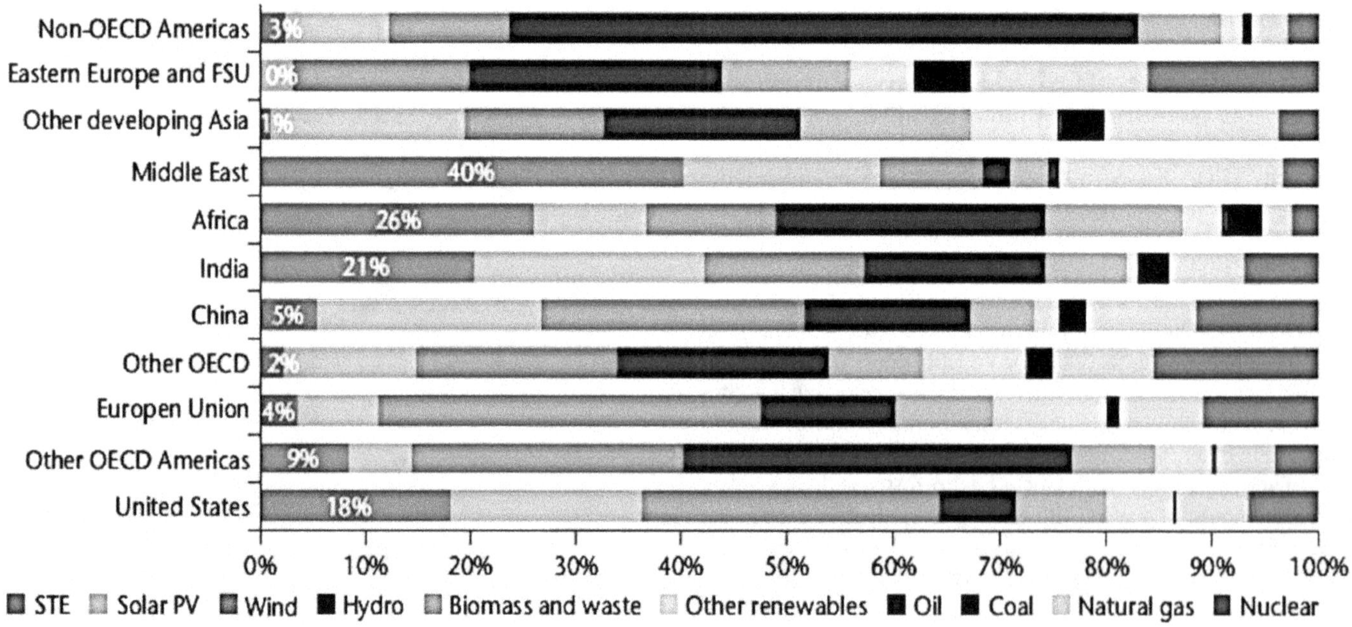

FIG. 3.23: IEA 2050 ROADMAP – GENERATION MIX IN %. (*Source* [75])

TABLE 3.7

THE RANKED LIST OF R&I ACTIVITIES (SOURCE [75])

	Ranked list of R&I activities
Ranking	**List of R&I Activities to reach the targets**
1	Activity 5: Improved central receiver molten salt technology
2	Activity 3: Parabolic trough with silicon oil
3	Activity 6: Next generation of central receiver power plants
4	Activity 1: Advanced linear concentrator Fresnel technology with direct molten salt circulation as heat transfer fluid and for high temperature thermal energy storage
5	Activity 2: Parabolic trough with molten salt
6	Activity 4: Solar tower power plant to commercially scale-up and optimize the core components of the open volumetric air receiver technology
7	Activity 8: Multi-tower central receiver beam down system
8	Activity 9: Thermal energy storage
9	Activity 10: Development of supercritical steam turbines optimised for the specifics of CSP applications
10	Activity 11: Development of advanced concepts for improved flexibility in CSP applications
11	Activity 12: Development and field test of CSP hybrid air Brayton turbine combined cycle sCO_2 systems
12	Activity 7: Pressurized air cycles for high efficiency solar thermal power plants

during peak evening times. It is expected that by 2020 there will be 20GW of CSP capacity installed globally, with an annual growth rate of 20%, produced by Solar Thermal Companies, Technology Providers, and Project Developers [77].

(1) "BrightSource Energy" [77] is a US-based Solar Thermal startup that focuses on the design, development and deployment of solar thermal power plants globally. Their technology was used in the largest CSP plant in the world, situated in Ivanpah, CA. A 377 MW system that went live in 2013, the system produces enough power for 140,000 California homes during peak hours. Other projects that are in construction phase are a 121 MW system in Israel and a 135 MW CSP plant with energy storage in China.

(2) "Agengoa Solar"[77] is a Spanish company researching and developing solar thermal technology in the 1980s. Abengoa Solar is involved in the design, manufacture and marketing of key components for solar facilities, which are used either for its own or third-party plants. The vast majority of their CSP plants operate in Spain and South Africa, with one plant in Chile and the UAE. In Spain all their CSP plants have a combined capacity of 492 MW. Each plant in Spain produces 50 MW sufficient for 25,700 households. Each plant offsets 31,400 tons of CO_2 being released each year.

(3) "Siemens" [77] is the largest green technology company in the world, although they are new when it comes to Solar Thermal technology. However, they are experienced in designing and developing steam boilers and turbine generators for CSP plants. Seimens involved with other CSP components, such as the design and development of parabolic troughs, central towers, and linear Fresnel concentrated solar power plants and hybrid plants.

(4) "Acciona" [77] is a Spanish company with a global presence and an engineering history dating back to the 1880's. Their core businesses are infrastructure and renewable energy. Acciona Energy works with five renewable energy technologies, namely wind, solar photovoltaic, solar thermal, hydroelectric and biomass. With over 8 913 MW of installed capacity, they have their own technology for the design, construction, operation and maintenance of CSP plants, with 314 MW in operation across six thermo-solar plants (five in Spain, and one in the US).

(5) "Solar Reserve" [77] is a global developer in utility scale solar power projects. They specialize in both Solar PV and Solar Thermal Plants, with over 482 MW of capacity installed as well as another massive 13 GW in the pipeline. The company also holds over 100 patents relating to CSP technology.

(6) "Trivelli Energia"[77] is a leading micro solar thermal technology company in Italy and Europe. They design and develop micro CSP plants used for process heating and cooling. In Europe, 70% of energy consumed in industry

is heat. Most of this heat is below 482 Fahrenheit, making CSP technology an ideal solution by providing clean, free and reliable heating from the sun! [77].

(7) "Abros Green" [77] is a German-based company that develops Solar PV and CSP project implementations internationally. They develop CSP plants that can be used in conjunction with fossil fuels such as natural gas, biomass or coal fired plants.

(8) "Torresol Energy" [77] was founded in 2007 and is based in Spain. Torresol Energy designs, develops and implements CSP projects. Torresol Energy claims to be the first to use salt as a thermal energy solution in their 17 MW Gemasolar plant located in Spain. The plant consists of two 650 Heliosats over a space of 195 hectares. The plant's thermal energy storage system provides power generation autonomy for up to 15 hours without sunlight. They also have two other CSP plants in Spain with a capacity of 50 MW.

A picture of the Torresol Energy Gemasolar plant in Fuentes de Andalucia near Seville is shown in reference [78]. This unique thermosolar power station in southern Spain can shrug off cloudy days energy stored when the sun shines lets it produce electricity even during the night. From the road between Seville and Cordoba, one can see its central tower lit up like a beacon by 2,600 solar mirrors, each 120 square metres (28,500 square feet) that surround it in an immense 195-hectare (480-acre) circle. "It is the first station in the world that works 24 hours a day, a solar power station that works day and night!" said Santago Arias, technical director of Torresol Energy, which runs the station. The mechanism is "very easy to explain," he said: the panels reflect the sun' s rays on to the tower, transmitting energy at intensity 1,000 times higher than that of the sun's rays reaching the earth. Energy is stored in a vat filled with molten salts at a temperature of more than 500 degrees C (930°F). Those salts are used to produce steam to turn the turbines and produce electricity. It is the station's capacity to store energy that makes Gemasolar so different because it allows the plant to transmit power during the night, relying on energy it has accumulated during the day [78].

"This type of station is expensive, not because of the raw material we use, which is free solar energy, but because of the enormous investment these plants require," Arias said [78]. The investment cost exceeds 200 million euros ($260 million). But "the day when the business has repaid that money to the banks (in 18 years, he estimates), this station will become a 1,000-euro note printing machine!" he said, recalling that oil prices have soared from $28 a barrel in 2003 to nearly $130. For now, the economic crisis has nevertheless cast a shadow over this kind of project: Spain is battling to slash its deficit as it slides into recession and has suspended aid to new renewable energy projects. Andalusia, Spain plant has been hard hit by the economic crisis with the country's highest unemployment rate at 31.23 percent. "We have three projects ready but stalled" because of the aid suspension, Arias said, admitting that in a difficult global economy the group has not managed to sell the Gemasolar technology abroad despite huge interest outside Spain [78].

Torresol Energy is a joint venture between the Spanish engineering group Sener, which holds 60 percent, and Abu Dhabi-financed renewable energy firm Masdar. "I use that energy as I see fit, and not as the sun dictates," Arias explained. As a result, the plant produces 60 percent more energy than a station without storage capacity because it can work 6,400 hours a year compared to 1,200-2,000 hours for other solar power stations, he said. "The amount of energy we produce a year is equal to the consumption of 30,000 Spanish households with an annual saving of 30,000 tons of CO_2" Arias said. Helped by generous state aid, renewable energies have enjoyed a boom in Spain, the world number two in solar energy and the biggest wind power producer in Europe, ahead of Germany [78].

The Torresol Energy Gemasolar thermasolar plant claims to be the first commercial-scale plant to apply central tower receiver and molten salt heat storage technology. The relevance of this plant lies in its technological uniqueness, since it opens up the way for new thermosolar electrical generation technology. The annual production of Gemasolar (110GWhe) is the equivalent of the energy generated in a conventional thermal plant burning 89,000 tons of lignite or the converted energy of 217,000 oil barrels [79].

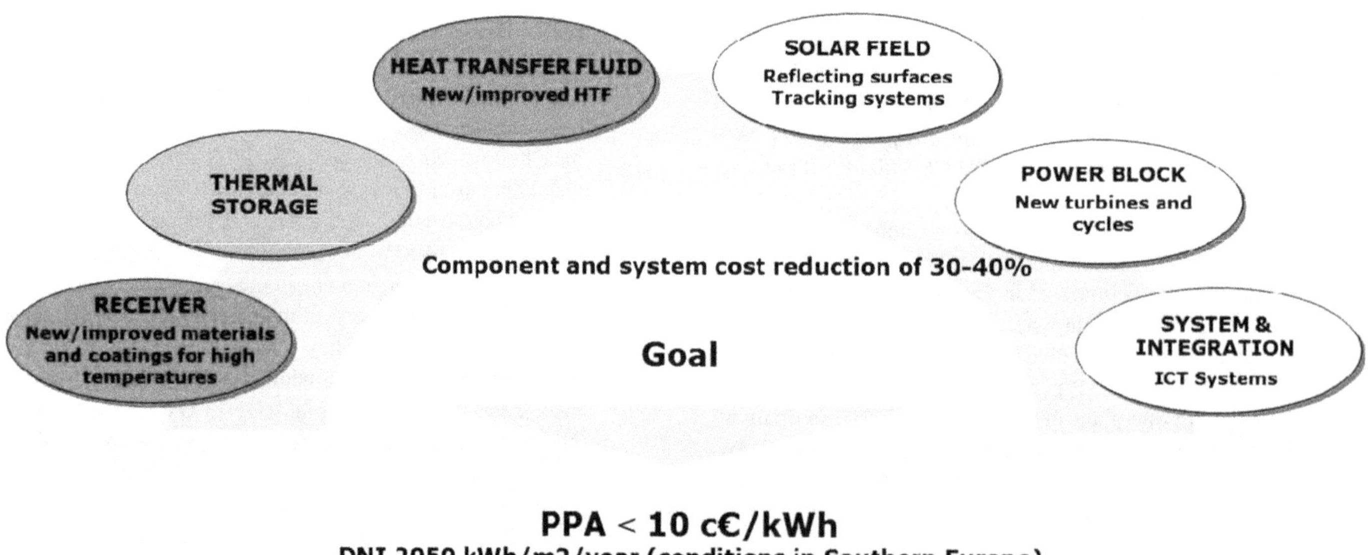

FIG. 3.24: MAIN ELEMENTS OF A CSP PLANT THAT HOLD POTENTIAL FOR COST REDUCTION [75]

FIG. 3.25: THE TORRESOL ENERGY GEMASOLAR THERMASOLAR PLANT IN FUENTES DE ANDALUCIA NEAR SEVILLE, SOUTHERN SPAIN [79].

The Olmedilla Photovoltaic Park in Olmedilla de Alarcón, Spain, is the world's largest photovoltaic plant which was completed in 2008. The Olmedilla Photovoltaic (PV) Park uses 162,000 flat solar photovoltaic panels to deliver 60 megawatts of electricity on a sunny day. It produces enough electricity to power more than 40,000 homes. The entire plant was completed in 15 months at a cost of about $530 million. Olmedilla was built with conventional solar panels, which are made with silicon and tend to be heavy and expensive. So-called "thin-film" solar panels, although less efficient per square meter, tend to be much cheaper to produce, and their the technology being tapped to realize the world's largest proposed PV plant, the Rancho Cielo Solar Farm in Belen, N. Mex., which is expected to cost $840 million, cover an area of 700 acres (285 hectares), and produce 600 megawatts of power [80], see figure 3.26

A larger solar power plant in the South Western Spain near Cadiz is currently being built in 2010 by Tentosul will have a peak capacity of 250 MW. This project is the first unsubsidized utility-scale solar project in Spain with an estimated cost of €275 million. The Spanish solar market represents about 5% of the global PV market and in recent years its growth has slowed down significantly compared to rest of Europe. This decrease in growth can be explained by the end of all subsidies to solar energy in 2012 as a result of a wider economic review by the Spanish government. According to the European Photovoltaic Industry Association (EPIA), Spain's national target for solar capacity is 8,367 MW by 2020. With over 5GW already installed, the EPIA estimates that Spain only needs to add 400MW of solar capacity every year to reach the necessary target. The EPIA also estimates that Spain should be able to add over 1500 MW of solar energy every year. According to The International Renewable Energy Agency (IRENA), Spain has a target to meet 3% of total energy demand from solar by 2020. The report goes to say that Spain has a target to generate 38% of its electricity from renewable energy by 2020 however by 2011 Spain already surpassed that target and more than 40% of its electricity is was being generated through renewable energy in that year. Spain has one of the highest levels of solar irradiance in Europe. With some regions receiving 2000 kWh per square meter annually, Southern Spain receives sunlight comparable to Northern Africa making this part of the country particularly suitable for PV deployment. According to EPIA, ground mounted solar farms account for about 80% of the total Spanish solar market and the rest of solar capacity is built for commercial and industrial use. Spain's residential solar market is only about 1% of the national solar market. Castilla La Mancha is the largest regional market with about 1000 MW solar capacity installed, Andalucia is the second largest with over 800 MW. Other regions which have significant amount of solar installations include Castilla y Leon, La Comunidad Valenciana, Extremadura and Murcia. All four of these regions have more than 300 MW of installed capacity each [81].

FIG. 3.26: THE OLMEDILLA PHOTOVOLTAIC PARK IN OLMEDILLA DE ALARCÓN, SPAIN. (*Source* [80])

Figure 3.27 is the 11 megawatt PS10 solar power tower produces electricity from the sun using 624 large movable mirrors called heliostats [82]. Solucar PS10 is the first solar thermal power plant based on tower in the world that generates electricity in a commercial way, created in 23 September 2007 [83], also see figure 3.13 [43]. Figure 3.28 shows three solar towers near Seville Spain, PS20, Eureka 5 and PS10 [82]; figure 3.29 shows PS20 and PS10 in Andalusia, Spain [82].

FIG. 3.27: SOLUCAR PS10 (SOURCE BY AFLORESM. [83])

FIG. 3.28: THREE SOLAR TOWERS NEAR SEVILLE SPAIN, PS20, EUREKA 5 AND PS10. (*Source* [82])

FIG. 3.29: PS20 AND PS10 IN ANDALUSIA, SPAIN. (*Source*:[82])

Figure 3.30 shows the Aerial view of the unit I, III, and IV of Abengoa Solar's Solnova Solar Power Station. The two towers and reflective mirrors in the background are the PS10 and PS20 solar power plants, also owned by Abengoa Solar. This region is also sometimes known as the Solar Platform and the picture figure 3.40 was taken on 7th September 2010 [82].

FIG. 3.30: THE FIRST THREE UNITS OF SOLNOVA IN THE FOREGROUND, WITH THE TWO TOWERS OF THE PS10 AND PS20 SOLAR POWER STATIONS IN THE BACKGROUND [82].

FIG. 3.31: THE 150 MW ANDASOL SOLAR POWER STATION. (*Source* [83])

The Andasol solar power station is a 150-megawatt (MW) concentrated solar power station and Europe's first commercial plant to use parabolic troughs located near Guadix in Andalusia, Spain see figure 3.31 [83]. The Andasol plant uses tanks of molten salt as thermal energy storage to continue generating electricity, irrespective of whether the sun is shining or not. Andasol is the first parabolic trough power plant in Europe. Andasol consists of 3 projects: Andasol-1 (completed 2008), Andasol-2 (completed 2009) and Andasol-3 (completed 2011). Each project generates approximately 165 GW-h each per year (a total of 495 GW-h for all three combined). The total cost of building the three projects was estimated to €900 million [83].

3.8.3 Top 10 Performing Countries for Solar Energy as of March 13th, 2017 [84]

International Energy Agency (IEA) carries out comprehensive programs of energy cooperation between 29 member states and the EC. According to IEA, for 2015, the global total cumulative PV installation is 227 GW worldwide. Most of this is grid-connected. This total amount of global PV produces more than 1.3 % of the world's electricity demand. IEA-PVPS, mem-

FIG. 3.32: SOLAR ENERGY PANELS. (*Source*: [84])

FIG. 3.33: A SOLAR PANEL FARM. (*Image Source* Pixabay [84])

ber countries account for 197 GW of the world's total with the remaining coming from all the rest. 2015 saw a massive growth in the solar energy market with 50.7 GW of additional installed capacity worldwide is about 26.5% increase in 2014. This trend seems to be a result of the ever declining cost of PV systems and drive for low carbon economies. In Europe, 3.5% of total electricity demand is now supplied by PV which is about 7% of peak demand. According to IEA 23 countries have at least 1 GW of cumulative PV capacity by the end of 2015 with 7 countries installing at least 1 GW in 2015 [84].

IEA's Photovoltaic Power Systems (IEA PVPS) Program is to get the best performance in the solar energy countries. This program's mission is to "enhance the international collaborative efforts which facilitate the role of photovoltaic solar energy as a cornerstone in the transition to sustainable energy systems." Participating countries for IEA PVPS are currently Australia, Austria, Belgium, Canada, China, Denmark, Finland, France, Germany, Israel, Italy, Japan, Korea, Malaysia, Mexico, the Netherlands, Norway, Portugal, Spain, Sweden, Switzerland, Thailand, Turkey, and the United States of America. The European Commission, "SolarPower" Europe, the Solar Electric Power Association, the Solar Energy Industries Association and the Copper Alliance are also members [84]. The IEA data is drawn mainly from national survey reports and information summaries supplied by representatives from each of the reporting countries, whose reports are available on their country websites. IEA also collects information from official member state governmental bodies and other reliable industrial sources. For those countries outside of the network, data is gathered from industrial associations such as The International Renewable Energy Agency (IRENA) and Renewable Energy Policy Network for the 21st Century (REN21) . Using the most recent report from IEA-PVPS the following list of the top ten countries for solar energy with cumulative PV capacities have been drafted, using results from IEA-PVPS.org [84].

First place for solar energy is China [84]:

In 2015 PV Cumulative Installed Capacity was 43.5 GW, for 2014 PV Cumulative Installed Capacity was 28.1 GW and in 2014 China occupied Second Position, so that the position change was +1. China is an IEA-PVPS Member. 2015 was a significant year for China, not only has it beaten the record for highest capacity in a single country but also taken the lead. China knocked Germany off the top spot this year with a total capacity of 43.5GW. China has installed between 2013 and 2014 a total of 21.55 GW of solar generation systems with 15.2 GW in 2015 alone. There is particular emphasis on PV distribution that began in 2014. China's ambition was to reach 35 GW by 2015! They now aim for 143 GW by the year 2020, an increase of 100 GW from previous announcements. Will this growth be sustainable for China? They certainly seem set to meet their ambitious 2020 target. China has implemented various schemes and incentives for the development of PV. They aim to develop utility scale PV, rooftop PV in city areas and micro-grids and off-grid applications in un-electrified regions of the country. With China's massive and dense population, not to mention notorious smog and pollution levels, it's great to see China doing so well. They have massively increased their PV capacity since 2009 to top the charts. China is after all, one of the world's largest manufacturers of solar panels [84].

Second place for solar energy is Germany [84]

In 2015 PV Cumulative Installed Capacity was 39.7 GW; in 2014 PV Cumulative Installed Capacity was 38.2 GW. In 2014 Germany's position was first. Germany is an IEA-PVPS Member. Germany lost its long-held lead in 2015 to the powerhouse that is China. The country experienced a market decline of new installations to 1.5 GW due to changing support schemes and tendering exercises. Germany now has 7.1% of its annual electrical demand supplied by PV. Germany has decided to opt for tenders for large scale PV installations from 2015 onwards. This is fairly significant because for the past three years, Germany has consistently added 7 GW of PV systems to the grid. This growth has been achieved because of a combination of several elements. These were a long term stability of support schemes, investor confidence and the appetite of residential, commercial and industrial building owners for PV installations. Customers are now forced to pay a significant percentage of the levy paid by consumers to finance renewable incentives in Germany. This includes self-consumed part of the PV market. In 2015 Germany's cumulative PV capacity consisted 18% of the world's total, which is pretty impressive. Germany actually has incentives for battery storage on PV systems, a stark contrast to most of Europe [84].

Third place for solar energy is Japan [84]

In 2015 PV Cumulative Installed Capacity was 34.4 GW and in 2014 PV Cumulative Installed Capacity was 23.3 GW in which year Japan occupied third position: Japan is an IEA-PVPS Member. Japan has managed to maintain its global ranking from 2014, not an insignificant feat. In the land of the rising sun, the rapid growth of the Japanese PV market lasted until 2014. This growth continued and the country reached around 11 GW, confirming Asia as the first world region for PV. Despite their maintained position for cumulative totals, Japan has also maintained its position in second place for installed capacity. Japan managed to install 10.8 GW in 2015, a slight growth rate compared to 2014, but a record-high year for the Japanese PV market. Interestingly, within the top ten countries, Asia-Pacific regions constitute around 50%. Japanese interest in renewable technology was greatly increased following the tragic accident at Fukushima. Japan has a national subsidy for residential storage batteries. Prefabricated housing builders are also proactive in promoting PV systems with storage batteries [84].

Fourth place for solar energy is USA [84]:

In 2015 PV Cumulative Installed Capacity was 25.6 GW and in 2014 PV Cumulative Installed Capacity was 18.3 GW. In 2014 US occupied fifth position and US is an IEA-PVPS Member. The United States of America managed to up the rankings in 2015 moving from fifth to fourth place. They also performed well for installed capacity in 2015 of 7.3GW, bringing them into third place. This compares to 4.75 GW and 6.2 GW in 2013 and 2014 respectively. This growth was mainly attributed to large utility-scale PV systems rather than rooftop installations. The US's total cumulative capacity means that they can provide 11% of the total PV capacity in the world. The US's growth has been helped by the EPA's final rulings on carbon emission reductions of 30% between 2020 and 2030. 2015 saw several countries, including Germany and the USA who have introduced competitive calls for tender to grant power purchase agreements (PPA) for PV. Also, compensation of net-metering was the cause of several debates in the US. This resulted in the establishment of either caps or small fees to be introduced in net-metering for PV. Globally, centralized PV now represents more than 60% of the market as of 2015. This has mainly been driven by China, the US and emerging PV markets [84].

Fifth place for solar energy: Italy [84]

In 2015 PV Cumulative Installed Capacity was 18.9 GW and in 2014 PV Cumulative Installed Capacity was 18.5 GW. In 2014 Italy occupied fourth position and is an IEA-PVPS member. Italy fell behind slightly in 2015, dropping from fourth in 2014 to fifth in 2015. Italy, Greece and Germany now produce 8, 7.4 and 7.1 percent of their annual electrical demand through PV alone. In 2015 Italy installed a modest 300-400 MW of PV systems in 2015. This contrasts to 9.3 GW in 2011, 3.6 GW in 2012 and 1.6 GW in 2013. This can be explained by the phase-out of the feed-in tariffs that are not granted anymore for new PV installations, leaving the market driven by the self-consumption scheme and additional tax rebates that are now in place. It is interesting to note that in 2015 several countries have surpassed the 1 % mark of PV contribution to electrical demand. Italy is one of them, in fact not just one but first amongst all. They now contribute close to 8% and 3.5% of the entire European PV contribution not to mention 8% of the total world cumulative PV contribution. Italy's average PV yield is around 1.3 kWh/kW and they installed a modest 300 MW of PV in 2015 [84].

The Italian Ministry of Economic Development (MISE) has issued a new decree to promote solar and renewable energies across the country's 20 minor islands, which include,

FIG. 3.34: A RESIDENTIAL PV SYSTEM IN ITALY. (*Source*: IMAGE: NOVEMBER 2, 2017 EMILIANO BELLINI, FLICKR/PUNTIN [86])

among others, the renowned Capri, Stromboli and Lampedusa islands. The decree sets the timeframe for the energy need of the islands to be covered with renewables by 2020. The new rules allow the installation of residential and commercial PV systems, as well as solar installations providing power to recharge-stations for electric vehicles. All PV and renewable energy power systems installed on the minor islands will be entitled to inject all or part of their power output to the islands' grids [85].

Every kWh of self-consumed or injected power will be remunerated with an additional tariff, set by Italian energy and gas regulator The Italian Regulatory Authority for Energy, Networks and the Environment (AEEGSI). The MISE said the rate will be commensurate to the cost of saved fuel for electricity generation. These incentives can't be cumulated with other Italian renewable energy incentive schemes such as the Conto Energia. The new legislation includes the improvement of the islands' grids. The program will be co-financed by the European Commission and the UC4 levy for the minor islands, which is paid for by energy consumers [85].

Italy installed 323 MW of solar in first nine months of 2017. The Italian solar market has grown by 13% so far this year and by 6% in the third quarter. New additions for September reached 32.2 MW [86].

Italy saw the addition of 323 MW of new PV systems in the first nine months of 2017, according to provisional numbers released by the Italian renewable energy association Anie Rinnovabili, which relies on data provided by the country's grid operator Terna. This result represents a 13% growth compared to the same period in 2016 and a 31% increase from the first nine months of 2015. In the third quarter of this year, new additions totaled 89.9 MW, up 6% from the same quarter of 2016. In September, approximately 32.2 MW of new PV systems were connected to the grid in Italy, up from 26.3 MW in August. September's result still gives the category for PV plants over 1 MW the largest share in this year's ranking with around 66.9 MW of newly installed power, followed by PV systems with a capacity between 4.5 kW and 6 kW (47.4 MW), PV systems ranging in size from 20 kW to 100 kW (44.9 MW), solar installations with a power of 10 kW to 20 KW (42.1 MW) and solar power generators with a capacity between 100 kW and 200 kW (32.5 MW). The region with the largest share so far this year is Lazio, where several new large-scale PV plants are located, with 81.4 MW, followed by Lombardia (41.9 MW), Veneto (30.5 MW) and Emilia Romagna (30.5 MW). According to the latest official statistic released by the energy agency Gestore dei Servizi Energetici (GSE), Italy had approximately 19.28 GW of installed PV capacity at the end of 2016. This means that the country should have reached approximately 19.6 GW of installed solar capacity at the end of September [86]. Figures 3.35 and 3.36 give graphical presentation of Italy as of 2014 and 2017 [87].

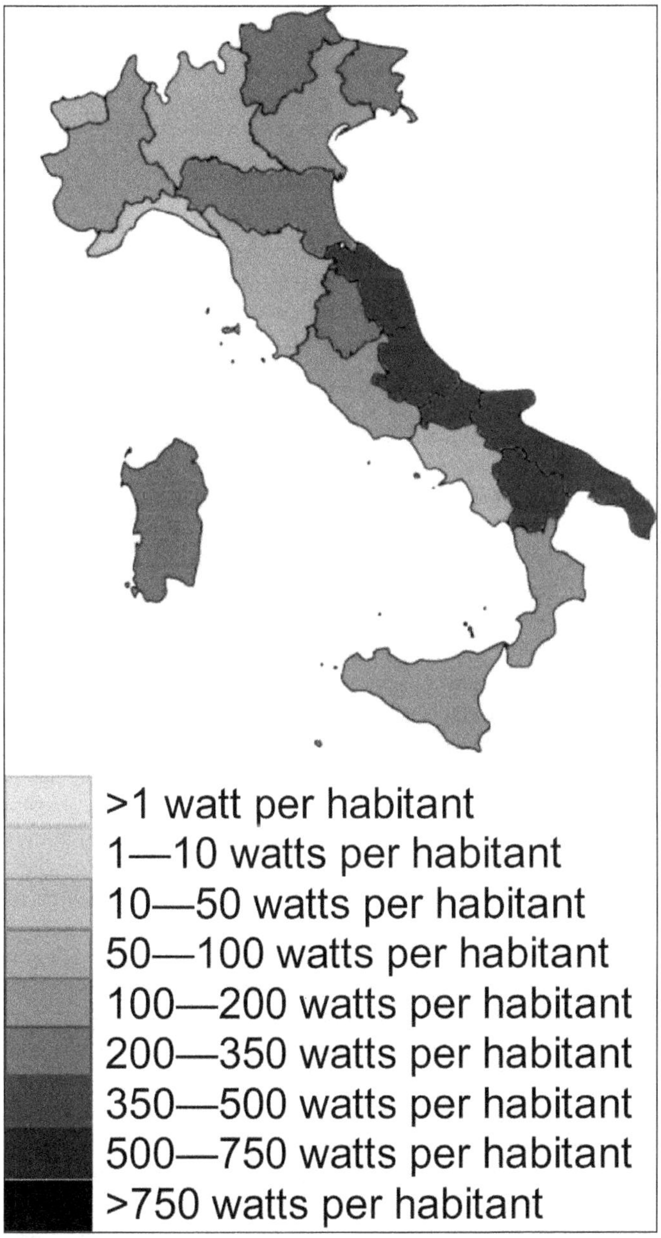

FIG. 3.35: ITALY PHOTOVOLTAIC CAPACITY IN 'WATTS PER CAPITA' BY REGION, 15 JUNE 2014 [87]

FIG. 3.36: ITALY'S SOLAR POWER POTEMTIAL_2017. (*Source* [87])

Italy sets 2030 solar target of 50 GW [88]

The Italian government has published a plan for climate and energy envisaging a central role for solar in the nation's energy mix, with PV expected to represent more than the half the power generation capacity from renewables by the end of the next decade. Italy's Ministry of Economic Development (MISE) issued the country's new National Integrated Plan for Climate and Energy 2030, which has provided more clarity on the renewable energy targets laid out in the energy strategy for the period 2020-2030 – the "Strategia Energetica Nazionale" – defined by the previous government in November 2017. According to the new plan, solar has seen its 2030 target for power generation revised up from 72 TWh to 74 TWh and it has also been given an installed capacity target – 50 GW – that was not present in the SEN, despite predictions around 30 GW would be needed to achieve the 72 TWh target. Italy has an installed PV capacity of almost 20 GW, meaning around 7 GW more will be required to hit the MISE target of 26.6 GW of PV capacity by 2025. Solar is expected to represent more than the half of Italy's renewables capacity, which is expected to reach 93.1 GW from around 54 GW currently. The second largest renewable energy source is expected to be hydropower – a traditional strength of the mountainous nation – with 19.2 GW, followed by wind (18.4 GW) and biomass (3.7 GW). The Italian government has fleshed out its plans for solar up to the year 2030 [88].

FIG. 3.37: OCTOPUS ENERGY BY EMILIANO BELLINI, JANUARY 11, 2019. (*Source* [88])

The Italian government has stipulated ground-mounted PV will have to be implemented with a minimal loss of agricultural land. "This new approach is aimed at supporting the installation of solar power generators mainly on buildings, sheds, car parks, service areas etc.," states the document. "However, the diffusion of large ground-mounted photovoltaic systems remains important for the achievement of the 2030 objectives, but unproductive areas not destined for other uses, or unused agricultural areas, will have to be used." The authorization process for developing large-scale PV and the availability of land appear set to remain significant obstacles for developers, PV magazine reported from Italy's main solar event, Key Solar, in Rimini in November, 2017. "On the authorization side, the government defines, together with the regions, clear rules that reduce [the] time frame and encourage the involvement of local realities, a decisive step to reach the 2030 targets," renewable energy consultant Gianni Silvestrini told PV magazine at the time. "The real challenge for this government will no longer be based on the definition of new incentives, but on the adaptation of the regulations and on the clarity of the authorization processes." Despite such challenges, Italy's private PPA segment saw an increasing number of deals for large-scale solar in recent months and the nation's first auction for large wind and PV projects is expected to be issued in the near future [88].

Sixth place for solar energy is UK [84]

In 2015 PV Cumulative Installed Capacity was 8.8 GW and in 2014 PV Cumulative Installed Capacity was 5.1 GW. In 2014 Italy occupied eighth position. Italy is an IEA-PVPS Member. The UK has historically been rather sluggish with PV uptake but not anymore. Moving up two places from 2014 to sixth place is pretty impressive. Although total installed capacity pales in comparison to sunnier, and frankly, much larger countries such as China or the US, the growth is still admirable. Improvements have been made in PV efficiency which is tandem with stronger government backing which has allowed the UK to become a poster child within the global solar market. Despite the cut in government funding of some incentives like "The Green Deal" (until it's rebirth in 2017) and Feed-in Tariffs (FiT's), had helped the UK turbo boost its PV market. The UK was Europe's largest market in 2015 with an install of 3.51 GW. This allowed the UK to beat Germany, which installed 1.5 GW, for increased PV installation in 2015 [84].

Seventh place for solar energy is France [84]

In 2015 PV Cumulative Installed Capacity was 6.6 GW and in 2014 PV Cumulative Installed Capacity was 5.7 GW with sixth position in 2014. France is an IEA-PVPS Member. France sadly dropped a position in 2015 despite stabilizing its market close to 0.9 GW installs. Behind the top seven, no country has installed more than 1 GW of PV systems in 2015. According to IEA, France installed 879 MW in 2015. Australia remains a strong market with 935 MW installed in 2015 but with an ongoing segmentation shift. Canada installed around 600 MW. France, like Germany, has also been reliant on PPA's. France has been using PPA's to medium-to-large scale PV systems for some years now. These include commercial installations. France's market used to be well behind in Europe and the world, it is now one of the best on the continent. A political decision was made several years ago to maintain the market at around 1 GW per year for the last few years. They also decided to increase it in the years to come. France's new additional capacity has been a bit sluggish in 2015 compared to previous years at 887 MW. This compares to 954 MW in 2014 and 1,120 MW in 2012. There is a conviction to increase these newly installed capacities in the years to come. Following 2015 United Nations Climate Change Conference, (COP 21), France has put a concerted effort into boosting its solar market by revising national PV installation targets. These are now 10.2 GW and will be between 18 to 20 GW by 2023. Will they achieve this? Only time will tell [84].

Eighth place for solar energy is Spain [84]

In 2015 PV Cumulative Installed Capacity was 5.4 GW and in 2014 PV Cumulative Installed Capacity was 5.38 GW. In 2014 Spain occupied seventh position. Spain is an IEA-PVPS Member. Spain has historically performed much higher than its current position prior to 2010, but has had a sluggish performance in recent years. They are not alone in Europe however. Some EU countries that previously grew quickly have now stalled or experienced small or significant reductions in total capacity. Spain has reported around 56 MW of new additions in a complex context of harsh regulations for consumers including "solar tax" on self-generated power. To most, this concept would be deemed a false economy, or worse, active sabotage of the PV market. Under Royal Decree 900/2015 regulatory framework for self-consumption was introduced. This established a maximum capacity for self-generation of electricity which must be equal or below a stated capacity. This specifies two tiers. Type 1 has a maximum capacity of 100 kW, where no compensation for any surplus is provided when fed to the grid. Type 2 has no limit on capacity. The surplus can be sold in the wholesale market directly through an intermediary. There is, however, tax due of 0.5 Euro's/MWh as well as a 7% tax on the electricity produced. According to the EIA, Regulation indicates that self-generated power above 10 kW is charged with a fee per kWh consumed as a "grid backup toll", commonly known as the "sun tax". Adding battery storage to the installation also implies an additional tax [84].

Spain is one of the top ten countries by solar photovoltaics installed capacity and the first country for concentrated solar power in the world. In 2018, the cumulative total solar power installed was 7,011 MW, of which 4,707 MW were solar PV installations and 2,300 MW were concentrated solar power.

FIG. 3.38: PV CAPACITY IN WATTS PER CAPITA BY AUTONOMOUS COMMUNITIES IN SPAIN IN 2013. (*Source*: [89])

In 2016, nearly 8 TWh of electrical power was produced from photovoltaics, and 5 TWh from CSP plants. During 2016 Photovoltaics accounted for 3% of total electricity generation and solar thermal an additional 1.9% [90].

Spain is one of the European countries with the most hours of sunshine. In the wake of the 2008 financial crisis, the Spanish government drastically cut its subsidies for solar power and capped future increases in capacity at 500 MW per year, with effects upon the industry worldwide. Between 2012 and 2016, new installations stagnated in Spain and with the controversial "sun tax" and intimidating regulation surrounding solar consumption introduced in 2015 were repealed in late 2018. 2019 is likely to see the industry beginning a dramatic rebirth following a series of positive developments meaning that growth in the sector would likely occur even without subsidies. Spain provides an interesting model of what happens to solar power development once grid parity has been surpassed by a fair margin in an industrialized country with good natural solar resources and a non-punitive regulatory environment [90].

As a legacy from Spain's earlier development of solar power, the country remains a world leader in concentrated solar power, accounting for almost a third of solar power installed capacity in the country, a much higher ratio than that for other countries as of 2017. Many large concentrated solar power stations remain active in Spain and may have provided some of the impetus for large CSP developments in neighboring Morocco. In 2017 Spain held large auctions for renewable energy capacity to be constructed by 2020: solar and wind projects each won 4 GW [90].

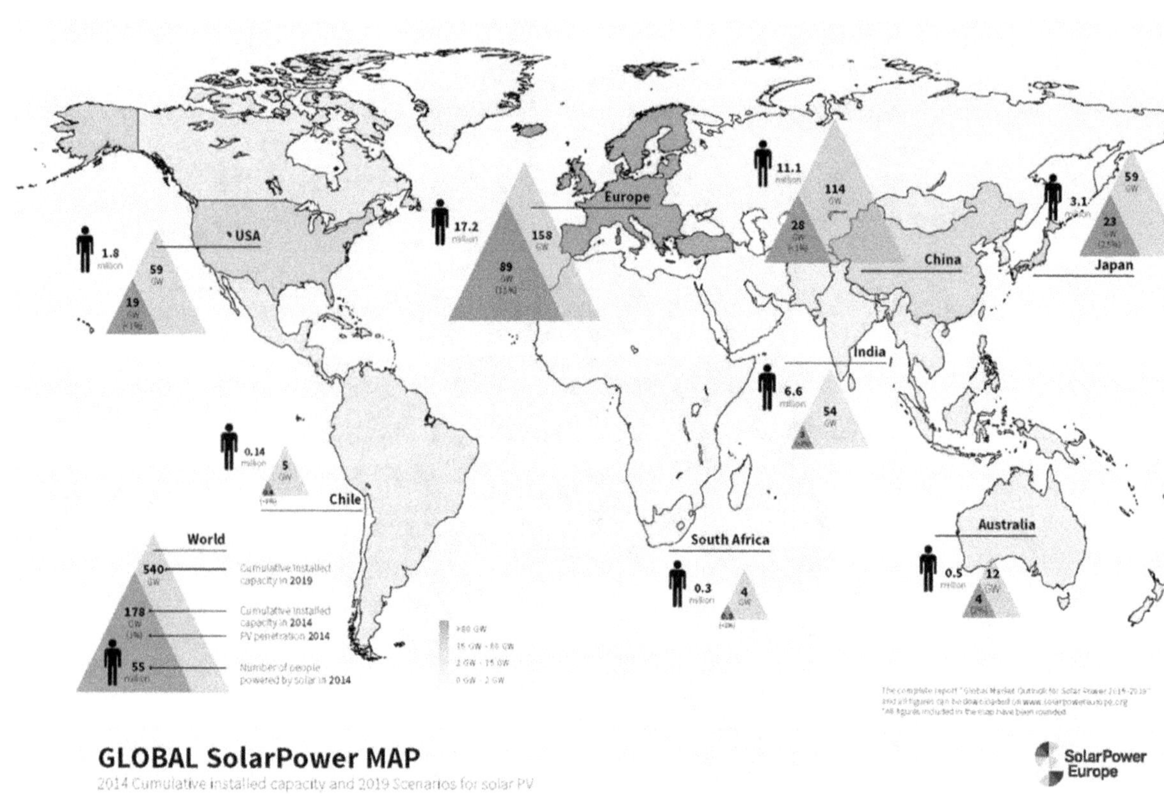

FIG. 3.39: GLOBAL PV INSTALLATIONS PER HABITANT. (*Source* Fig. 3, pg 20 of 60 [91])

The global extent of Solar Energy is depicted in figure 3.39 [91] and the amount of solar energy in vogue in Europe in figure 3.40 [91]

Ninth place for solar energy: Australia [84]

In 2015 PV Cumulative Installed Capacity was 5.1 GW and in 2014 PV Cumulative Installed Capacity was 4.1 GW. In 2014 Australia occupied ninth position. Australia is an IEA-PVPS member. Australia was maintaining its position from 2014, and is ninth in the world for total installed PV capacity. This is mainly due to its mature and strong PV market around 1022 MW new installations in one year. Their PV market constitutes around 2% of the global market for new PV, which is pretty impressive. They also make up around 2% of the world's total cumulative PV capacity, currently 228 GW worldwide. In previous years Australia has shown a nice growth of 811 MW in 2013 and 863 MW in 2014. According to IEA, Australia has "more than 5.1 GW of PV systems installed and commissioned, mainly in the residential rooftops segment (more than 1.5 million buildings now have a PV system; an average penetration of 19% in the residential sector, with peaks up to 50%), with grid-connected applications." Australia's growth in PV in 2015 is solely because of three projects under the solar flagships program that were commissioned. These were the solar farms at Nyngan (134MW), Broken Hill (64MW) and Moree (70MW). Their utility scale growth was around 287 MW with distributed application increasing 12% to 709MW this year. IEA explains Australian Government support programs impacted significantly on the PV market in recent years. The 45 000 GWh Renewable Energy Target (RET) (a quota-RPS system) consists of two parts – the Large-scale Renewable Energy Target (LRET) and the Small-scale Renewable Energy Scheme (SRES) [84].

Tenth place for solar energy: India [84]

In 2015 PV Cumulative Installed Capacity was 5 GW and in 2014 PV Cumulative Installed Capacity was 0.8 GW. In 2014 India was not in the top 10 position. India is an IEA-PVPS Member. In 2015 India got promotion to top 10. Formerly nowhere to be seen, India has smashed expectations and elevated itself into the world's top ten for PV capacity. India could become one of the global PV market leaders in the coming years. Next, to India, Pakistan seems promising with several hundreds of MW installed. India has managed to progress nicely to around 2GW new installs in 2015. This is a significant leap forward that could lead to an even larger development in the coming years. These are exciting times for the sub-continent. India has more than 1 Billion inhabitants and has, for years, experienced severe power shortages. The Indian PV market seems to have responded to this call to boost India's PV capacity to 2.1 GW in 2015 from 779 MW in 2014. IEA explains, the PV market in India is driven by a mix of national targets and support schemes at various legislative levels. The Jawaharlal Nehru National Solar Mission aims to install 20 GW of grid-connected PV systems by 2022 and an additional 2 GW of off-grid systems, including 20 million solar lights. Some states have announced policies targeting large shares of solar photovoltaic installations over the coming years. Finally, 2 GW of off-grid PV systems should be installed by 2017. India unveiled a new target of 100 GW in 2014 for the country. 60 GW of this to be centralized with the remaining 40 GW to be supplied by rooftop PV installations. India's central government has provided massive support for PV which should lead to good growth and significant increases in installations in the years to come [84].

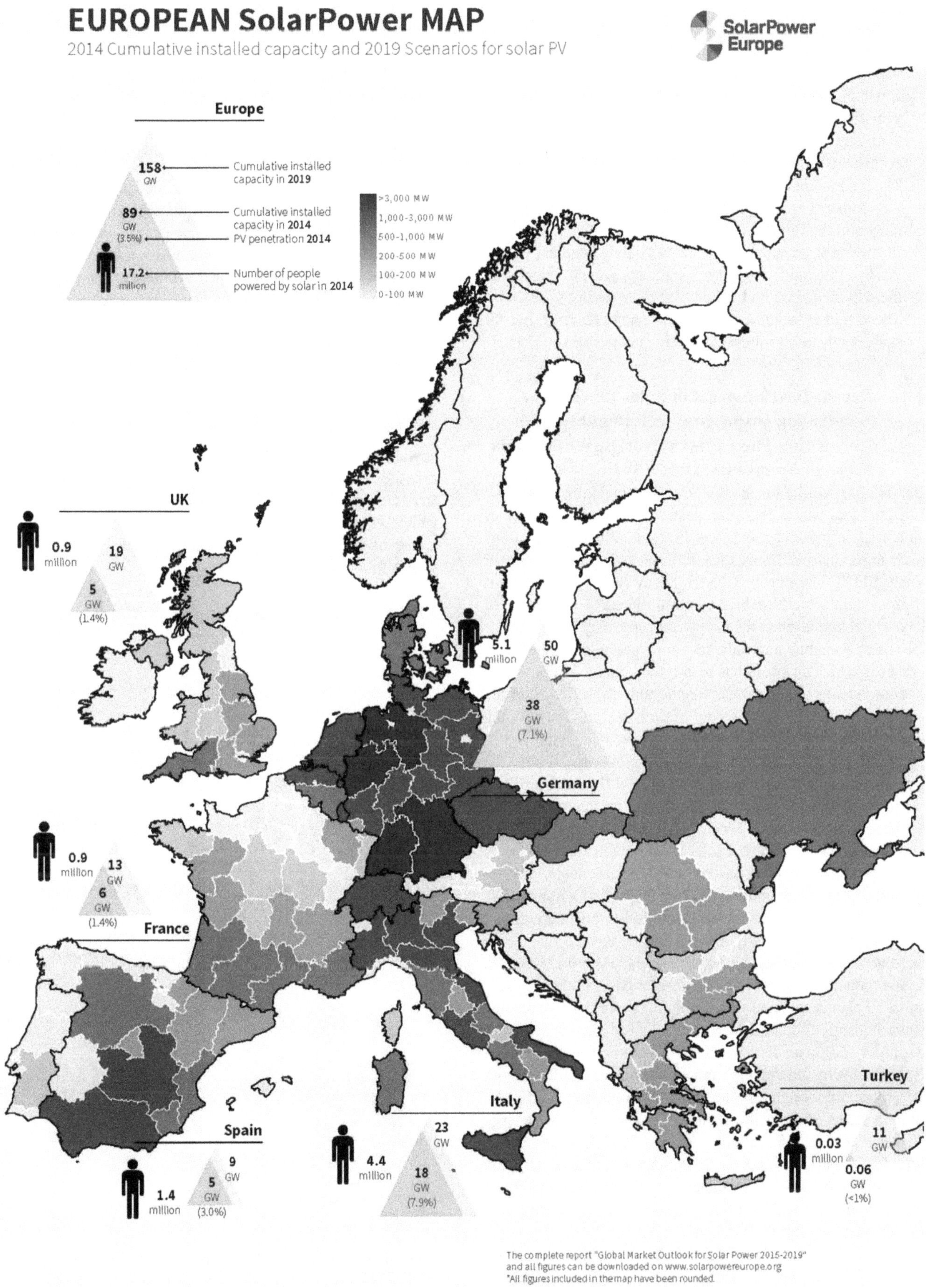

FIG. 3.40: EUROPEAN PV INSTALLATIONS PER HABITANT, 2013. (*Source*: pg 24 of 60 [91])

Conclusion [84]

- India is committed to tackle power shortages which has woken the PV market in India
- It will be interesting to see China's progress in the coming years.
- Germany, the US and the UK featured near the top and will continue to be there.
- The PV market is looking pretty healthy globally and many governments are committed to sustain growth in PV installations in the future.
- In the latest developments, ISTAT, Italy's National Statistics Institute, said the country's economy grew 0.4 percent in the second quarter and 1.5 percent over the last year, 2017. Those figures would be modest in many countries, but for Italy, it's the most robust economic growth since 2011 [92].

3.8.4 Recent Developments in Solar Investment Arbitration Disputes: Czech Republic and Italy Enjoy Their First Victories, While Spain Seeks to Annul First Defeat [93]

International Arbitration Focus Team in November 2017 had over the past four years, three European States hit with a wave of investment claims from solar power investors alleging violations of the Energy Charter Treaty ("ECT") and Bilateral Investment Treaties ("BIT"). The Energy Charter Treaty (ECT) is an international agreement which establishes a multilateral framework for cross-border cooperation in the energy industry [93].

The Czech Republic and Italy have now prevailed in their first solar cases in which an award has been issued. Spain, after initially succeeding in two cases, has been found liable for breaches of the fair and equitable treatment standard but it is seeking to annul the adverse award. The Czech Republic recently won the first of seven investment treaty cases brought by investors over reforms to the country's solar power sector introduced in 2011. In an award by a 2 to 1 majority, dated October 11, 2017, The International Centre for Settlement of Investment Disputes (ICSID) tribunal held that a tax on the electricity output of solar power plants did not violate legitimate expectations of investors and was reasonable, proportionate and non-arbitrary under the Czech Republic-Germany BIT. The remaining six United Nations Commission on International Trade Law (UNCITRAL [94]) cases pending against the Czech Republic are based on the same controversial State measures. In 2016, Spain managed to prevail in two solar cases under the ECT of almost 30 claims against it. After obtaining two positive results, Spain was recently found liable in its third solar case. In an award dated May 4, 2017, an ICSID tribunal unanimously determined that Spain had "crossed the line" and violated its obligations under the ECT to accord foreign investors fair and equitable treatment by upending a series of financial incentives in the renewable energy sector that had been offered under Spain's 2007 legislation. The Tribunal found that, starting from 2008 Spain began gradually dismantling this favorable regime and eventually deprived the investor-claimants of their €126 million investment in the construction of three thermo-solar plants in Spain. [93]

In May 2017, the claimants sought enforcement of the award in the United States District Court for the Southern District of New York, and quickly obtained an order granting enforcement. However, Spain filed a motion to vacate the Order, alleging, inter alia, that it was an "impermissible attempt" to circumvent the U.S. Foreign Sovereign Immunities Act of 1976. On November 13, 2017, the Southern District granted the motion and vacated the Order. Moreover, Spain has also challenged the award itself, seeking annulment on the grounds of alleged due process failings. The ICSID ad-hoc committee on annulment was constituted on October 23, 2017 [93].

Finally, Italy won the first of nine proceedings started by several foreign energy companies under the ECT. These claims were brought following reforms to the Italian power sector that reduced feed-in tariffs available to solar investors. [Despite Italy's withdrawal from the ECT of January 1, 2016, the "sunset clause" provides investors with a twenty year window to bring claims relating to investments made before January 2016 filed for annulment of the unfavorable ICSID award in May 2017. It differs from the other eight claimants in that it founded its claim on an earlier, non-retroactive version of the controversial reforms passed in 2013. The other claims against Italy were brought in quick succession following a similar 2014 reform, which did have retroactive effect [93]

Although these awards tend to suggest that investors face an upward battle in successfully bringing claims against host States, it is too soon to identify any significant trend. Indeed, there is a large number of pending cases to be decided in the near future, which will shed further light on this issue and, in particular, on the nature of State measures that are perceived to be unlawful under investment treaties [93].

3.8.5 Solar Power Complex in Morocco [95]

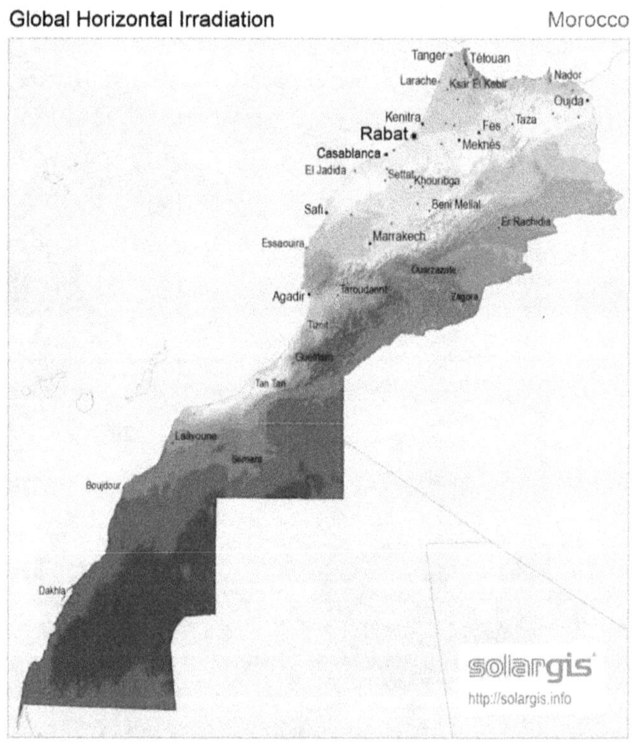

FIG. 3.41: SOLAR RADIATION MAP: GLOBAL HORIZONTAL IRRADIATION MAP OF MOROCCO, SOLARGIS OLAR RESOURCES IN MOROCCO_WIKI. (*Source* [95])

Solar power in Morocco is enabled by having one of the highest rates of solar insolation, about 3,000 hours of sunshine per year and up to 3,600 hours in the desert, see figure 3.51 [95]. Morocco has launched one of the world's largest solar energy projects costing an estimated $9 billion. The aim of the project is to create 2,000 megawatts of solar generation capacity by the year 2020. Five solar power stations are to be constructed, including both *photovoltaic* and *concentrated solar power* technology. The Moroccan Agency for Solar Energy (MASEN), a public-private venture, has been established to lead the project. The first plant will be commissioned in 2015, and the entire project in 2020. Once completed, the solar project will provide 38% of Morocco's annual electricity generation [95].

Morocco is the only African country to have a power cable link to Europe (2,100 MW) and aims to benefit from the €400bn ($573.8bn) expected to come from the ambitious pan-continental "Desertec Industrial Initiative". Desertec plan sketches the possibility of a sustainable supply of power to Europe, the Middle East and North Africa (EU-MENA) as proposed by Trans-Mediterranean Renewable Energy Cooperation (TREC). DESERTEC EU-MENA Map, figure 3.42 [95] is a sketch of possible infrastructure for a sustainable supply of power to Europe, the Middle East and North Africa (EU-MENA) proposed by TREC). The red squares represent the total surfaces needed for solar collectors of Concentrating Solar Thermal Power (CSP) plants to provide the present electricity demands, for the world (18.000 TWh/a, 300x300 km^2), for Europe (EU 3.200 TWh/a, 125x125 km^2) and for Germany or MENA (Middle East and North Africa, about 600 TWh/a, 55x55 km^2). The square labeled "TRANS-CSP Mix EUMENA 2050" indicates the surface needed for solar collectors to realize DESERTEC in EU-MENA (according to the TRANS-CSP scenario by DLR). This way the needs for seawater desalination and about two-thirds of the rising electricity consumption in the MENA region could be provided and about 17 percent of the European electricity consumption in 2050 (2,940 TWh/a, 120x120 km in total). In reality numerous CSP-Plants will be spread in the deserts of EU-MENA and around the globe, see https://en.wikipedia.org/wiki/Solar_power_in_Morocco#/media/File:DESERTEC-Map_large.jpg). Theoretical space needed for solar power plants to generate sufficient electric power in order to meet the electricity demand of the world, Europe (EU-25) and Germany (de) respectively are shown in figure 3.42 [93].

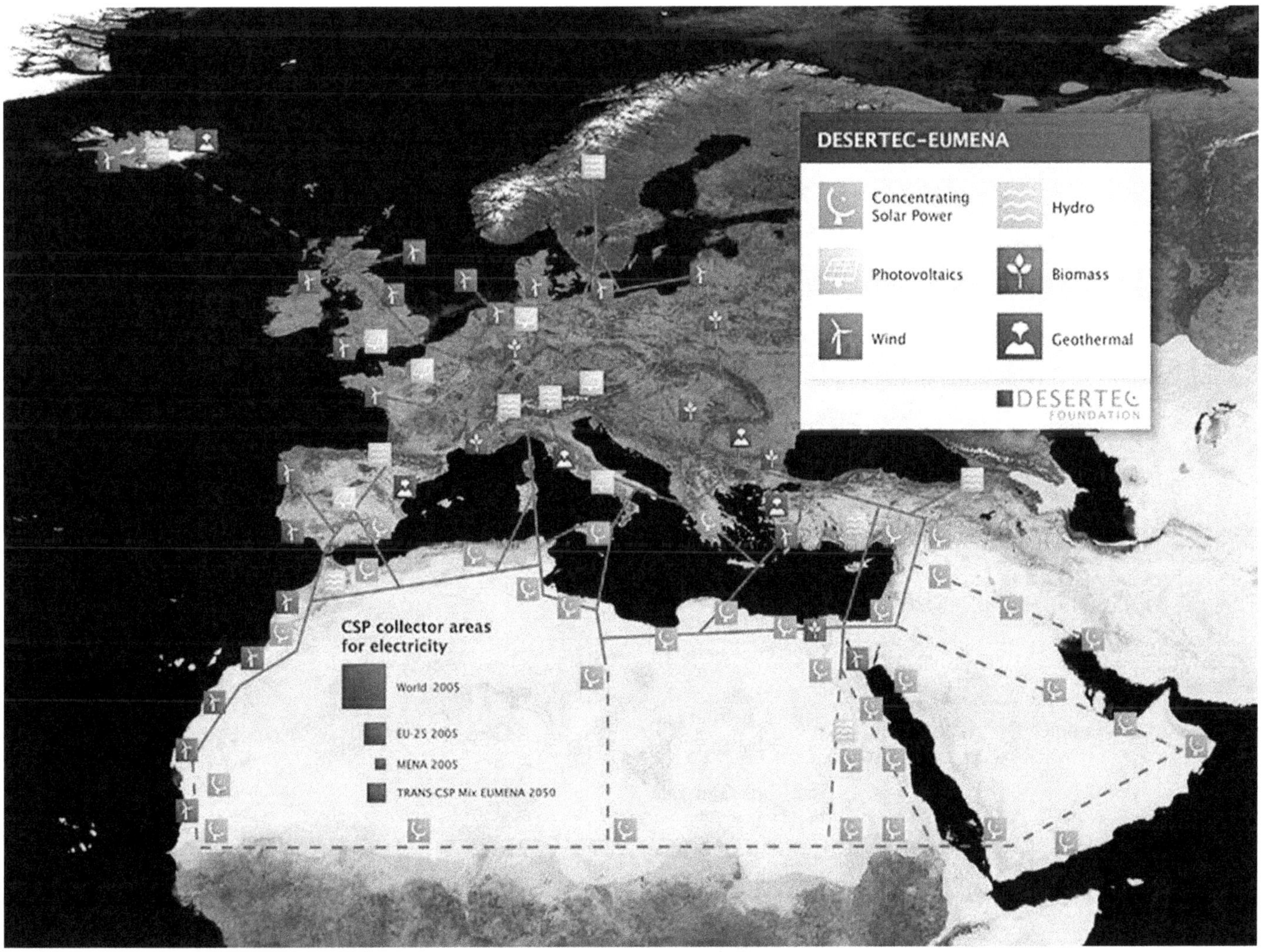

FIG. 3.42: MAP ACCORDING TO THE DESERTEC EU-MENA SOLAR POWER PLAN [95]

The Moroccan Agency for Solar Energy invited expressions of interest in the design, construction, operation, maintenance and financing of the first of the five planned solar power stations, the 500 MW complex in the southern town of Ouarzazate, that includes both PV and CSP. Construction officially began on 10 May 2013. The project is divided in 3 phases: a 160MW concentrated solar power project, a 200MW parabolic mirror plant, and a 150MW solar trough plant. The 160 MW first phase, *Noor I*, was brought online in February 2016. The project was awarded to a consortium led by Saudi Arabia's ACWA Power, which sells the electricity produced for $0.19/kWh, and co-financed by the World Bank and the European Investment Bank. the end of 2018. These two phases provide another 580 MW and cover 6,000 acres [95].

Ouarzazate Solar Power Station (OSPS), also called Noor Power Station (نور, Arabic for light) is a solar power complex located in the Drâa-Tafilalet region in Morocco, 10 kilometers (6.2 mi) from Ouarzazate town, in Ghessat rural council area. At 510 MW, it is the world's largest concentrated solar power plant. With an additional 72 MW photovoltaic system the entire project is planned to produce 582 MW at peak when finished and is being built in three phases and in four parts. The total project is expected to cost $9 billion. The plant will be able to store solar energy in the form of heated molten salt, allowing for production of electricity into the night. Phase 1 comes with a full-load molten salt storage capacity of 3 hours. Morocco Unveils A Massive Solar Power Plant In The Sahara. The Noor I power plant is located near the town of Ouarzazate, on the edge of the Sahara. It's capable of generating up to 160 megawatts of power and covers thousands of acres of desert, making the first stage alone one of the world's biggest solar thermal power plants. Noor II, commissioned in 2018 and Noor III, commissioned in January 2019, store energy for up to eight hours. It will cover an area of 2,500 hectares (6,178 acres). The project is being developed with the help of the Spanish consortium TSK-Acciona-Sener and is the first in a series of planned developments at the Ouarzazate Solar Complex by the Moroccan Agency for Solar Energy (MASEN). As of December 2015, the plant was expected to begin commercial operation in June 2017. Water consumption for the Ouarzazate Noor complex is estimated at 2.5 to 3 million m3 per year for one wet- cooling project (Noor I) and two dry cooling projects (Noor II and III), due to the need to clean the reflectors regularly. Noor IV will be a 72 MW Photovoltaic power station with the total investment in this project to be 750 millions MAD [96].

TABLE 3.8

MOROCCO RENEWABLE ENERGY PROJECTS NOW TO 2030 (SOURCE: [95])

Name	Output	Location	Technology	Status	Notes
Ain Beni Mathar	400MW	Ain Bni Mathar	Parabolic trough		Also runs on fossil fuels (Gas)
Noor I	160 MW	Drâa-Tafilalet	Parabolic trough		Part of Ouarzazate Solar Power Station
Noor II	200 MW	Drâa-Tafilalet	Parabolic trough		Part of Ouarzazate Solar Power Station
Noor III	150 MW	Drâa-Tafilalet	solar power tower		Part of Ouarzazate Solar Power Station
Noor IV	70 MW	Drâa-Tafilalet	Photovoltaics		Part of Ouarzazate Solar Power Station
Sebkhate Tah	500MW				
Foum Al-Oued Boujdour Noor Midelt Noor PV II	500MW 100 MW 800 MW 800MW	Taroudant, Kalâat Seraghna, Bejaâd, Guercif, El Hajeb		Planned	
Noor-Tafilalet	120 M	Zagora, Erfoud and Missour		Planned	
Noor Atlas	800MW	Tata/Tahla/Tan Tan/ Outat El Haj/Ain Beni Mathar/ Boudnib/Bo uanane/ Boulemane	Photovoltaics		
Noor Argana	200 MW	Boumalne/Tinghir/ Errhamna/Essaouira	Photovoltaics		

FIG. 3.43: NOOR III SOLAR TOWER OF THE OUARZAZATE POWER STATION AT DUSK.JPG. (*Source*: Picture by Marc Lacoste, November 6, 2018 [96])

3.8.6 Solar Power Generation in South Africa [97]

Solar power in South Africa includes photovoltaics (PV) as well as concentrated solar power (CSP). Solar Africa provides clean energy solutions to South African residential complexes, small businesses & property developers. The southern African region, and in fact the whole of Africa, has sunshine all year round. The annual 24-hour global solar radiation average is about 220 W/m2 for South Africa compared with about 150 W/m2 for parts of the USA, and about 100 W/m2 for Europe and the United Kingdom [97].

In 2016, South Africa had 1,329 MW of installed solar power capacity. Installed capacity is expected to reach 8,400 MW by 2030. In 2014 several solar farms were commissioned, including the 96 MW Jasper Solar Energy Project, one of Africa's largest photovoltaic power stations providing enough solar power for 30,000 homes. As of 1 January 2016 the South African government gave a tax incentive through the South African Revenue Service for the installation of photovoltaic solar energy generation systems. Despite this aggressive tax incentive, South African companies are slow to adopt grid-connected photovoltaic solar systems due to the lack of public dialogue from the government concerning photovoltaic solar energy. Through its holding in Eskom, the South African government is dependent on Eskom's revenue stream to avoid a bailout with public funds. In addition to the government tax incentive for photovoltaic solar energy generation, certain municipalities have further incentives for residential and commercial customers, such as a feed-in tariff or net metering. As of January 2015 a total of 593 MW produced from PV plants were connected to the grid. The Department of Energy estimates that 23% of this target can be met through solar water heating and Eskom is therefore actively encouraging consumers to switch to solar water heating [97].

TABLE 3.9

GROWTH OF PHOTOVOLTAICS IN SOUTH AFRICA SINCE 2011 (SOURCE: [97])

Year	Installations in MWp		Production in GWh
	Cumulative capacity	Added capacity	
2011	1	–	–
2012	41	40	–
2013	122	81	67
2014	957	800	1183
2015	965	8	2185
2016	1329	364	2347
2017	1479	150	3095

According to the IMF's 2019 World Economic Outlook report, six of the world's fastest-growing economies are in Africa. Over the next three decades, Africa's population is expected to double to 2.4 billion. By 2050, about one in four people in the world will be African and the continent's economy could be worth $29 trillion, more than four times its current GDP (in PPP terms). Africa's large market and economic potential underlie the growing efforts by China, India, Europe, North America and the Middle East to deepen trade, investment and economic ties with the continent. Currently, China leads the pack in terms of trade, investment, and business with Africa. Solar energy is the fastest and cheapest way to light up Africa's thousands of villages which are so remote and too much spread apart that it would take decades and billions of investment dollars to take electricity to all corners of African continent [98].

3.8.7 Solar Power Generation in Chile, South America [99]

Northern Chile has very high solar incidence in the world. In October 2015 Chile's Ministry of Energy announced its "Roadmap to 2050: A Sustainable and Inclusive Strategy", which plans for 19% of the country's electricity to be from solar energy, 23% wind power, and 29% hydroelectric power [99]. Solar potential in Chile is given in figure 3.44.

The time line of Solar power is recent and given below for the period from 2014 to 2016. In 2016 a Solar Pack was won in an electricity auction by bidding $29.1/MWh which was considered a record low price. During the period 2014-2015 The Atacama 1 solar complex was proposed as a 110 MW solar thermal electric plant (the first in Latin America) and a 100 MW photovoltaic plant. The solar thermal plant was to include 17.5 hours of thermal storage. These technologies complement each other to supply clean and stable energy 24 hours a day. The complex is located in the commune of María Elena, Segunda Región. Construction of the solar thermal electric plant commenced in 2014 and the plant is scheduled to begin operating in the second quarter of 2017. Construction of the photovoltaic plant commenced in January 2015 and the plant began operating in June 2016 with 160 MW of panels, the largest solar plant in Chile at the time [99], figure 3.55 shows irradiance and installed capacity of Chile [99].

The 100-megawatt (MW) Amanecer Solar CAP, a photovoltaic power plant located near Copiapó in the Atacama Desert, which was the largest in Latin America at that time, capable of

TABLE 3.10

GROWTH OF PHOTOVOLTAICS IN SOUTH AFRICA SINCE 2011 (SOURCE: [97])

Installations in MWp		Production in GWh
Cumulative capacity	Added capacity	
100	100	188
200	100	590
300	100	676

FIG. 3.44: SOLAR POTENTIAL IN CHILE [99].

generating 270 gigawatt-hours (GWh) of electricity per year, was inaugurated in June 2014. The 70 MW photovoltaic Salvador Solar Park went online in November 2014, to produce 200 GWh of electricity per year. The plant is located 5 kilometers south of El Salvador, in the Atacama region. It is one of the first in the world to supply competitively priced solar energy to the open market without government subsidy. The 60 MW photovoltaic Lalackama I plant went online in 2014 to produce 160 GWh of electricity per year. The nearby 18 MW Lalackama II plant went online in May 2015 capable of generating approximately 50 GWh per year. Both plants feature photovoltaic inverters designed and manufactured by Elettronica Santerno, an Italian company [99].

The 141 MW photovoltaic Luz Del Norte (Light of the North) plant, located northeast of Copiapó in the Atacama region, uses more than 1.7 million cadmium telluride modules. The first two blocks of this project were connected to Chile's central power grid in October 2015. The 79 MW Pampa Norte PV solar plant began operating in April 2016 at a site southwest of Taltal in Chile's Antofagasta Region. It uses approximately 258,000 polycrystalline silicon photovoltaic modules and is capable of generating more than 200 GWh per year. The 97 MW Carrera Pinto photovoltaic plant began is located 60 kilometres from the city of Copiapó in the Atacama Region. The first 20 MW of the plant was connected to the grid in early January 2016, with the remaining 77 MW connected in August 2016. The plant is capable of generating over 260 GWh per year. The 246 MW El Romero single axis tracking solar photovoltaic plant began operating in November 2016 at Vallenar in the Atacama region, with a 493 GWh annual average output. It was the largest solar farm in Latin America when it opened. It uses 776,000 polycrystalline silicon photovoltaic modules. The solar irradiance has been measured at 853 W/m² [99].

In October 2017, within the framework of the Renewable Energy Solutions Program, the Deutsche Energie Agentur GmbH (dena RES programme) at the Tierra Atacama Hotel, a photovoltaic system with 155.88 kWp in combination with a battery storage system was put into operation. The off-grid solar system was built by the German EPC (Engineering, procurement, and construction) service provider Kraftwerk Renewable Power Solutions. Qinous's battery system with 335 kWh lithium-ion battery storage and 180 kW inverter power enables the diesel generators to be switched

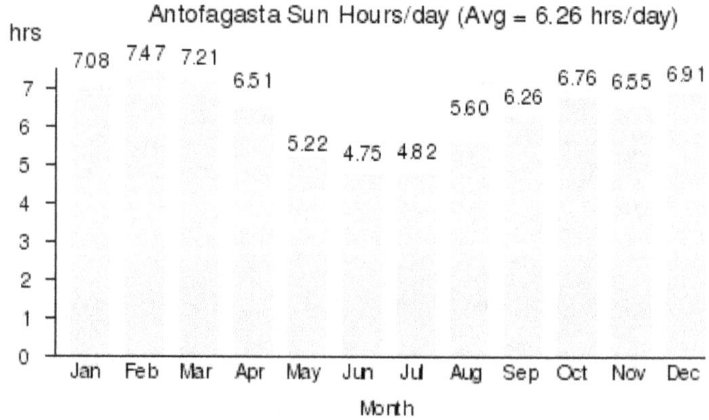

FIG. 3.45: IRRADIANCE AND INSTALLED CAPACITY OF CHILE. (*Source*: [NREL cited in 99].

TABLE 3.11

CHILE PHOTOVOLTAICS CAPACITY (MW_p) (SOURCE: [99])

End of the year	Installed capacity operating	Under construction	Presented for environmental assessment
2011	0	1	<769
2012	2	2.5	4,109
2013	6.7	128	8,184
2014	362	983	12,559

off completely during the day because the battery system can take over the power supply instead of the diesel generators. The annual diesel requirement can thus be reduced by up to 60%. As early as 2013, the hotel was named the most sustainable tourism company in Chile following the commissioning of the first small PV diesel hybrid plant [99].

3.9 NOMENCLATURE AND UNIT CONVERSIONS

bar Unit of pressure equal to 100 kilopascals
CLFR Compact Linear Fresnel Reflector
CRS Central Receiver System
CSP Concentrating Solar Power
DE Dish Engine
GEF Global Environmental Facility
GW Gigawatt (109 Watts)
GWh Gigawatt-hour (109 Wh)
Ha/MW Hectares per MW
HTF Heat Transfer Fluid
IEA International Energy Agency
ISCCS Integrated Solar Combined Cycle System
kJ/kgK kiloJoule per kilogram and Kelvin
kW kiloWatt (103 Watts)
kWh/m^3 kiloWatt-hour per cubic meter
LFR Linear Fresnel Reflector
m meter
m^2 square meter
mbar millibar
mrad milliradian
MW Megawatts (106 Watts)
MWe Megawatts electric (106 Watts)
MWth Megawatts thermal (106 Watts)
NSO Nevada Solar One
O&M Operation and Maintenance
PCM Phase Change Material
PTC Parabolic Trough Collector
PV Photovoltaics
R&D Research and Development
SEGS Solar Electricity Generating Systems
SHP Solar Heat and Power
SPG Solar Power Group
STE Solar Thermal Electricity
TWh TeraWatt hour (1012 Wh)
W/mK Watts per meter and Kelvin

3.10 REFERENCES

1. Becker M., Meinecke W., Geyer M., Trieb F., Blanco M., Romero M., Ferriere A. (2002), "Solar Thermal Power Plants". In: "The future for Renewable Energy 2: Prospects and Directions". Eurec Agency. Pub. James & James Science Publishers Ltd., London, UK. pp. 115-137.
2. Herring G. "Concentrating solar thermal power gains steam in Spain, as momentum builds for major projects in the US, North Africa, the Middle East, Asia and Australia". Photon International, December 2009, 46-52.
3. Nitsch, J. and Krewitt, W. "Encyclopedia of Energy", Elesevier Inc., Volume 5, 2004.
4. Philibert C. (2004). International Energy Technology Collaboration and Climate Change Mitigation, Case Study 1: Concentrating Solar Power Technologies, OECD/IEA Information Paper, Paris. COM/ENV/EPOC/IEA/SLT(2004)8.
5. IEA (2010) Energy Technology Perspectives 2010 – Scenarios and strategies to 2050. ISBN 978-92-64-08597-8.
6. Romero M., Zarza E. (2007) "Concentrating Solar Thermal Power". In: Handboook of Energy Efficiency and Renewable Energy. F. Kreith and Y. Goswami (Eds.) Chapter 21. pp. 1-98. CRC Press Taylor & Francis Group, Boca Raton, Florida.
7. Kearney A.T. (2010) "Solar thermal electricity 2025". STE industry roadmap for the European Solar Thermal Electricity Association (ESTELA). Available at: www.atkearney.com.
8. Winter C.J., Sizmann R.L., Vant-Hull L.L. (Eds) (1991), "Solar Power Plants", Springer-Verlag, Berlin, ISBN 3-540-18897-5.
9. Kodama T. (2003), "High-temperature solar chemistry for converting solar heat to chemical fuels", *Progress in Energy and Combustion Science* **29,** 567–597.
10. Rabl, A. Active Solar Collectors and Their Applications. New York, Oxford University Press, 1985: pp. 59-66. ISBN: 0-19-503546-1.
11. Welford W.T., Winston R. (1989) "High Collection Non-Imaging Optics". New York; Academic Press.
12. Biggs F., Vittitoe C. (1979). "The Helios model for the optical behavior of reflecting solar concentrators". Report Sandia National Labs. SAND76-0347. March 1979.
13. Mills D. (2004) "Advances in solar thermal electricity technology". *Solar Energy* 76 (2004) 19–31.
14. Geyer M. (2002), Panel 1 Briefing Material on Status of Major Project Opportunities. The Current Situation, Issues, Barriers and Planned Solutions. International Executive Conference on Expanding the Market for Concentrating Solar Power (CSP) – Moving Opportunities into Projects; 19 – 20 June 2002; Berlin, Germany.
15. DeMeo E.A., Galdo J.F. (1997), "Renewable Energy Technology Characterizations", TR-109496 Topical Report, December 1997, U.S. DOE-Washington and EPRI, Palo Alto, California.
16 Price H., Luepfert E., Kearney D., Zarza E., Cohen G., Gee R., Ma-honey R. (2002), "Advances in Parabolic Trough Solar Power Technology", Int. J. Solar Energy Eng., Vol. 124, pp. 109-125.

17. Fernandez-Garcia A., Zarza E., Valenzuela L., Perez M. "Parabolic-trough solar collectors and their applications". Renewable and Sustainable Energy Reviews 14 (2010) 1695–1721.
18. Rabl, A. Active Solar Collectors and Their Applications. New York, Oxford University Press, 1985: pp. 59-66. ISBN: 0-19-503546-1.
19. Eck, M. Zarza, E. Eickhoff, M. Rheinländer, J. Valenzuela, L. Applied research concerning the direct steam generation in parabolic troughs. Solar Energy, no. 74, 2003: pp. 341-351.
20. Kearney, D.W and Cohen, G.E. (1997) "Current experiences with the SEGS parabolic trough plants". In: Proceedings of the 8th International Symposium on Solar Thermal Concentrating Technologies. Vol. 1. Cologne, Germany, 1996. Becker, M.; Böhmer, M. eds. Heidelberg, Alemania, C.F. Müller, 1997: pp. 217-224.
21. Goebel O. "Shams One 100 MW CSP Plant in Abu Dhabi"; Proceedings SolarPACES 2009 (CD). Ref. manuscript: 15523; Berlin, Germany; 15-18 September 2009. Ed. DLR, Stuttgart, Germany. ISBN 978-3-00-028755-8.
22. Romero M. (2004)."Solar Thermal Power Plants". In: "Report on research and development of energy technologies". Edited by IUPAP working group on energy; October 6, 2004. pp 96-108.
23. Horn M, Führing H, Rheinländer J. "Economic analysis of integrated solar combined cycle power plants. A sample case: The economic feasibility of an ISCCS power plant in Egypt". Energy 2004;29:935–1011.
24. Relloso S., Delgado E. "Experience with molten salt thermal storage in a commercial parabolic trough plant. Andasol-1 commissioning and operation. Proceedings SolarPACES 2009 (CD). Ref. manuscript: 11396; Berlín, Germany; 15-18 September 2009. Ed. DLR, Stuttgart, Germany. ISBN 978-3-00-028755-8.
25. Falchetta M., Mazzei D., Crescenzi T., Merlo L. "Design of the Archimede 5 MW molten salt parabolic trough solar plant". Proceedings SolarPACES 2009 (CD). Ref. manuscript: 11608; Berlin, Germany; 15-18 September 2009. Ed. DLR, Stuttgart, Germany. ISBN 978-3-00-028755-8.
26. Laing, D., Steinmann, W.-D., Tamme, R., and Richter, C., 2006, "Solid Media Thermal Storage for Parabolic Trough Power Plants", Solar Energy, 80(10), pp. 1283-1289.
27. Laing D., Bahl C., Bauer T., Lehmann D., Steinmann W.D. "Thermal energy storage for direct steam generation". Proceedings SolarPACES 2009 (CD). Ref. manuscript: 12055; Berlin, Germany; 15-18 September 2009. Ed. DLR, Stuttgart, Germany. ISBN 978-3-00-028755-8.
28. Zarza, E., Valenzuela, L., Leon, J., Hennecke, K., Eck, M., Weyers, H. D., and Eickhoff, M., 2004, "Direct steam generation in parabolic troughs: Final results and conclusions of the DISS project", Energy, 29(5-6), pp. 635-644.
29. Kalogirou S.A. "Solar thermal collectors and applications". Progress in Energy and Combustion Science. Volume 30, Issue 3, 2004, 231-295.
30. Mills D.R and Morrison G.L.(1999). Compact linear Fresnel reflector solar thermal powerplants. Solar Energy. V68, pp 263 – 283.
31. Bernhard R., Hein S., de Lalaing J., Eck M., Eickhoff M., Pfänder M., Morin G., Häberle A. (2008). Linear Fresnel Collector Demonstration on the PSA, Part II – Commissioning and first Performance Tests, Proceedings of the 14th Solar Paces Symposium, Las Vegas, USA. http://www.nrel.gov/docs/gen/fy08/42709CD.zip.
32. Hautmann G., Selig M., Mertins M., "First European Linear Fresnel power plant in operation – Operational experience&outlook". Proceedings SolarPACES 2009 (CD). Ref. manuscript: 16541; Berlin, Germany; 15-18 September 2009. Ed. DLR, Stuttgart, Germany. ISBN 978-3-00-028755-8.
33. Dersch J., Morin G., Eck M., Häberle A.: Comparison of Linear Fresnel and Parabolic Trough Collector Systems – System Analysis to determine break even Costs of Linear Fresnel Collectors. Proceedings SolarPACES 2009 (CD). Ref. manuscript: 15162; Berlin, Germany; 15-18 September 2009. Ed. DLR, Stuttgart, Germany. ISBN 978-3-00-028755-8.
34. Romero M., Marcos M.J., Téllez F.M., Blanco M., Fernández V., Baonza F., Berger S. (2000), "Distributed power from solar tower systems: A MIUS approach", Solar Energy, 67 (4-6) 249-264.
35. Kolb G.J. (1998). Economic evaluation of solar-only and hybrid power towers using molten-salt technology, Solar Energy. 62, 51-61.
36. Falcone P.K. (1986), "A handbook for Solar Central Receiver Design", SAND86-8009, Sandia National Laboratories, Livermore, (USA). http://prod.sandia.gov/techlib/access-control.cgi/1986/868009.pdf.
37. Sánchez M., Romero M. (2006) "Methodology for generation of heliostat field layout in central receiver systems based on yearly normalized energy surfaces", Solar Energy, 80, pp 861-874.
38. Lipps F.W., Vant-Hull L.L. (1978), "A Cellwise method for the optimization of large central receiver systems" Solar Energy 20, 505-516.
39. Kistler B.L. (1986), "A User's Manual for DELSOL3: A Computer Code for Calculating the Optical Performance and Optimal System Design for Solar Thermal Central Receiver Plants". Sandia Report, SAND-86-8018. Sandia National Laboratories, USA. http://prod.sandia.gov/techlib/access-control.cgi/1986/868018.pdf.
40. Mavis, C.L. (1989), "A description and assessment of heliostat technology", SAND87-8025, Sandia Nat. Labs., January 1989.
41. Romero M., Conejero E. and M. Sánchez (1991) "Recent experiences on reflectant module components for innovative heliostats", Solar Energy Materials 24, 320-332.
42. Monterreal, R., Romero, M., García, G. and Barrera, G. (1997)," Development and testing of a 100 m^2 glass-metal heliostat with a new local control system". Solar Engineering 1997, pp. 251-259, Eds. D.E. Claridge and J.E. Pacheco, ASME, New York, 1997. ISBN: 0-7918-1556-0.
43. Osuna R., Fernández V., Romero M., Sanchez M. (2004). "PS10: A 11-MW solar tower power plant with saturated steam receiver", Proceedings 12th SolarPACES International Symposium, 6-8 October 2004, Oaxaca, Mexico. S3-102. CD-Rom. Eds. C. Ramos and J. Huacuz. ISBN: 968-6114-18-1.
44. Silberstein E., Magen Y., Kroyzer G., Hayut R., Huss H. (2009) "Brightsource solar tower pilot in Israel's Negev operation at 130 bar @ 530°C Superheated Steam". Proceedings SolarPACES 2009 (CD). Berlín, Germany; 15-18 September 2009. Ed. DLR, Stuttgart, Germany. ISBN 978-3-00-028755-8.
45. Schell S. (2009). "Design and evaluation of eSolar's heliostat fields". Proceedings SolarPACES 2009 (CD). Berlín, Germany; 15-18 September 2009. Ed. DLR, Stuttgart, Germany. ISBN 978-3-00-028755-8.
46. Romero M., Buck R., Pacheco J.E. (2002), "An Update on Solar Central Receiver Systems, Projects, and Technologies.", Int. J. Solar Energy Eng., Vol. 124, pp. 98-108.
47. Becker M. and Vant-Hull L.L. (1991). "Thermal receivers". In "Solar Power Plants", Winter C.J., Sizmann R.L., Vant-Hull L.L. (Eds) , Springer-Verlag, Berlin, pp. 163-197. ISBN 3-540-18897-5.
48. Agrafiotis C.C., Mavroidis I., Konstandopoulos A.G., Hoffschmidt B., Stobbe P., Romero M., Fernández-Quero V. (2007) "Evaluation of porous silicon carbide monolithic honeycombs as volumetric receivers/collectors of concentrated solar radiation". Solar Energy Materials & Solar Cells 91 (2007) 474–488.
49. Palero S., Romero M., Castillo J.L. (2008) "Comparison of Experimental and Numerical Air Temperature Distributions Behind a Cylindrical Volumetric Solar Absorber Module". Journal of Solar Energy Engineering 130, 011011-1-8.
50. Marcos M.J., Romero M., Palero S. (2004) "Analysis of air return alternatives for CRS-type open volumetric receiver". Energy, 29, 677–686.
51. Hoffschmidt B., Fernandez V., R. Pitz-Paal, M. Romero, P. Stobbe, F. Téllez (2002). "The Development Strategy of the HitRec Volumetric Receiver Technology - Up-Scaling from 200kWth via 3MWth up to 10MWel –". 11th SolarPACES International Symposium on Concentrated Solar Power and Chemical Energy Technologies. September 4-6, 2002. Zurich, Switzerland. pp. 117-126. ISBN: 3-9521409-3-7.

52. Romero M., Marcos M.J., Osuna R. and Fernández V. (2000), "Design and Implementation Plan of a 10 MW Solar Tower Power Plant based on Volumetric-Air Technology in Seville (Spain)". SOLAR ENGINEERING 2000-Proceedings of the ASME International Solar Energy Conference, Madison, Wisconsin, June 16-21, 2000. Ed.: J.D. Pacheco and M.D. Thornbloom, ASME, New York, 2000. ISBN: 0791818799.
53. Tellez F., Romero M., Heller P., Valverde A., Reche J.F., Ulmer S., Dibowski G. (2004). "Thermal Performance of "SolAir 3000 kWth" Ceramic Volumetric Solar Receiver", Proceedings 12th SolarPACES International Symposium, 6-8 October 2004, Oaxaca, Mexico. S9-206. CD-Rom. Eds. C. Ramos and J. Huacuz. ISBN: 968-6114-18-1.
54. Kribus A. (1999). "Future directions in solar thermal electricity generation". In: Solar thermal electricity generation. Colección documentos CIEMAT. CIEMAT, Madrid, Spain. pp. 251-285. ISBN: 84-7834-353-9.
55. Grasse W., Hertlein H.P., Winter C.J. (1991), "Thermal Solar Power Plants Experience" . In "Solar Power Plants", Winter C.J., Sizmann R.L., Vant-Hull L.L. (Eds), Springer-Verlag, Berlin, pp 215-282. ISBN 3-540-18897-5.
56. Radosevich L.G., Skinrood A.C.(1989), "The power production operation of Solar One, the 10 MWe solar thermal central receiver pilot plant", J. Solar Energy Engineering, 111, 144-151.
57. Pacheco J.E., Gilbert R. (1999), "Overview of recent results of the Solar Two test and evaluations program". In Renewable and Advanced Energy Systems for the 21st Century RAES'99 April 11–15, 1999 — Maui, Hawaii, pp. RAES99-7731, Eds.R. Hogan, Y. Kim, S. Kleis, D. O'Neal and T. Tanaka; ASME, New York, 1999. ISBN: 0-7918-1963-9.
58. Epstein M, Liebermann D., Rosh M, Shor A. J (1991), "Solar testing of 2 MW (th) water/steam receiver at the Weizmann Institute solar tower", Solar Energy Materials, Vol. 24, pp. 265-278.
59. Fernandez-Quero V., Osuna R., Romero M., Sanchez M., Ruiz V., Silva M., (2005) EURECA: Advanced Receiver For Direct Superheated Steam Generation In Solar Towers, As An Option For Increasing Efficiency In Large Low Cost Direct Steam Generation Plants. Proceedings of the 2005 Solar World Congress ISES-2005. 6-12 August, Orlando, Florida. Ed. By D.Y. Goswami, S. Vijayaraghaveng, R. Campbell-Howe. Pub. American Solar Energy Society, Boulder, Colorado, USA .ISBN 0-89553-177-1.
60. Tyner CE., Pacheco J.E. (2009). "eSolar's power plant architecture". Proceedings SolarPACES 2009 (CD). Berlín, Germany; 15-18 September 2009. Ed. DLR, Stuttgart, Germany. ISBN 978-3-00-028755-8.
61. Kelly, B., Singh, M. (1995) "Summary of the Final Design for the 10 MWe Solar Two Central Receiver Project," Solar Engineering: 1995, ASME, Vol. 1, p. 575.
62. Pacheco, J. E., H. E. Reilly, G. J. Kolb, C. E. Tyner (2000), "Summary of the Solar Two Test and Evaluation Program". Proceeding of the Renewable Energy for the New Millennium, Sydney, Australia, March 8-10, 2000. pp.1-11.
63. Ortega J.I., Burgaleta J.I., Tellez F. (2006) "Central Receiver System (CRS) solar power plant using molten salt as heat transfer fluid". Proceedings of the 13th International Symposium on Concentrated Solar Power and Chemical Energy Technologies. June 20, 2006, Seville, Spain. M. Romero, D. Martínez, V. Ruiz, M. Silva, M. Brown (Eds.) (2006). Pub. CIEMAT. ISBN: 84-7834-519-1.
64. Burgaleta J.I., Arias S., Salbidegoitia I.B. (2009) "Operative advantages of a central tower solar plant with thermal storage system". Proceedings SolarPACES 2009 (CD). Ref. manuscript: 11720; Berlin, Germany; 15-18 September 2009. Ed. DLR, Stuttgart, Germany. ISBN 978-3-00-028755-8.
65. Mancini T., Heller P., Butler B., Osborn B., Schiel W., Goldberg V., Buck R., Diver R., Andraka C., Moreno J. (2003), "Dish–Stirling Systems: An Overview of Development and Status", Int. J. Solar Energy Eng., Vol. 125, pp. 135-151.
66. Lopez C., Stone K., (1992), "Design and Performance of the Southern California Edison Stirling Dish," Solar Engineering, Proc. of ASME Int. Solar Energy Conf., Maui, HI, ISBN 0-7918–762-2, pp. 945–952.
67. Keck T., Heller P., Weinrebe G., (2003). "Envirodish and Eurodish — system and status". Proc. ISES Solar World Congress, Göteborg, Sweden, June 2003. ISBN: 91-631-4740-8.
68. Diver, R.B., (1987), J. Solar Energy Engineering, 109 (3), 199-204.
69. Moreno J. B., Modesto-Beato M., Rawlinson K. S., Andraka C. E., Showalter S. K., Moss T. A., Mehos M., and Baturkin V., (2001), "Recent Progress in Heat-Pipe Solar Receivers," SAND2001-1079, 36th Intersociety Energy Conversion Engineering Conf., Savannah, GA, pp. 565–572.
70. Stine W., Diver R.B (1994), "A Compendium of Solar Dish/Stirling Technology", report SAND93-7026; Sandia National Laboratories, Albuquerque, New Mexico.
71. IEA (2010) Technology Roadmap – Concentrating Solar Power. Available free at http://www.iea.org.
72. Pitz-Paal R., Dersch J., Milow B., Ferriere A., Romero M., Téllez F., Zarza E., Steinfeld A., Langnickel U., Shpilrain E., Popel O., Epstein M., Karni J. (2005). ECOSTAR Roadmap Document for the European Commission; SES-CT-2003-502578. Edited by: Robert Pitz-Paal, Jürgen Dersch, Barbara Milow. Deutsches Zentrum für Luft- und Raumfahrt e.V., Cologne, Germany. February 2005. http://www.vgb. org/data/vgborg_/Forschung/roadmap252.pdf.
73. Energy and Power Generation Handbook - Established and Emerging Technologies, edited by K. R. Rao, ASME Press, 2011, New York, NY
74. www.eera-set.eu/wp-content/uploads/JP-CSP_Position-Paper_March-2017.pdf
75. https://setis.ec.europa.eu/system/files/set_plan_-_csp_initiative_implementation_plan.pdf
76. https://setis.ec.europa.eu/system/files/implementation_of_the_csp_targets.pdf
77. https://www.disruptordaily.com/8-csp-companies-look-2018
78. https://phys.org/news/2012-03-thermosolar-power-station-spain-night.html
79. https://wordlesstech.com/torresol-energy-gemasolar/
80. https://solarenergypowerplants.blogspot.com/2010/10/olmedilla-photovoltaic-park.html
81. https://www.worldenergy.org/data/resources/country/spain/solar/
82. https://en.wikipedia.org/wiki/Solar_power_in_Spain
83. https://en.wikipedia.org/wiki/Andasol_Solar_Power_Station
84. https://interestingengineering.com/top-10-performing-countries-for-solar-energy
85. https://www.pv-magazine.com/2017/05/08/italy-issues-new-regulation-to-promote-solar-and-renewables-in-non-interconnected-islands/
86. https://www.pv-magazine.com/2017/11/02/italy-installed-323-mw-of-solar-in-first-nine-months-of-2017/
87. https://en.wikipedia.org/wiki/Solar power in Italy
88. https://www.pv-magazine.com/2019/01/11/italy-sets-2030-solar-target-of-50-gw/
89. https://commons.wikimedia.org/wiki/File:Watts_per_capita_Spain.svg
90. https://en.wikipedia.org/wiki/Solar power in Spain#cite note-epia-2014-31
91. https://issuu.com/aie eur electrical contractors/docs/global market outlook 2015 -2019 lr/14
92. http://www.xinhuanet.com/english/2017-08/24/c 136550296.htm
93. https://dzta7c9ve7na4.cloudfront.net/wp-content/uploads/sites/2/2017/12/01090324/Alert-International-Arbitration-30-11-17.pdf
94. United Nations Commission on International Trade Law (Please hyperlink this to https://en.wikipedia.org/wiki/ United Nations Commission on International Trade Law)
95. https://en.wikipedia.org/wiki/Solar power in Morocco
96. https://en.wikipedia.org/wiki/Ouarzazate Solar Power Station
97. https://en.wikipedia.org/wiki/Solar power in South Africa
98. https://www.smallstarter.com/browse-ideas/why-the-usa-and-africa-are-losing-out-on-trade-and-investment-opportunities/
99. https://en.wikipedia.org/wiki/Solar power in Chile.

CHAPTER

4

SOLAR ENERGY APPLICATIONS IN INDIA

Deepak Yadav and Rangan Banerjee

4.1 ACRONYMS

MNRE	Ministry of New and Renewable Energy
ISA	International solar alliance
COP21	Conference of the Parties 21
JNNSM	Jawaharlal Nehru National Solar Mission
PV	Photovoltaics
kWh	Kilowatt Hour
GW	Gigawatts
HTSS	Hydrothermal-Solar Scheduling
PLF	Plant Load Factor
CSP	Concentrated Solar Power
MW	Megawatts
NITI	National Institute for Transforming India, Aayog
NTPC	National Thermal Power Corporation
NVVN	National Vidyut Vyapar Nigam
SECI	Solar Energy Corporation of India
CEL	Central Electronics Ltd.
DCR	Domestic Content Requirement
SECI	Solar Energy Corporation of India
WTO	World Trade Organization
DISCOMs	Distribution Companies
CAPEX	Capital Expenditure
OPEX	Operational Expenditure
MSEDCL	Maharashtra State Electricity Distribution Company
DG	Diesel Generators
VAh	Volt Ampere hour
SBI	State Bank of India
CREDA	Chhattisgarh Renewable Energy Development Agency
PHC	Primary Health Care
CFA	Central Financial Assistance
NABARD	National Bank for Agriculture and Rural Development
SNA	State Nodal Agencies
CPSUs	Central Public Sector Utilities
LFR	Linear Fresnel Reflector
PPA	Power Purchase agreement

4.2 INTRODUCTION

India has a population of 1.1 billion people (1/6th of the world population) and accounts for less than 5% of the global primary energy consumption. The average electricity consumption in India has increased from 704 kWh/person/year in 2008-09 to 1149 kWh/person/year in the year 2017-18. However, this is significantly lower than in developed countries. For eg., the average electricity consumption in the US was 12,830 kWh/person/year in 2016. In India, most states have peak and energy deficits. In 2008-09, the average deficit was about 8.2% for energy and 12.6% for the peak. These deficits reduced in 2017-18. The average deficit now is about 0.8% for energy and 1.1% for peak power [1]. In 2018, India achieved 100% electrification of villages but not all households. It is estimated that out of 181 million rural households, 22% do not have access to electricity [2]. India's development strategy is to provide energy access to all households.

The scarcity of fossil fuels and climate change problem has resulted in an increased emphasis on renewable energy sources. India has a dedicated ministry focussing on renewables (Ministry of New and Renewable Energy, MNRE). MNRE oversees renewable energy installation and generation in the power sector. On a global scale, India took the lead in forming the International solar alliance (ISA) on the sidelines of COP21 in Paris in 2015 [3]. ISA is an alliance of 121 sunshine-rich countries that lie completely or partially between the tropic of cancer and tropic of Capricorn. The primary objective of ISA is to enable efficient utilization of solar energy in the world.

Figure 4.1 shows the share of renewable and solar power in the energy mix of the Indian power sector from 2001 to 2018. It can be seen that the share of renewable energy has witnessed a steady growth over the past decades. The main sources of renewable energy in the present supply mix are wind, solar, small hydro and biomass systems. As on 31st July 2018, the installed capacity of grid-connected renewables is more than 71 GW. India intends to have 175 GW of renewable energy installation by 2022 [4]. Out of this, it is expected that solar installations will achieve 100 GW.

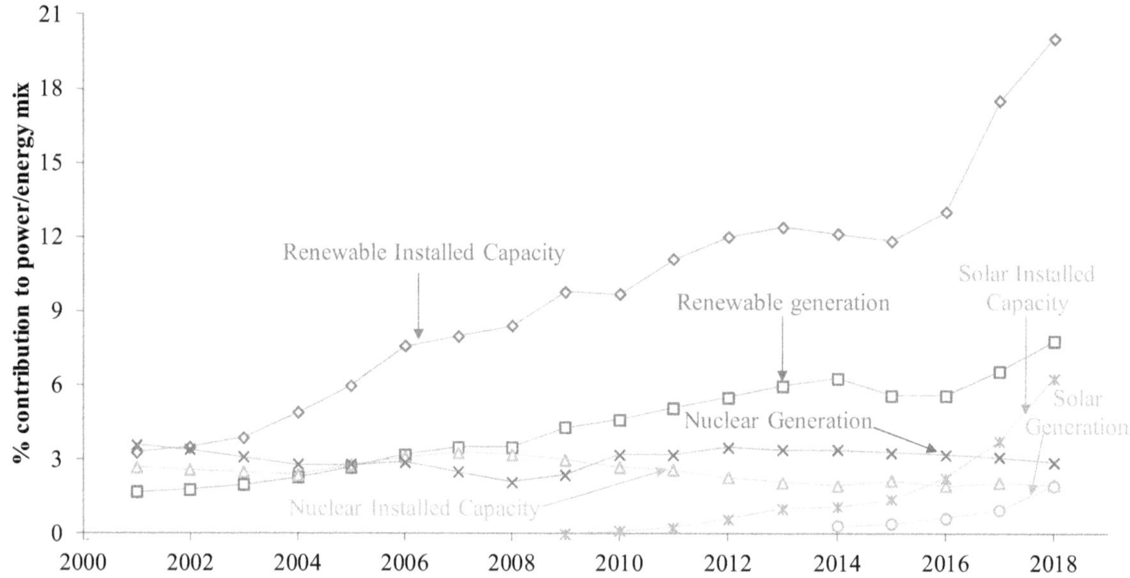

FIG. 4.1: SHARE OF RENEWABLE AND SOLAR POWER IN THE ENERGY MIX OF THE INDIAN POWER SECTOR [6].

The integration of renewable energy sources in the electric grid will be a challenge in the future due to the intermittent nature of power generation. This can be partially addressed by using algorithms like hydrothermal-solar scheduling (HTSS). HTSS assesses the impact of large-scale solar integration with a hydrothermal system, in terms of electricity mix, generation schedules and the annual generation costs [5].

The share of installed capacity and power generation from different sources in India is shown in Figure 4.2 [7]. It can be seen that the share of solar power in the installed capacity is 6.6%. However, the share of solar power in the electricity mix is only 2% due to lower plant load factor (PLF). The typical PLF of solar plants in India is 18-20%.

In 2010, India launched the Jawaharlal Nehru National Solar Mission (JNNSM) as a part of its climate change mission to develop cost-effective solar power solutions. The national solar mission has provided an impetus to the growth of solar installations in India. The cumulative installed capacity has increased from 12.8 MW in 2009 to 21,651 MW in 2018. This capacity addition has been primarily due to large-scale centralized plants. The solar-rooftop systems have not grown to the expected capacity. It is observed that out of the 21.6 GW of installed capacity, only 2.5 GW are rooftop systems while the rest are centralized power plants.

Most of India enjoys excellent solar radiation. Almost the entire country has insolation greater than 1900 kWh/m^2/year with about 300 days of sunshine. Figure 4.3 shows a map with the solar insolation ranges for different parts of the country. The cumulative installed capacity of solar power in a few significant states and six neighbouring countries (Afghanistan, Bangladesh, Myanmar, Nepal, Pakistan, and Sri Lanka) are also indicated in Figure 4.3.

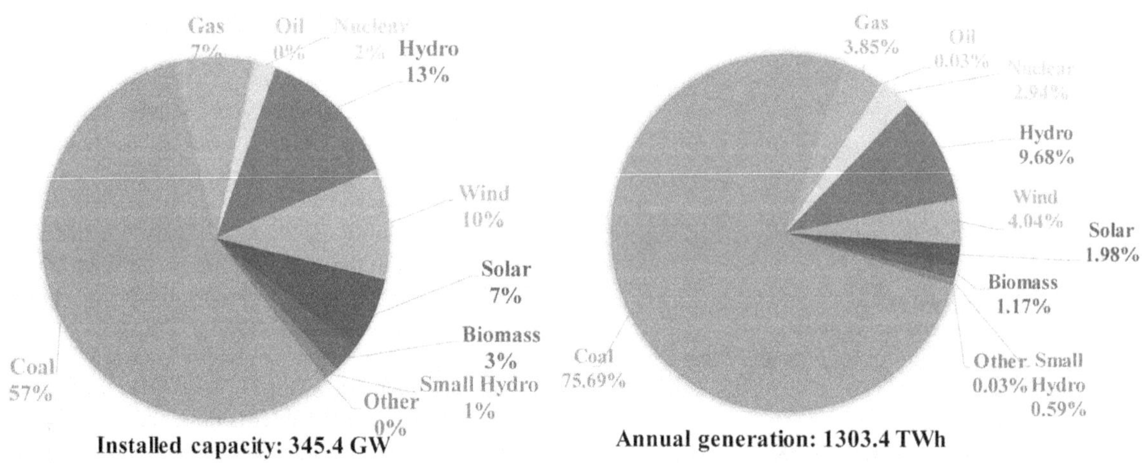

FIG. 4.2: SHARE OF POWER GENERATION IN INDIA (2017-18) [6].

FIG. 4.3: AVERAGE GLOBAL HORIZONTAL INSOLATION FOR INDIA AND NEIGHBOURING COUNTRIES [8].

4.3 APPLICATIONS OF SOLAR ENERGY

Figure 4.4 shows the pathways for end uses of solar energy. For solar energy applications, the two distinct routes are solar photovoltaics (PV) and solar thermal. The energy outputs of solar thermal technology are heat and power. In India, power generation from the PV route has witnessed rapid growth. In 2018, the cumulative installed capacity of PV systems exceeded 21000 MW while the operational capacity of concentrated solar power (CSP) plants in only 228.5 MW. Apart from power generation, solar thermal systems are indigenously available for water heating, drying, cooking, and cooling. However, except for solar water heating systems, solar thermal applications have had limited success.

4.4 JAWAHARLAL NEHRU NATIONAL SOLAR MISSION

Before the launch of the solar mission, the total installed capacity of solar power plants in India was only 12.8 MW. The interest in solar power generation was generated by the launch of the national solar mission on January 11, 2010. The implementation of the national solar mission was divided into three phases. The first phase was from 2010-2013, the second phase extended to 2017 and the third phase was from 2017-2022. The roadmap for achieving the large scale deployment of the solar system is shown in Table 4.1.

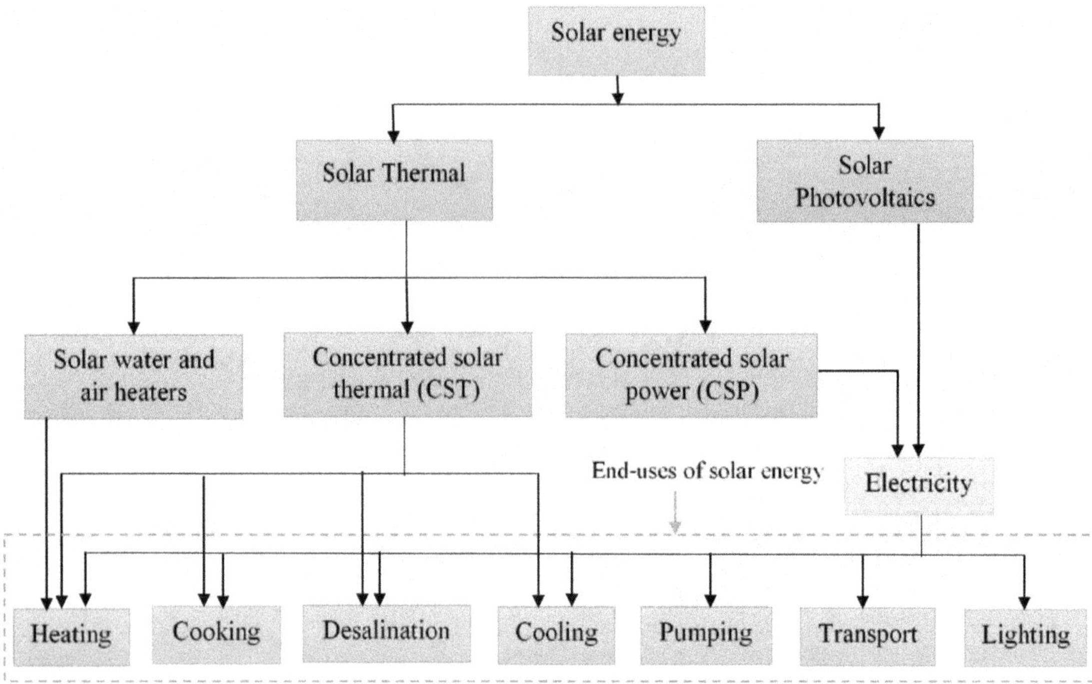

FIG. 4.4: PATHWAYS FOR END-USES OF SOLAR ENERGY.

The objectives of the solar mission were as follows [9]:

- To create a policy framework for the deployment of 20 GW of solar power by 2022.
- To create an indigenous manufacturing capacity for solar PV and thermal systems.
- To promote off-grid solar power by deploying 2 GW by 2022.
- To deploy 15 million m^2 of solar collectors by 2017 and 20 million m^2 by 2022.
- To deploy 20 million solar lighting systems for rural application by 2022.

Subsequently, the targets were revised in 2015 to deploy 100 GW by 2021-22 [10]. The revised annual targets are shown in Table 4.2. These targets will be met by installing 40 GW of rooftop and 60 GW of ground-mounted utility-scale power plants. Beyond 2022, NITI (National Institute for Transforming India) Aayog, a government 'Think Tank', shows the possibility of achieving 479 GW of solar power by 2047 [4].

TABLE 4.1

ROADMAP FOR DEPLOYING SOLAR SYSTEMS IN INDIA [9]

Sr. No.	Application segment	Target for phase 1 (2010-13)	Target for phase 2 (2013-17)	Target for phase 3 (2017-22)
1	Solar collectors	7 million m^2	15 million m^2	20 million m^2
2	Off grid solar applications	200 MW	1,000 MW	2,000 MW
3	Utility grid power, including roof top	1,000-2,000 MW	4,000-10,000 MW	20,000 MW

TABLE 4.2

REVISED TARGETS FOR DEPLOYING SOLAR SYSTEMS IN INDIA [10]

Category	Year-wise targets (MW)							
	2015-16	2016-17	2017-18	2018-19	2019-20	2020-21	2021-22	Total
Rooftop	200	4,800	5,000	6,000	7,000	8,000	9,000	40,000
Ground-mounted	1,800	7,200	10,000	10,000	10,000	9,500	8,500	57,000
Total	2,000	12,000	15,000	16,000	17,000	17,000	17,500	97,000*

*3,743 MW were commissioned till 2014-15

In India, the solar power plants are allotted based on the reverse bids invited by the nodal agency. The implementation of the first phase (2010-2013) was done by a subsidiary of National Thermal Power Corporation (NTPC), NTPC Vidyut Vyapar Nigam Ltd. (NVVN). Subsequently, the Ministry of New and Renewable energy (MNRE) set up the Solar Energy Corporation of India (SECI) [11] to act as the nodal agency for implementing the solar mission.

4.5 GRID-CONNECTED PV STATIONS

The Indian PV industry started in 1976 with the research and development efforts of Central Electronics Ltd. (CEL). In the 1980s, public sector undertakings like CEL, Bharat Heavy Electricals and Udaya Semiconductors manufactured solar cells in India. Since 1992, the manufacture of solar cells was liberalised. At present, the solar cell and module manufacturing is predominantly driven by large private sector companies like Jupiter solar, Indosolar, and Tata Power Solar Systems Limited [12].

Figure 4.5 shows the trend of increase in cell and module manufacturing capabilities in India. Till August 2017, the operating capacity of module manufacturing in India was 5506 MW. However, the solar cells manufacturing capacity was only 1667 MW[12]. This is because the module manufacturers typically sell modules that use imported solar cells.

Before the launch of the national solar mission, the domestic manufacturing companies used to export solar cells to European countries. However, the launch of the solar mission and cost reductions in China after 2010, lead to a surge in solar import. In the financial year 2017-18, India imported 89% of its solar requirement from China alone [14]. Therefore, the domestic cell manufacturing capacity has not grown to the extent of PV capacity added in the country. The indigenous cell manufacturing capability is under-developed due to unavailability of raw materials, lack of technical know-how, lack of large-scale demand for domestically manufactured cells, under-developed supply chain and unskilled workforce [13,15].

Figure 4.6 shows the trend of variation in net solar exports and imports in India. It can be seen that the revenues due to solar exports exceeded imports till 2011 and India was a net exporter of solar PV systems. From the beginning of 2011 to the end of 2012, the price of PV panels in China dropped from 1.8$/W to 0.65$/W [13]. Therefore, from 2011 onwards, the quantum of imports from China has increased significantly. The domestic manufacturing industry has not been able to compete with China in terms of costs. Therefore, India is now a net importer of solar systems. In 2017-18, solar import from China was worth 3,837 Million $. This was 5% of India's annual import from China and 0.82% of India's overall import bill.

The Indian government intends to revive the indigenous solar manufacturing capability. It has planned to achieve this objective by a three-pronged approach: the imposition of anti-dumping duty on imported solar cells and modules, creating demand through domestic content requirement (DCR) and supporting manufacturers by providing multiple subsidies [16]. In July 2018, India imposed a two-year safeguard duty on cells imported from China and Malaysia. The levy is 25% for the first year starting July 30. For the two subsequent six-month periods, the levy is 20% and 15% respectively [17]. The solar energy corporation of India (SECI) has invited bids for 5 GW/year solar manufacturing capability linked with power purchase agreements (PPAs). The maximum tariff for these bids was capped at 2.93 Rs./kWh for 25 years. However, only one manufacturer (Azure power [18]) bid for the auction [19]. Subsequently, in October 2018, the government postponed the manufacturing linked bids for the fourth time to ensure that these comply with the WTO norms [20].

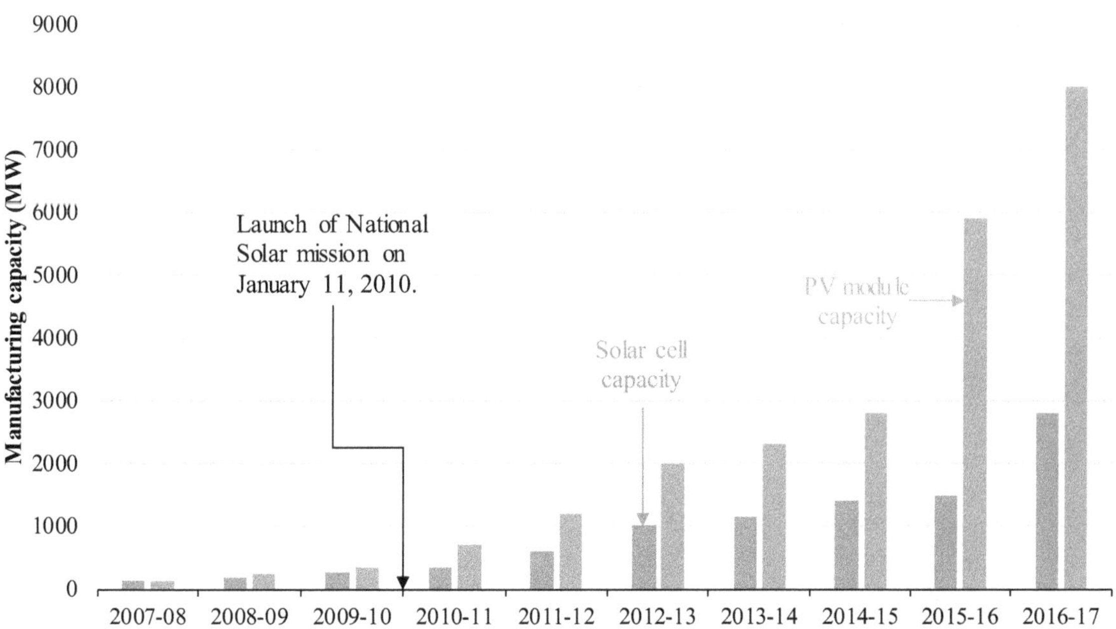

FIG. 4.5: SOLAR CELL AND PV MODULE MANUFACTURING CAPACITY IN INDIA [13].

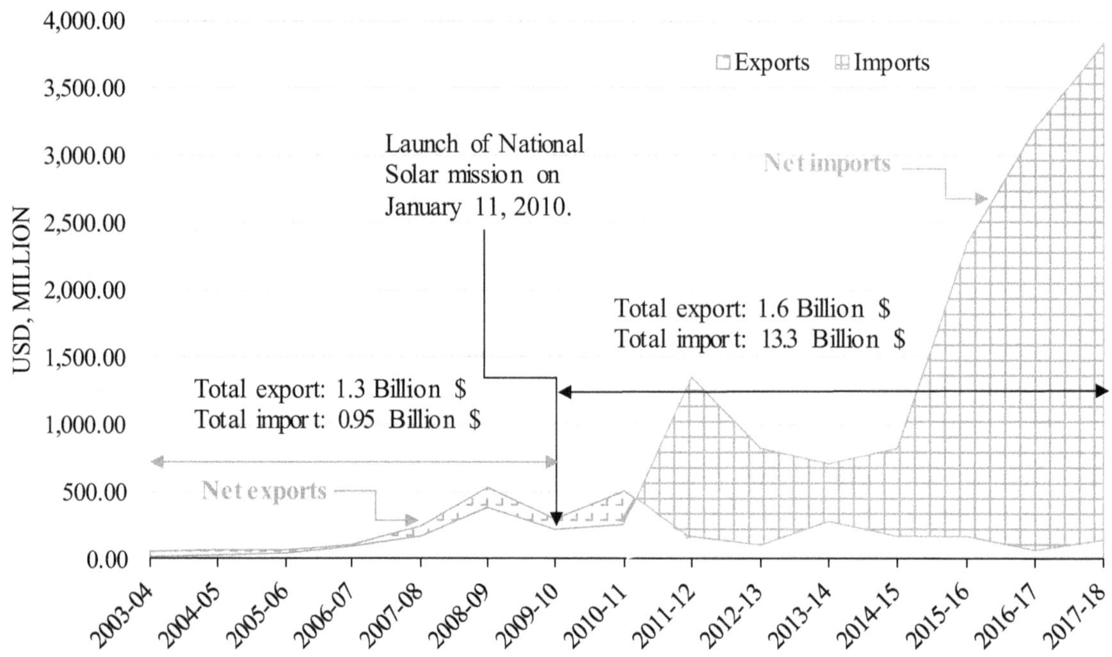

FIG. 4.6: INDIAN SOLAR PV IMPORTS AND EXPORTS [15].

In India, the solar plants are allotted based on tenders issued by SECI. These tenders are well received by the private sector and aggressive bids have been obtained. Figure 4.7 shows the trend of variation in the tariff bids and the cumulative installed capacity of solar power plants in India. Over the years, the tariff bids have significantly reduced due to the low price of panels imported from China. The lowest bid of Rs. 2.44/kWh was obtained for the development of a 200 MW solar plant at Bhadla Solar park, Rajasthan.

In 2014, the government announced setting up at least 25 Solar parks and Ultra Mega Solar Parks. The solar parks have all clearances, transmission networks, water access, road connectivity etc. Solar parks offer project developers 'plug-and-play' model with all facilities in exchange for a usage fee. These parks have been of particular interest to the international project developers who benefit from a pre-existing power evacuation facility and avoiding complex land acquisition process in India [21]. The average project size in solar parks is 110 MW. The large project size has led to market concentration among the project developers. The top 10 developers took 60% of the projects sanctioned in 2017-18[21].

SECI regularly auctions the blocks with these solar parks for installing PV plants. This scheme was developed to facilitate the

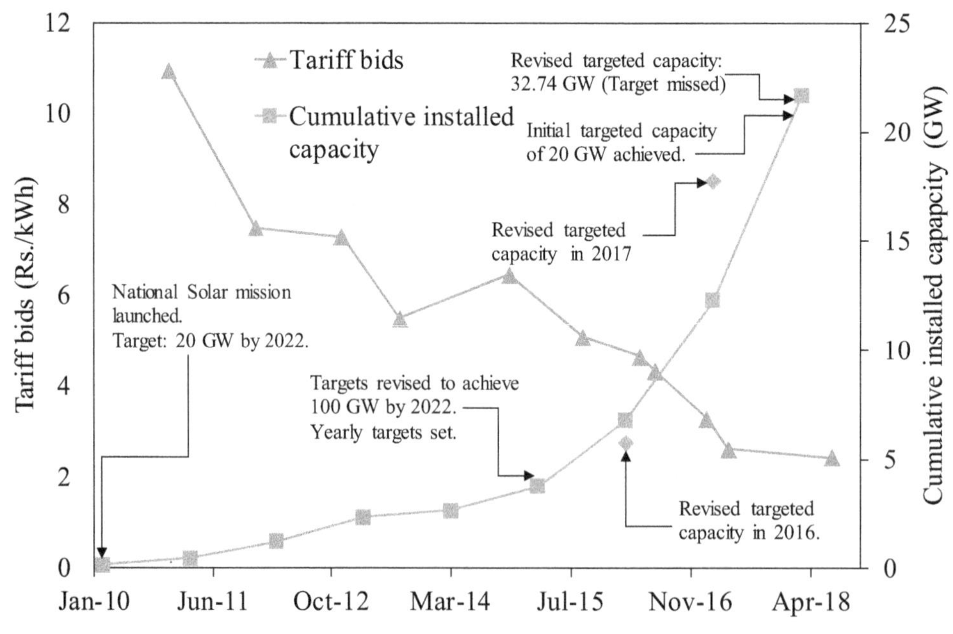

FIG. 4.7: TARIFF BIDS AND CUMULATIVE INSTALLED CAPACITY OF PV PLANTS IN INDIA.

TABLE 4.3

LIST OF ULTRA MEGA SOLAR POWER PLANTS IN INDIA [22]

Sr. No.	Plant name	Capacity commissioned (MW)	Approved capacity (MW)	Tariff (Rs./kWh)
1	Ananthapuramu-I Solar Park, Andhra Pradesh	1000	1500	2.71-3.57
2	Kurnool Ultra Mega solar park, Andhra Pradesh	1000	1000	4.63-5.13
3	Rewa Ultra Mega solar, Madhya Pradesh	750	750	2.97
4	Bhadla solar park, Rajasthan	746	2255	2.44 - 4.36
5	Pavagada solar park, Karnataka	600	2000	2.82-4.79
6	Kamuthi Solar power project, Tamil Nadu	648	648	5.1-7.01
	Upcoming Solar Parks			
7	Dholera Solar Park, Gujarat	–	5000	–
8	Kadapa Ultra Mega Solar Park, Andhra Pradesh	–	1000	–
9	Odisha Solar Park, Odisha	–	1000	–

development of solar parks with a capacity exceeding 500 MW. Till June 2018, the government has approved 45 solar parks with a total capacity of 26449 MW [22]. The government now intends to install 40 GW solar power by 2019-20 in these solar parks. Table 4.3 lists the various solar parks in the country. The approved capacity, capacity commissioned, and the tariff bids are also indicated. A few solar parks in Dholera, Gujarat and Odisha are under approval phase and will be auctioned after receiving the necessary clearances.

Adani Green Energy commissioned one of India's largest PV plant at Kamuthi, Tamil Nadu [23]. The plant has a capacity of 648 MW and was built with an investment of Rs. 45.5 Billion. The plant covers an area of 2500 acres and was completed within a record eight months. The tariff for 313 MW commissioned before 31st March 2016 was Rs. 7.01/kWh. The tariff for the remaining 335 MW capacity added beyond this date was Rs. 5.1 /kWh [24]. Figure 4.8 shows a picture of the solar power plant.

It can be seen that the solar tariff has reached a record low of Rs.2.44 /kWh. However, the state distribution companies (DISCOMs) are not keen on purchasing solar power even at the lowest price of Rs. 2.44/kWh. The DISCOMs are mandated to ensure that a fixed percentage of their energy consumption is from renewable sources. This is termed as renewable purchase obligation (RPO).

FIG. 4.8: 648 MW SOLAR PLANT AT KAMUTHI, TAMIL NADU [23].

The solar RPO target for 2018-19 is 6.75%. This has been revised and for the years 2021-22, the solar RPO target is 10.5%. Thus, the increasing solar RPO targets for DISCOMs was expected to drive the growth of solar installations in India. However, it is observed that 16 states and union territories achieved less than 60% of their RPOs in 2016-17 [25].

The DISCOMs in many states have signed long-term PPAs with the thermal power stations. These PPAs have a fixed cost component and a variable cost. The fixed cost has to be paid even if no electricity is drawn from the thermal station. For eg., A state pays Rs. 1.60/kWh (fixed cost) and Rs. 1.65/kWh (variable cost) [26]. Since the fixed cost has to be paid, the effective price of solar power even at the lowest bid of Rs. 2.44/kWh, becomes higher {Rs. 4.04(2.44+1.60)/kWh} than the amount that thermal stations charge {Rs. 3.25(1.60+1.65)/kWh} for fossil-based electricity.

This has created a situation where there are no buyers of renewable energy. If SECI, the nodal agency for solar power implementation in India, accepts the solar bids and no PPAs are signed for solar power, then it has to pay Rs. 40 lakhs/year/MW to the project developers. This had led a condition where SECI is not inviting bids for solar installations. As a result of this, 4000 MW of solar bids have been scrapped and very PPAs are now being signed. The situation has further aggravated because the Government has levied 25% tariff on solar imports with the intent of boosting domestic manufacturing. This has caused a further increase in solar tariffs.

The Government plans to address these issues by amending the Electricity Act, 2003. According to the proposed amendments, a penalty of Rs. 1-5/kWh will be levied on the DISCOMs in case of non-compliance with the renewable PPAs [27].

4.6 ROOFTOP SOLAR

The Government of India has announced an ambitious target of setting up 40 GW solar power on rooftops by 2022. The road-map for achieving this has been announced, and annual targets for each state has been allotted [28]. Maharashtra, Uttar Pradesh, Tamil Nadu, Gujarat, and Rajasthan are expected to have a total cumulative installation of 18 GW by 2022. The Government plans to achieve the rooftop target by installing solar PV systems in residential, commercial, Government and industrial sectors. The solar potential of various sectors in a city can be obtained by using high-granularity land use data and GIS-based image analysis [29].

There are two business models for installation of solar rooftop systems in India: CAPEX (Capital Expenditure) and OPEX (Operational Expenditure) models. The CAPEX model is a self-funding model where the consumer owns, finances, operates, maintains and consumes the energy generated by the PV system. In the OPEX model, a project developer is responsible for the installation and maintenance of the PV system. The developer enters into a long-term power purchase agreement with the host who consumes the power generated by the rooftop system. It is estimated that about 3/4th of the solar rooftop systems follow CAPEX model.

In India, a few states follow the net metering policy. Net energy is the difference between the energy consumed and energy supplied to the grid. In this policy, the excess energy produced by the rooftop systems during day time is fed to the electric grid. The owner is billed only for the net energy used. India, however, does not have a uniform policy for net metering systems. Net metering is mandatory in many states but is unregulated in states like Andhra Pradesh, Tamil Nadu and West Bengal. These states have legally non-binding guidelines [30]. This is a critical issue for

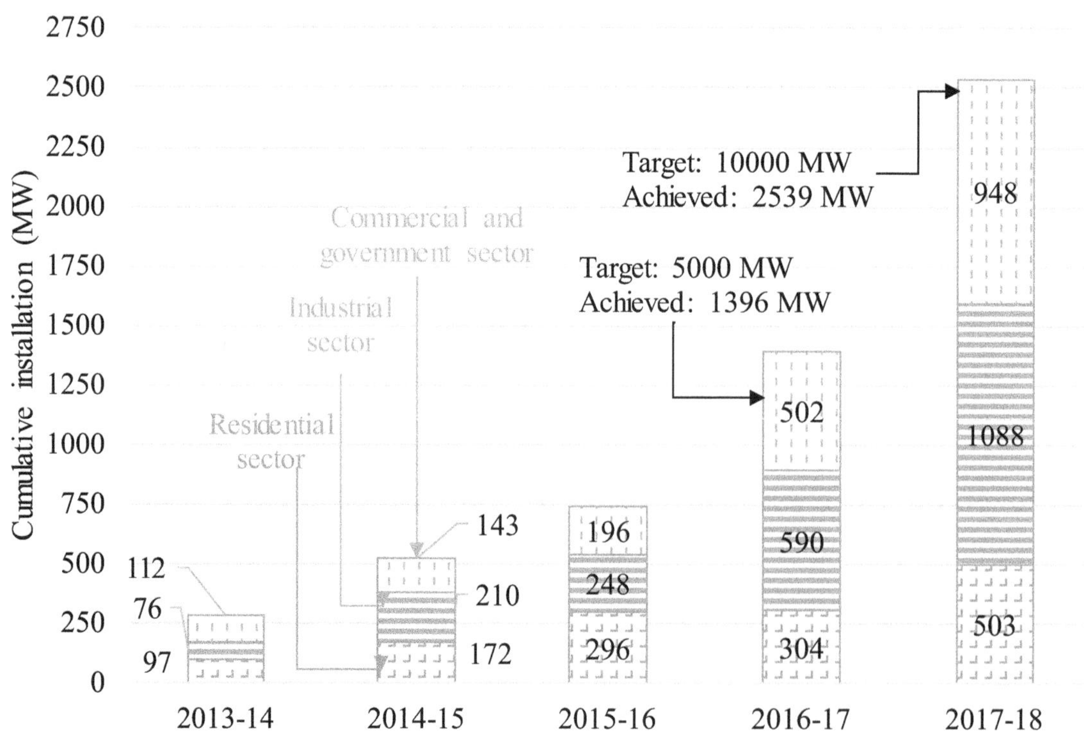

FIG. 4.9: CUMULATIVE INSTALLED CAPACITY OF ROOFTOP SOLAR PLANTS IN INDIA [32].

residential consumers who might not use solar power during day time when the production is higher. A few states have imposed a limit on the amount of power that can be fed into the grid, which further aggravates the situation [31].

Figure 4.9 shows the distribution of the cumulative installed capacity of solar rooftop systems in India [32]. It can be seen that there has been a steady growth in rooftop installations. However, the installed capacity is significantly lower than the annual targets set in 2015 [32]. Out of the annual installation target of 10 GW by 2017-18, only 25% has been achieved. This implies that by 2017-18, only 6.3% of the targeted capacity of 40 GW by 2022 has been added in the country.

The industrial and commercial sectors drive the growth in solar rooftop systems. The solar rooftop system is typically 17% cheaper than the average industrial tariffs [33]. Therefore, for industrial and commercial consumers, solar electricity is viable even without the net metering policy. The growth has been sluggish in the residential sector. The slow growth in the residential sector is due to cheap electricity for homeowners with low consumption, high capital cost and lack of financing options [30].

Most of the DISCOMs in India are state-owned. These DISCOMs typically charge a higher rate for high-end consumers and industries at Rs. 9-10/kWh and provide subsidised electricity to low-end consumers and farm sectors at Rs. 2.7-6/kWh. As a result of high tariffs, the industries and high-end consumers want to reduce their electricity bill by installing solar rooftop systems. DISCOMs, however, are not in favour of promoting solar rooftop systems because it could hurt their business prospects. The state-owned DISCOMS are under severe financial stress, and the total loss in 2017-18 was Rs. 173.52 Billion [34]. As a consequence of this, there is a clash between DISCOMS and high-end consumers. In fact, Maharashtra State Electricity Distribution Company (MSEDCL) plans to levy a surcharge of Rs. 1.25/kWh on solar prosumers [35]. To address these issues, the Government has formed SRISTI (Sustainable Rooftop Implementation for Solar Transfiguration of India) to bring solar prosumers and DISCOMs on a single platform [36]. The amendments in the Electricity Act, 2003 are also expected to address these issues.

To boost solar rooftop installations, the state governments have come up with a decentralized solar policy. A few states offer an additional subsidy to the rooftop installations. For eg., in Madhya Pradesh, the tariff bids for solar rooftop installation has reached Rs. 1.38/kWh for 8.6 MW grid-connected rooftop system [37,38]. This tariff is with the central and state governments subsidies of 23% and 20% respectively. The benchmark cost of grid-connected rooftop projects ranges from Rs. 53-60/W_p depending on the scale of installation [39].

The developments for rooftop installations in the residential, industrial and commercial sectors are discussed in the following sections:

4.6.1 Residential sector

The solar rooftop system in the residential sector is designed to generate electricity for households. The capacity typically ranges from 1 kW-5 kW depending on the connected load and area available for solar installation. The systems can either be grid-connected or stand-alone. A stand-alone system has battery backup while the grid-connected system can be installed with/without batteries. The battery capacity depends on the power outage frequency and duration. Net energy policy in the residential sector is likely to be successful in states that ensure a reliable power supply. In energy-deficient states, the consumer is likely to install a battery backup for offload periods. As seen in Figure 4.9, the cumulative installed capacity in the residential sector is only 503 MW. This capacity is projected to reach 5 GW by 2022 [40].

4.6.2 Industrial Sector

The typical capacity of solar installations in the industrial sector ranges from 100 kW-200 kW. This capacity is attractive for large scale developers. Typically, the solar power generated in industries is spontaneously consumed to meet lighting and air-conditioning loads. The inductive loads are usually fed grid power [40]. As seen in Figure 4.9, the industrial sector has the largest contribution to the solar rooftop installations in India. It is expected that about 44% of the remaining target will be met by solar installations in the industrial sector alone [33]. The solar installations in the industry are eligible for the net metering policy.

4.6.3 Commercial and government sector

The solar systems in the commercial sector include installations at petrol pumps, ATMs and telecom towers. The solar installations in government sectors are on canal-top, within establishments like army, ordnance factory, public sectors utilities (PSUs) etc. These installations are discussed briefly in this section.

4.6.3.1 Petrol pumps Petrol pumps have a continuous operation. However, the grid supply is either intermittent or not available at many remote locations. These petrol pumps, therefore, depend on diesel generators (DG) for meeting the power demand. The

TABLE 4.4

LIST OF SOLAR INSTALLATIONS AT PETROL PUMPS [40,42]

Parameters	Site 1	Site 2	Site 3	Site 4	Site 5	Site 6
Owner	IOCL	IOCL	Essar	IOCL	HP	IOCL
Location	Shadnagar, Andhra Pradesh.	Kodada, Andhra Pradesh.	Aurangabad, Maharashtra.	Panipat, Punjab	Panipat, Punjab	Chehal, Punjab
Solar capacity (kW)	2	2	3	3	6	2
Battery capacity (Ah)	180	180	180	180	150	150
No. of batteries	8	8	16	15	16	16

TABLE 4.5

LIST OF SOLAR INSTALLATIONS IN TELECOM TOWERS [40]

Parameters	Site 1	Site 2	Site 3
Operators	Vodafone	Single Operator	Reliance, BSNL, Vodafone
Location	Soldha, Haryana	Kolar, Karnataka	Sundarban, West Bengal
Power source	Grid, Diesel and Solar	Grid and Solar	Grid, Solar, Wind and Diesel
Total Load (kW)	1.2	1	3
Grid availability (hours/day)	10	9	6
Battery capacity VAh	48	48	48
	600	6000	3600
DG rating (kVA)	15	-	20
DG operation (hours/day)	Not used	Not used	2

cost of DG power is about Rs. 17/kWh [40]. Solar systems offer a viable and clean alternative to the DG sets. The typical loads at a petrol station are pumps, lighting, and cooling systems. This power requirement can be met by installing solar systems with battery backup. Table 4.4 lists some of the solar installations at petrol pumps in India. Indian Oil Corporation Limited (IOCL), Hindustan Petroleum (HP), and Bharat Petroleum Corporation Limited (BPCL) have solar-powered 25%, 9% and 7% of their respective petrol pumps in the country [41].

4.6.3.2 Telecom towers Similar to petrol stations, telecom towers are exposed to intermittent grid supply or are located in a remote location that has no access to grid electricity. The government has also mandated that 75% and 33% mobile towers in rural and urban areas respectively have to be powered by a mix of grid and renewable energy systems by December 2020 [43]. In this context, solar-powered systems become important to ensure mobile connectivity throughout India. Table 4.5 lists a few hybrid/solar powered telecom towers in India.

4.6.3.3 Miscellaneous commercial applications In India, solar PV systems have been installed in Automated Teller Machines (ATMs), hospitals, schools, and universities. State Bank of India (SBI) has installed solar panels on 1,200 ATMs and targets to have 10,000 solar powered ATMs by 2020. SBI has also installed a rooftop system on 150 buildings and has targeted another 100 rooftop projects by 2020 [44].

In the health sector, the Chhattisgarh Renewable Energy Development Agency (CREDA) has installed off-grid solar systems in 570 out of 800 primary health care (PHC) systems in the state. The solar-powered PHCs in Chhattisgarh have improved the quality of healthcare. For eg., the PHCs have reported a 59% increase in outpatient services, a 78% increase in deliveries and a 45% improvement in laboratory services [45]. Till June 2016, solar rooftop installations in hospitals and institutes were 2.4 MW and 8.3 MW respectively [39].

4.6.3.4 Defence sector The solar potential in military cantonments and ordnance depot are 5000 MW and 950 MW respectively. The government has planned to install 300 MW solar power in various establishments under the ministry of defence. Till 31st December 2017, the total installed capacity of solar PV systems in the defence sector was 22 MW. The total allotted capacity is 357 MW [46]. The solar tariff is Rs. 4.5/kWh.

4.6.3.5 Solar power plants on canal banks and tops The government has approved the installation of 100 MW grid-connected solar PV systems on canal banks and canal tops. It is expected that 50 MW will be installed on canal tops and 50 MW on canal banks. Till 31st December 2017, the total installed capacity on canal top and banks was 18 MW and 31 MW respectively [46].

4.6.3.6 Solar power plants in Central Public Sector Utilities (CPSUs) The government intends to install 1000 MW of grid-connected solar PV systems in CPSUs and government organizations. Till 31st December 2017, about 765 MW capacity has been commissioned [46].

4.7 OFF-GRID SOLAR PHOTOVOLTAIC SYSTEMS

The off-grid applications of PV systems include solar lamps, home lighting, and pumping systems. The off-grid solar systems are generally used in rural areas where people have low purchasing power. Therefore, the government is providing central financial assistance (CFA) for off-grid and decentralized power generation. The state nodal agencies (SNA) are responsible for implementing and disbursing 30% CFA received from the government. There is a separate implementing agency, National Bank for Agriculture and Rural Development (NABARD), for distributing solar pumps and lighting systems. NABARD provides 40% CFA on the benchmark cost of solar pumps and lighting systems.

In India, solar lighting systems were developed in the 1970s. Figure 4.10 indicates the cumulative installations of off-grid solar systems. It can be seen that solar street lights and home lighting systems had steady growth over the years. Since 2014, the government has been promoting solar pumping systems for irrigation. The government has sanctioned more than 2.4 lakh pumps. Out of the sanctioned capacity, about 1.47 lakh pumps have already been installed. There has also been a strong emphasis on solar home lighting systems. The government has already distributed 2.32 million solar lamps and lanterns in the rural area and targets to distribute an additional 7 million solar lamps to school going children.

Figure 4.11 shows the state-wise distribution of off-grid solar power in India. These numbers are available from state nodal agencies. The off-grid solar power by other agencies like NABARD is also indicated. The total installed capacity till 31st December 2017 was 182 MW.

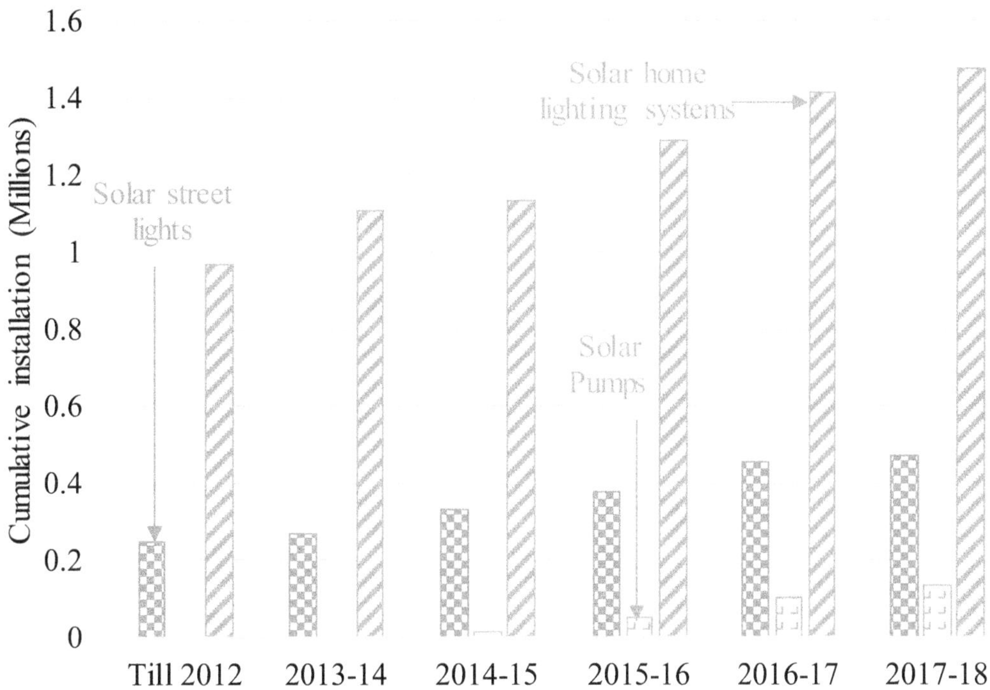

FIG. 4.10: CUMULATIVE INSTALLATIONS OF THE SOLAR STREET LIGHT, PUMPS, AND HOME LIGHTING SYSTEMS.

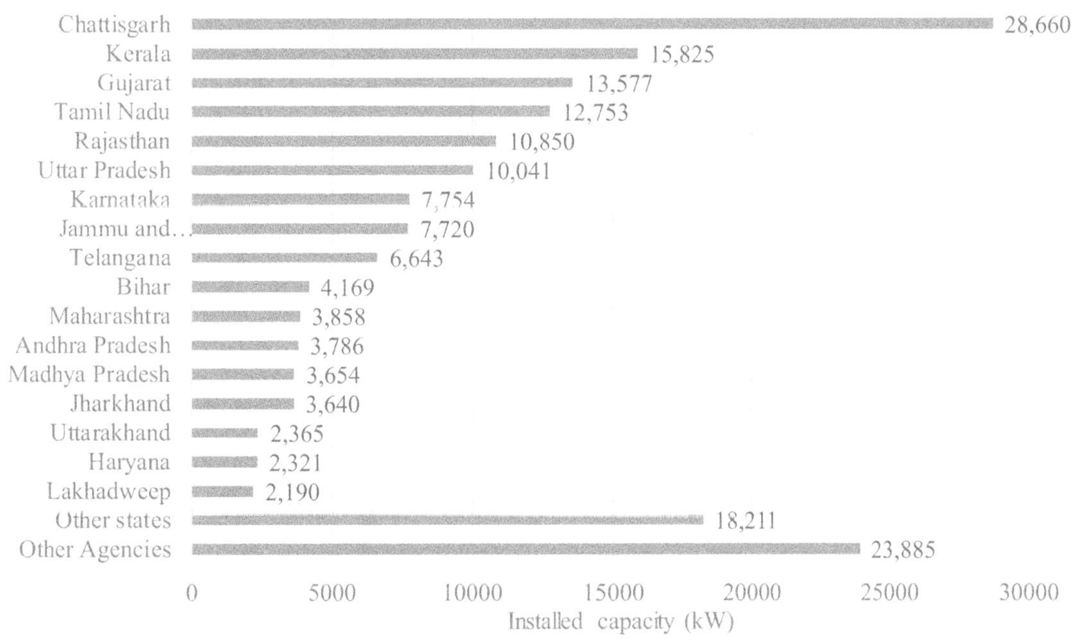

FIG. 4.11: STATE-WISE DISTRIBUTION OF OFF-GRID SOLAR POWER.

4.8 VILLAGE ELECTRIFICATION AND ISOLATED LOAD MANAGEMENT USING SOLAR PV

Microgrids are small-scale power-grids that can provide a reliable power supply to inhabitants of remote locations where grid electricity is either intermittent or not available. Microgrids, therefore, are important for enabling rural electrification. Depending on the resource availability, the renewable energy based mini-grids are powered by solar PV, wind, micro-hydel and biomass systems.

The government of India has launched a national policy for renewable energy based micro- (<10 kW) and mini- (>10 kW) grids. Till 2022, the ministry intends to deploy at least 10000 renewable energy based mini-grids with a minimum installed capacity of 500 MW. Till July 2018, under the solar off-grid and

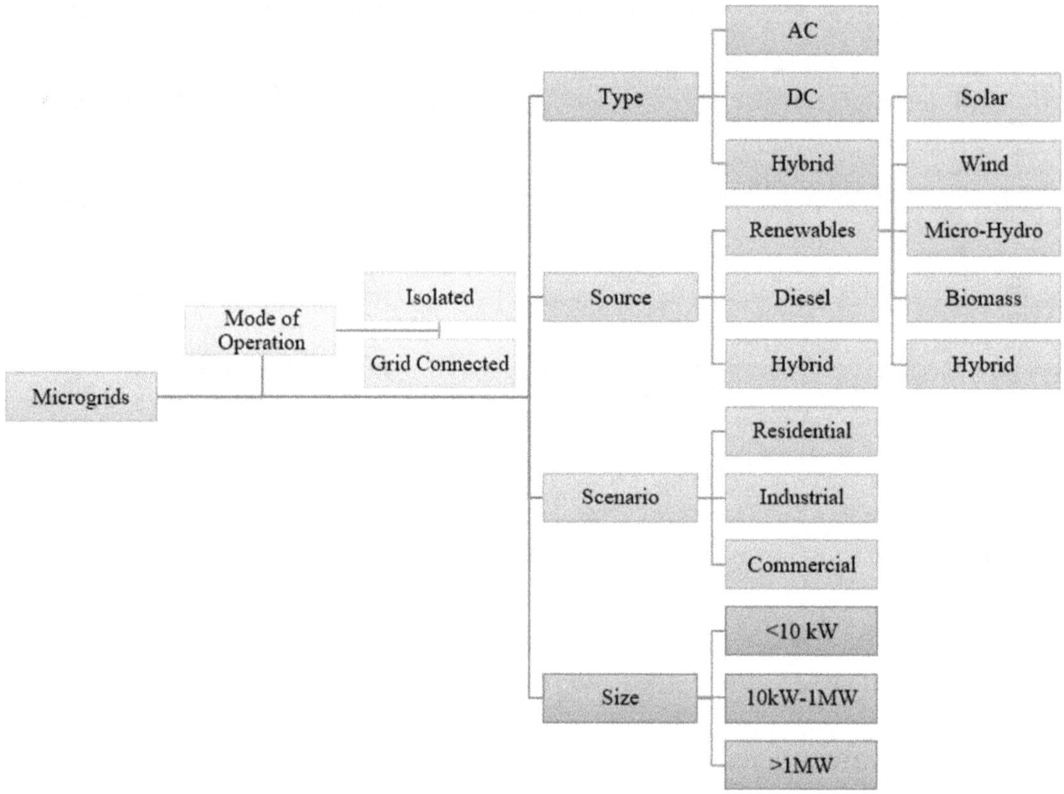

FIG. 4.12: CLASSIFICATION OF MICRO-GRIDS.

decentralized applications program, India has installed 63 solar-based microgrids totalling to 1899 kW [47].

Figure 4.12 shows the classification of micro-grids. On a broad level, the Microgrids are classified based on the mode of operation (Isolated/Grid connected), type of power output (AC/DC/Hybrid), source of power (Renewable/Hybrid), use (residential/commercial) and size (kW/MW). The source of electricity typically depends on geographic location and resource availability. A few locations might have a hybrid configuration to boost grid reliability.

The concept of PV-based mini-grids was first demonstrated at Sagar Island, West Bengal and Lamni Village, Chattisgarh [48]. Based on the initial success, several successful solar PV distributed generation systems were installed in a few islands. West Bengal Renewable Energy Development Agency (WBREDA)

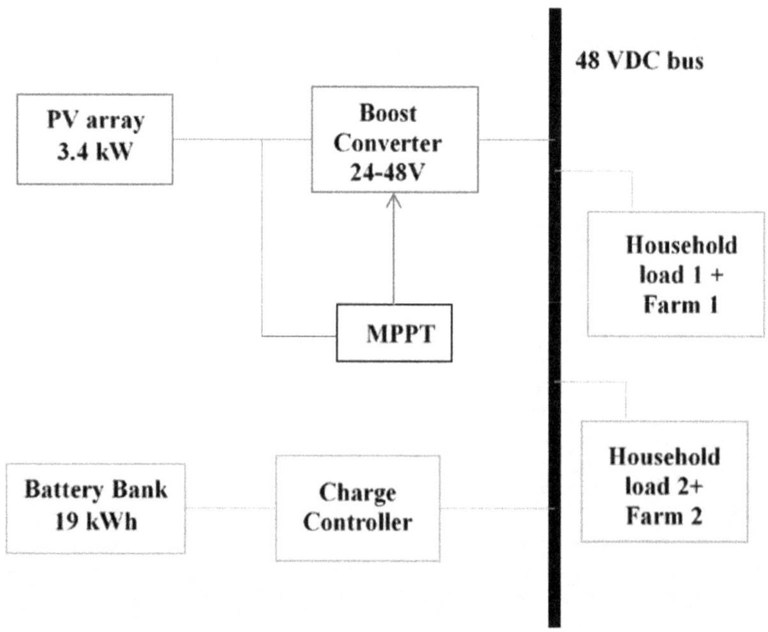

FIG. 4.13: SCHEMATIC OF AN ISOLATED RURAL LOAD WITH PHOTOVOLTAIC PLANT [51].

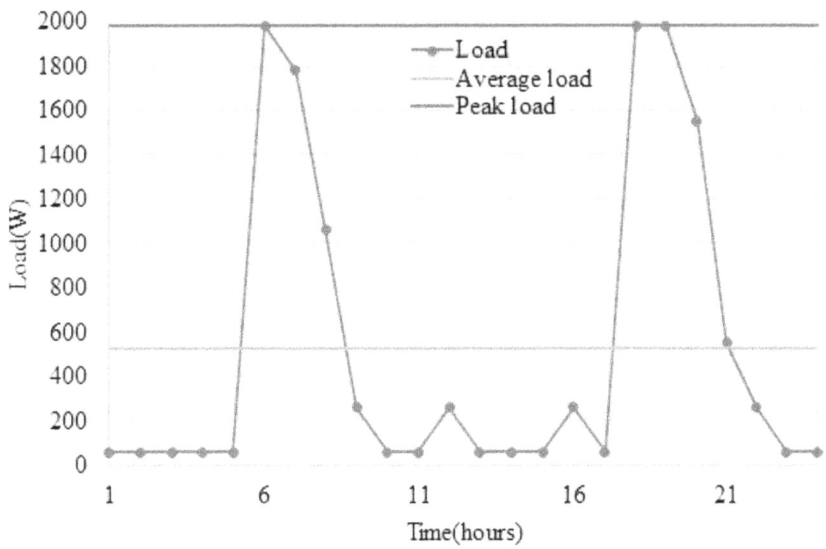

FIG. 4.14: LOAD PROFILE TAKEN OVER A DAY [51].

has installed nine solar mini-grids in the Sunderbans. The system has a peak rating of 345 kW and meets the energy requirement of 1750 consumers. A 5.5 kW PV-based system was installed in the Neechli Babhan Village, Rajasthan. The system has a capital cost of Rs. 2.5 million and provides electricity to 80 households [49]. The electricity output is stored in batteries that provide 24-hour reliable power. The project developers availed a 30% subsidy from the government on the capital cost and collected Rs. 0.28 million from the users as a connection charge. The users have a wireless prepaid meter and pay a tariff of Rs. 31.25/kWh. The cost of power generation is Rs. 27/kWh.

Figure 4.13 shows a schematic of the PV power system, along with the voltage values. Typically, Microgrids have interconnected loads, distributed generation and an energy storage device like the battery. Figure 4.14 shows the load profile of a typical PV-based isolated power system. It is seen that the average load is 0.5 kW, while the peak load is 2 kW. The total energy consumption in a day is 12.2 kWh. One of the main issues in an isolated system is matching the supply with the demand. The demand variability and its effect on system size (PV plant, batteries) in an isolated power system can be optimized by pinch analysis and design space approach [50].

Several attempts have been made to improve the capacity factor and the economic viability of isolated grids by linking with a constant/base load. In many of the island grids, solar photovoltaic systems have also been hybridised with the existing diesel micro-grids. In 2006, BHEL installed a 50 kW (peak) PV system along with a diesel engine generator and battery storage (two 75 kVA diesel generators) at Bangaram island in Lakshwadeep [52]. The system provides a 24-hour power supply to the island. Large scale (>100 kW) mini-grid commissioned in Sundarban, West Bengal have hybrid PV-wind and PV-Biomass gasifier systems [48]. A similar case is of Bhamane village in Karnataka that has a PV plant hybridized with a micro-hydel system. The system provides a reliable power supply to 32 households consisting of 160 people. Table 4.6 lists a few solar PV based Microgrids in India.

TABLE 4.6

SAMPLE ISOLATED SOLAR PV POWER PLANTS IN INDIA [40]

Location	PV capacity (kW)	Inverter capacity (kVA)	Battery capacity (Ah)	Battery voltage (V)	Connected load (kW)	Number of households
Chadoi, Odisha	0.4	-	200	24	-	13
Rajganga Hamelt, Odisha	0.4	-	200	24	-	14
Darewadi, Pune	9.36	12.5	600	48	-	40
Rajanganga, Odisha	6	6	500	48	-	34
Kanaka, Odisha	5	5	600	48	-	39
Baguli, Odisha	2.5	2	500	48	-	32
Nurda village, Jharkhand	28	20	1200	120	9.5	350

TABLE 4.7

LIST OF SOLAR PV-DRIVEN RO PLANTS IN INDIA [53]

Sr. No.	Technology	Developer	Capacity (LPD)
1	Kotri, Rajasthan	Manthan	5000
2	Sinodiya, Rajasthan	Manthan	5000
3	Mordikala, Rajasthan	Prayatna Sansthan and Manthan	5000
4	Khoda and Behrampur, U.P.	SauryaEner Tech	7000
5	Silowata, Rajasthan	PrayatnaSansthaan	5000

4.9 SOLAR DESALINATION SYSTEMS

Solar photovoltaic systems can be used to power the reverse osmosis (RO) plants. The RO plants can be used to obtain potable water either from brackish water or sea water. A typical solar-RO plant consists of PV modules, batteries, charge-controllers, pumps, inverter, storage tanks, sand-filter, and RO system. The PV-driven RO plants are especially popular in villages around Sambhar lake, Rajasthan that suffer from drought and saline groundwater. These villages have water councils that collect a one-time deposit of Rs. 500 and charge monthly maintenance fee of the system. Table 4.7 lists a few PV driven RO plants in India.

Barefoot College, Kotri in 2006 installed the first solar-RO system in India. The system cost was Rs. 1.5-1.6 Million. The solar RO system installed at Silowata and Mordi cost Rs. 3.5 Million in 2012. It is observed that the major maintenance cost of the PV-based system is due to the frequent replacement of the RO membrane. The RO membrane typically costs Rs. 30,000-35,000 [53].

4.10 SOLAR WATER HEATING SYSTEMS (SWHS)

Solar water heating systems (SWHS) are used for heating water for household consumption. In India, more than 90% of solar water heaters are installed in the urban area. The penetration levels are 2% in urban areas and 0.1% in the rural areas. It is observed that the residential sector accounts for 80% of the total solar water heater demand. Other potential users are hotels and hospitals.

Figure 4.15 shows an estimation of the impact of electric water heaters on the load curve for Pune in Maharashtra. It can be seen that electric water heating contributes significantly to the morning peak of the utility. Solar water heaters can effectively reduce morning peak. Some municipalities like Bangalore, Thane, and Kalyan are making solar water heaters compulsory for new housing units, hotels, hospitals. The estimated potential for solar water heating in India is 60 million m^2. This will lead to annual electricity savings of 12.2 billion kWh [54].

Figure 4.16 shows the trend of growth in the cumulative installed capacity of solar water heaters in India. Before 2006, the solar water heaters market was dominated by the Flat plate collectors (FPCs), and there were no installations of Evacuated tube collectors (ETCs). To boost the growth of solar hot water systems, the scheme 'Accelerated Development and Deployment of Solar Water Heating System in Domestic, Industrial and Commercial Sectors' was started in 2008. The government set a target of installing 1.4 Million m^2 of solar collectors during 2008-2010 and allotted Rs. 495 Million for the scheme. The cumulative installed capacity of solar water heaters increased from 2.15 Million m^2 in 2008 to 3.5 Million m^2 by the time national solar mission was launched in 2010 [56]. By 2015, the actual installation of solar hot water systems was about 9.54 million m^2 [57]. Out of this, the total installed capacity of evacuated tube collectors (ETC) was 5.91 million m^2. The rest was based on FPC technology [57]. ETCs are now more popular than the FPCs.

Solar thermal water heaters are manufactured, installed and serviced by indigenous manufacturers. Till 22nd November 2014, there were 63 Bureau of Indian Standards (BIS) approved manufacturers of flat plate collectors (FPC), and 55 MNRE approved manufacturers (assemblers) of evacuated tube collector (ETC) systems in the country. However, since the launch of the national solar mission, the indigenous manufacturers are struggling to compete with cheap imports from China.

TABLE 4.8

INDUSTRIAL APPLICATIONS OF SOLAR FLAT PLATE SYSTEMS

Sr. No.	Location	Application	Comments
1	Panchmahal dairy, Godha	20,000 lpd 80° C preheating boiler feed water	236 flat plate collectors storage 20,000 litres savings 110 litres/day furnace oil
2	Synthokem Labs Pvt. Ltd., Hyderabad	10,000 lpd boiler feed water preheating	16% reduction in energy cost, 2 years payback
3	Kangaroo Industries, Ludhiana	Electroplating 100 lpd	-
4	GFTCL Kakinada	120,000 lpd Boiler feedwater preheating	1309 collectors, Payback period 5 years

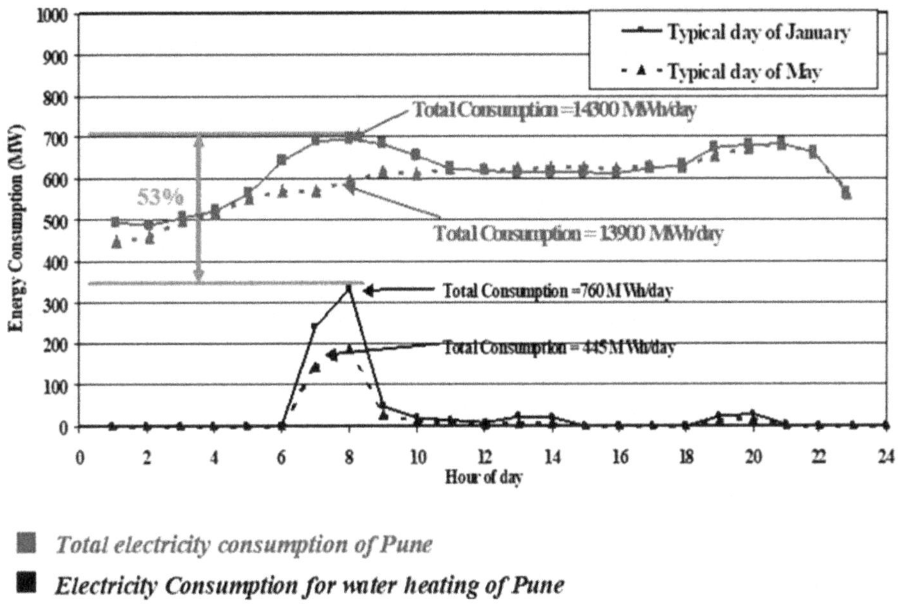

FIG. 4.15: LOAD CURVE REPRESENTING ENERGY REQUIREMENT FOR WATER HEATING [55].

The MNRE benchmark cost of evacuated tube collectors (ETC) and flat plate collectors (FPC) are Rs. 8000-8500 and Rs. 10000-11000 per m^2 respectively. A typical 250 litres per day household system costs about Rs. 25,000 (US 347$) in India. Since solar water heaters have commercialized, MNRE has discontinued the subsidy on these systems with effect from 1st October 2014.

Many industries have low-temperature process heat requirements (hot water, steam) which are ideal for the application of solar thermal systems. In India, there have been several installations of flat plate and evacuated tube collectors for preheating boiler feedwater and low-temperature process heating. Table 4.8 provides a few examples of industrial solar flat plate water heating systems in the country.

Unlike solar water heaters, flat plate collector based solar air heaters are not very popular, and the total installed capacity in India is only 11,900 m^2 [57]. There are only 8 manufacturers of flat plate collectors based air heaters that are empanelled with the MNRE.

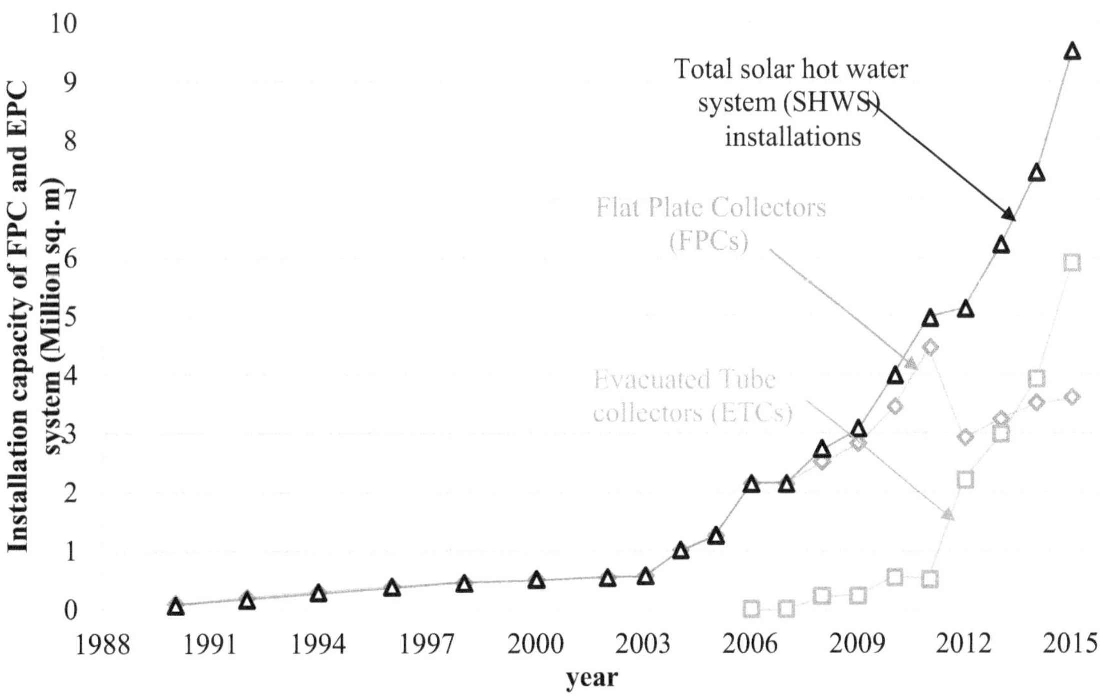

FIG. 4.16: GROWTH OF SOLAR WATER HEATING SYSTEMS IN INDIA [57].

4.11 CONCENTRATED SOLAR THERMAL (CST) SYSTEMS

India consumed 204 million tons of oil in 2016-17, which is 4.81% of the global consumption. A significant amount (~40%) of this is consumed in industries. CST systems can reduce the use of fuel oil in textile, chemical, food, and leather industries. Several indigenous manufacturers are developing solar concentrating systems that generate steam either through parabolic trough concentrators (KIE Solatherm and Thermax), Fresnel paraboloid dish (Clique), Scheffler (Gadhia solar) or Compact Linear Fresnel Reflectors (KG Design Services, Coimbatore). MNRE has empanelled 34 manufacturers/suppliers of CST systems.

To promote CST systems, the government has targeted cumulative installation of 90,000 m^2 under the scheme of "Off-grid and Decentralized CST technologies for Community Cooking, Process Heating and Space Cooling Applications in Industrial, Institutional and Commercial Establishments" [58]. Figure 4.17 shows the state-wise distribution of cumulative installed capacity of CST systems in India. It is observed that most of the projects in India have central financial assistance (CFA) from MNRE while a few projects

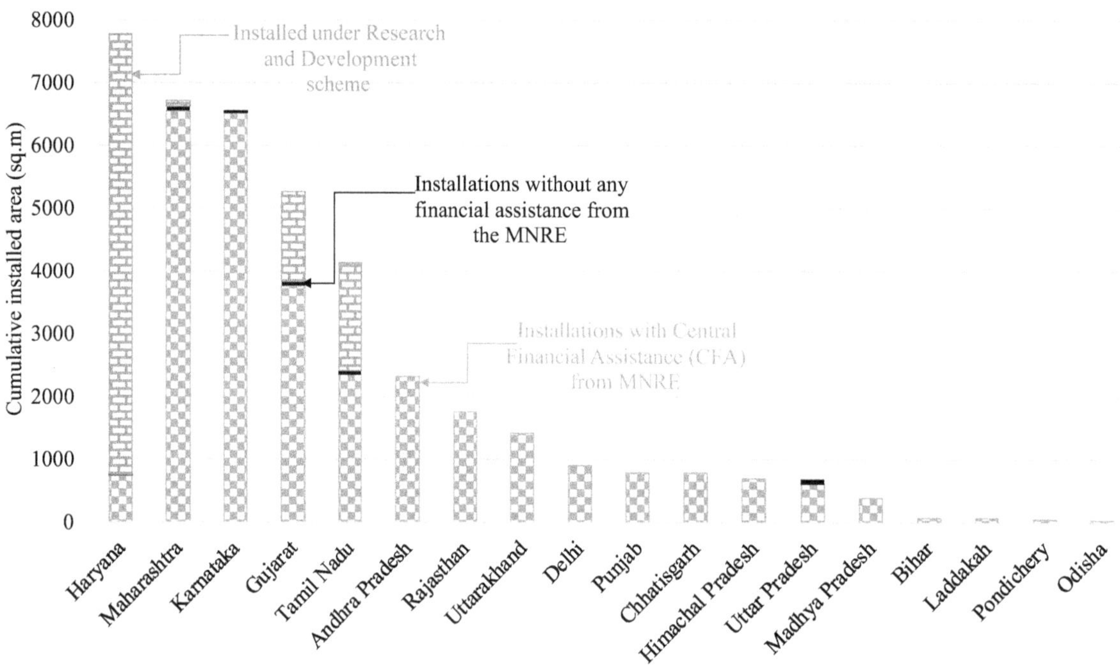

FIG. 4.17: STATE-WISE INSTALLED CAPACITY OF CST SYSTEMS IN INDIA (TILL DECEMBER 2014).

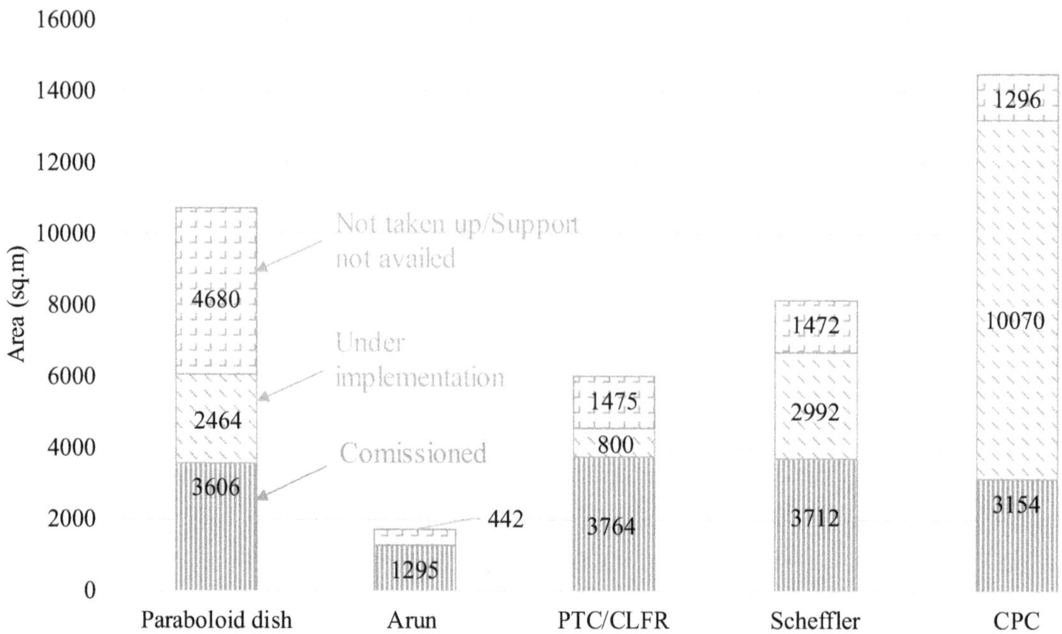

FIG. 4.18: TECHNOLOGY-WISE STATUS OF THE PROJECT [60].

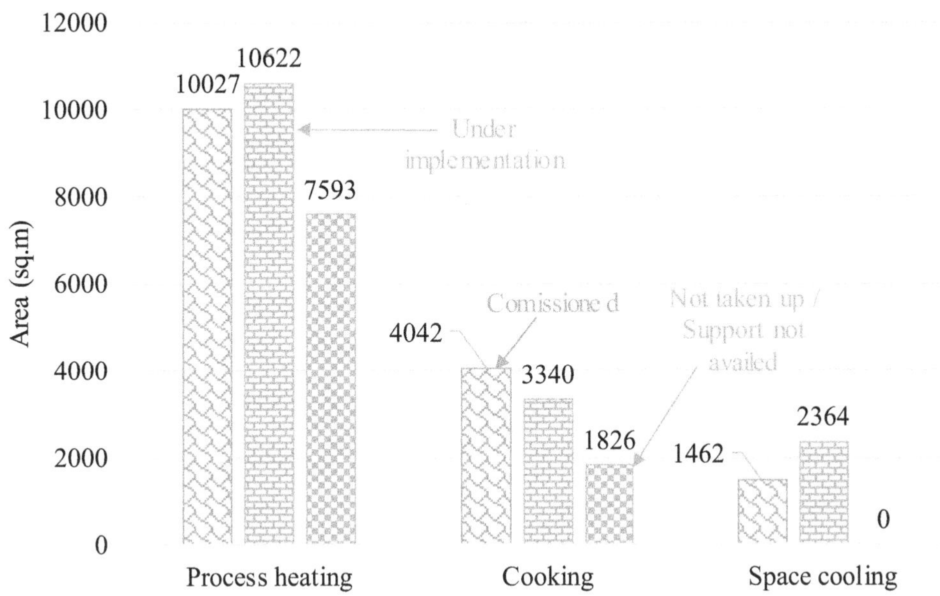

FIG. 4.19: PROCESS-WISE STATUS OF THE PROJECT [60].

are installed under research and development scheme. Most of the research projects are installed at The National Institute of Solar Energy (NISE), an autonomous research institute located in Gurgaon, Haryana.

There are very few CST projects without any assistance from the MNRE. This is because the cost of CST systems in India is very high. A few components like high-efficiency low-iron glass are imported which increases the system cost. The benchmark cost of CST technologies ranges from 7000-2000 Rs./m^2 depending on the type of tracking system. MNRE provides 30-60% subsidy on the benchmark cost of the CST systems. The amount of subsidy depends on the state in which the system is installed. There are also additional benefits for UNDP-GEF projects [59].

Despite the high cost, the government has made efforts to increase the penetration of CST technologies in the industry. The UNDP-GEF project on 'Market development of concentrating solar technologies for process heat applications' was initiated in April 2012. The target was installing 45,000 m^2 CST system by 2017. Figures 4.18 and 4.19 show the technology and process wise installation of CST systems in India under this scheme.

From Figures 4.18 and 4.19, it is observed that compound parabolic concentrator (CPC) is the most popular technology while process heat has been the widely accepted application of CST technology. There are a significant number of projects that have not taken up, or where no support has been availed. It is seen that cooking, space cooling and process heating are the major applications of CST technology. Therefore, a detailed discussion on these applications of CST technology is done in the following sections:

4.11.1 Solar thermal cooking systems

In India, the first solar cookers were developed by the National Physical Laboratory in 1954 with a circular parabolic reflector focussing on to a cooking pot. Subsequently, box type solar cookers were developed in 1961. It is estimated that about 0.64 million solar cookers have been installed in the country [61]. MNRE has empanelled 29 suppliers for manufacturing and selling of box and dish type solar cookers.

In the early 1990s, the solar dish cooker for community application (Scheffler cooker) was developed and manufactured in India. The aperture area of the dish was about 7 m^2. Figure 4.20 shows a schematic of the Scheffler cooker. The automatically tracked parabolic reflector is located outside the kitchen and reflects the incident solar energy through an opening in its mouth wall. A secondary reflector is used to focus the solar radiation on to the cooking vessel. Till March 2017, the total installed capacity of solar reflectors for cooking was 4042 m^2. Another 3340 m^2 are under the implementation phase [60].

CST systems are suitable for community kitchens and hostels. Table 4.9 lists some of the large community installations of solar cooking in India. The cooking system installed at Tirumala Tirupati Devasthanam, Andhra Pradesh is considered to be one of the world's largest solar cooking system. It consists of 106

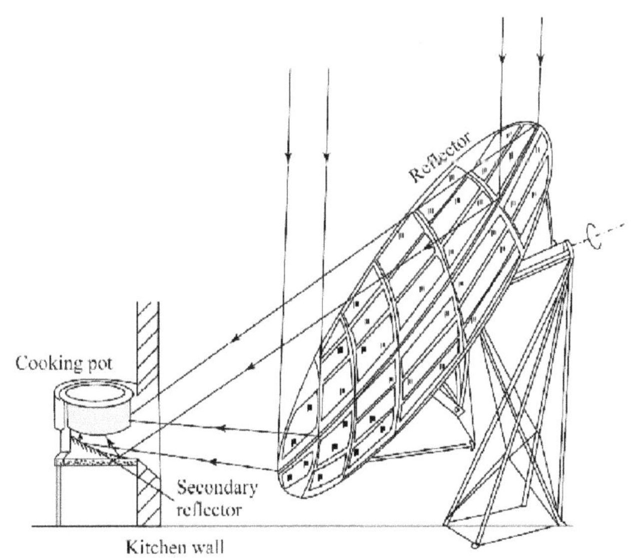

FIG. 4.20: SCHEMATIC OF A SCHEFFLER COOKER [62].

TABLE 4.9

LARGE SOLAR COOKING INSTALLATIONS IN INDIA

Sr. No.	Location	Cooking capacity people/day	Collector area
1.	Sri Sai Baba Sansthan, Shirdi, Maharashtara	40,000	73 reflective dishes of 16 m^2 each (1168 m^2)
2.	Tirupalli, Andhra Pradesh	15000	106 collectors of 9.2 m^2 each (975 m^2)
3.	Tamil Nadu agricultural university, Tamil Nadu	1000	10 dishes of 25 m^2 each (250 m^2)
4.	Brahmakumaris Ashram, Talleti, Mount Abu	10,000	800 m^2
5.	Brahmakumaris, Gurgaon	2000	265 m^2
6.	Auroville, Pondicherry, Tamil Nadu	1000	15m diameter solar bowl concentrator

automatically tracking parabolic concentrators that generate 4 tonnes of steam per day at 10 kg/cm^2, 180° C. steam. The system was installed in 2002 with a capital cost of Rs. 11 million. The government provided a 50% subsidy for this plant. It is estimated that the system saves 118,000 litres of diesel per year.

In 2016, ten dishes of 25 m^2 each (total area 250 m^2) were installed at Tamil Nadu agricultural university by A.T.E enterprises [63], Pune [60]. The total cost of the project was Rs. 8.2 million. These dishes have heat storage and cook food for 1000 students even during non-sunshine hours. The system saves 130 kg of LPG per day. The simple payback period of the system is three years.

4.11.2 Solar thermal cooling systems

CST systems can be used for space cooling and refrigeration. The cooling demand in various sectors is at a peak during the time of the day when solar energy is available. India is blessed with abundant solar energy that can be coupled with industrial vapour absorption machines (VAM) to provide comfort cooling.

Table 4.10 lists some of the CST based cooling systems in India. The solar system installed at Mahindra Vehicles, Pune consists of 70 Scheffler dishes with a total area of 1120 m^2. The Scheffler dishes produce steam at 8.5 bar, 167 °C, which is used to operate a 120 TR double-effect vapour absorption machine (VAM). uring monsoons, the Scheffler dishes yield low-grade heat in the form of hot water at 70 °C. This hot water is utilized in the LPG vaporization tank. The system results in electricity savings of 200 MWh/year and mitigates 140 tons CO_2/year. The monetary benefits are Rs. 1.5 million/year. This implies a payback period of 8.3 years.

4.11.3 Solar thermal systems for industries

CST systems find application in various industries for meeting the industrial process heat demands. A system of imported line focussing parabolic concentrators for producing process steam at 150°C and 100 kg/hr was installed at a silk factory in Mysore in 1987 [64]. Scheffler dishes have been used to generate saturated steam at 5 bar (20 dishes of 12.64 m^2) in a demonstration project supported by the Ministry of New and Renewable Energy at the Global Hospital and Research Centre, Mount Abu [65]. The overall system efficiency was about 28%.

TABLE 4.10

CONCENTRATED SOLAR THERMAL (CST) BASED COOLING SYSTEMS IN INDIA [59]

Sr. No.	Location (Year of commissioning)	Capacity (type of VAM)	Solar technology (total area)	Solar output	Solar cost (Million Rs.)
1	Civil hospital, Thane, Maharashtra (2011)	212 TR with Double effect	Scheffler, 2040 m^2	Steam at 7 bar, 150 °C	39.9
2	NPCIL, Kota, Rajasthan (2013)	100 TR with Triple effect	PTC, 641 m^2	Water at 17 bar, 200 °C	24.5
3	Honeywell Tech., Hyderabad (2013)	100 TR with Triple effect	PTC, 821 m^2	Water at 17 bar, 200 °C	21.3
4	Mahindra Vehicles, Pune, Maharashtra	120 TR with Double effect	Scheffler, 1120 m^2	Steam at 8.5 bar, 167 °C	12.5
5	Turbo Energy Ltd., Chennai (2011)	50 TR with Double effect	Dish, 338 m^2	Water at 15 bar, 180 °C	8

TABLE 4.11

INDUSTRIAL APPLICATIONS OF CONCENTRATED SOLAR THERMAL SYSTEMS

Sr. No.	Plant Location	Industry use	Fuel Saved	Solar technology (total area)	Solar output	Solar cost (Million Rs.)
1	Hindustan Vidyut Products Ltd., Haryana	Vulcanization of cables	Diesel	Scheffler, 320 m^2	Water at 10 bar, 85°C	5.5
2	ITC hotel, New Delhi	Laundry	Natural Gas	Dish, 338 m^2	Steam at 8 bar, 175 °C	10
3	ITC hotel, New Delhi	Laundry	Natural Gas	Scheffler, 128 m^2	Steam at 8 bar, 120 °C	3.5
4	Chitale dairy, Pune, Maharashtra	Pasteurization of Milk	Furnace Oil	Dish, 320 m^2	Steam at 4.5 bar, 150 °C	7.2
5	SKF Technologies, Mysore, Karnataka	Phosphating tank	Diesel	PTC, 256.4	Pressurized water at 130 °C	7

A Fresnel paraboloid dish (ARUN160) with two-axis tracking has been developed by Clique [66] and installed at Mahananda dairy, Latur in Maharashtra. This is a 160 m^2 dish that is used to replace steam fed from a 1 tonne/ hour boiler. Figure 4.21 shows a schematic of its integration into the dairy utility system. The boiler can be used as a backup for days where there is inadequate energy from the solar system, resulting in a reliable heat supply system without any need for storage.

Table 4.11 lists a few industrial application of CST systems. The solar system installed at Hindustan Vidyut Products Ltd., Haryana consists of 20 Scheffler dishes with a total area of 320 m^2. The industry uses heat for vulcanization of cables. In this process, the cable-core is dipped in hot water at 85 °C for 18 to 24 hours. Before installation of the solar thermal system, the industry used a diesel boiler for heating the thermic fluid to 125°C. This thermic fluid was circulated in the coils that heated the tank to maintain the desired temperature. Now, the solar thermal system has been integrated with the diesel boiler to meet the heating requirement of the industry. The total capital cost of the system was Rs. 5.5 million. MNRE provided a grant of Rs. 1.65 million. The solar system saves 270 litres of diesel/day.

4.11.4 Solar thermal desalination

Solar stills have been traditionally used for obtaining potable water from saline or brackish water. These systems, however, suffer from low efficiency (<35%) and low productivity

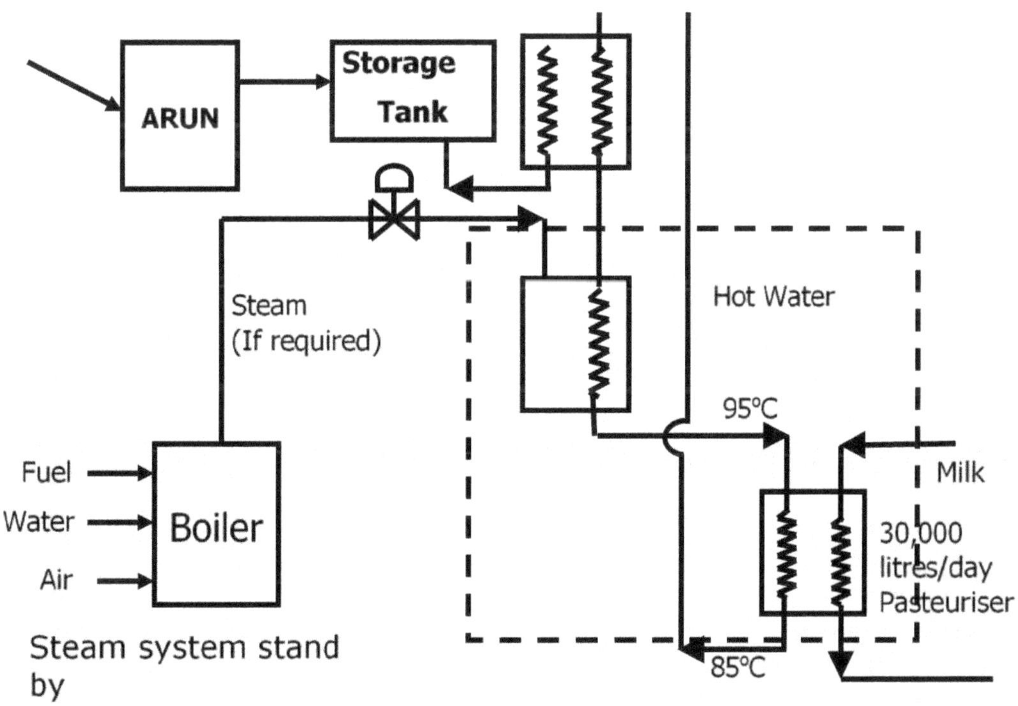

FIG. 4.21: SOLAR SYSTEM FOR A DAIRY IN MAHARASHTRA.

(3-4 L/m²/day) [53]. Concentrating solar collectors are developed as an alternative to solar stills. These systems heat water to boiling temperature for producing steam. The steam is then condensed in a condenser to obtain potable water. The solar collectors can also be coupled with multiple effect flash and distillation chambers to achieve higher efficiency.

In 2013, KGDS renewable energy Pvt. Ltd., commissioned an LFR (Linear Fresnel Reflector) system coupled with multi-effect desalination (MED) unit to produce 6000 LPH of desalinated water at Ramanathapuram, Tamil Nadu [67]. The total collector area is 1404 m². Currently, the group is installing a solar desalination system that will provide 120 TPD of potable water at NTPC thermal power station, Tamil Nadu.

4.12 SOLAR THERMAL POWER GENERATION

Before the launch of National solar mission, India had limited experience in solar thermal power generation. There was a 50 kW line-focusing parabolic trough collector (PTC) demonstration plant (imported) at the NISE, Gwalpahari (near Delhi) by the Ministry of New and Renewable Energy. Due to maintenance problems, the unit was derated and then combined with a biomass gasifier system. An integrated solar combined cycle plant with a 35 MW solar component (based on line focussing parabolic collectors) was planned at Mathania in Rajasthan in the early 1990s. However, this plant was not installed due to problems in obtaining multilateral funding. Demonstrations units based on imported paraboloid dishes have been installed near Hyderabad (20 kW rating) and at Vellore (10 kW rating).

The launch on the national solar mission in 2010 revived interests in the CSP technology. The phase 1 of the mission had a target capacity of 1000 MW. The idea was to provide equal emphasis on CSP and PV technologies, and there was 50:50 allocations for these technologies. NVVN, the nodal agency for phase 1, invited proposals for setting-up 500 MW CSP plants under the reverse bidding mechanism. Subsequently, NVVM signed PPAs with 7 companies for installing CSP plants with a total capacity of 470 MW.

Apart from these seven projects, three projects with a capacity of 10 MW each were selected under the migration scheme taking the total approved capacity to 500 MW. All projects were planned to be commissioned by May 2013. However, none of the project developers were able to commission the plants within the prescribed time-frame. The first CSP plant under the national solar mission was commissioned by Godawari Green Energy Ltd. in June 2013.

The tariffs, quoted in the reverse bidding of CSP plants, ranged from 10.49-12.24 Rs./kWh. These tariffs were 25-31% lower than the CERC benchmarked tariff of 15.31 Rs./kWh [56]. The global tariffs at that time ranged from 12.42 Rs./kWh in South Africa to 18.90 Rs./kWh in Spain. Only Morocco, with CSP tariff of 10.21 Rs./kWh, had a value lower than those quoted in India [68]. Therefore, it was observed that the bidders quoted unviable tariffs with the only intent of obtaining orders. As a result of this, only three plants with a cumulative capacity of 225 MW have been commissioned till 2018. The other 4 bidders are looking to exit the solar thermal business by selling the plants.

In addition to these large-scale projects, there are three pilot demonstration of CSP technology in India. Acme India commissioned the first pilot-scale CSP plant in April 2011 [69]. The plant

TABLE 4.12

CSP PLANTS IN INDIA [56]

Sr. No.	Plant	Size and technology	Tariff	Current Status
1	Godawari Green, Jaisalmer, Rajasthan	50 MW, PTC	12.20	Commissioned in June 2013.
2	Reliance power, Jaisalmer, Rajasthan	125 MW, LFR	11.97	Commissioned in November 2014.
3	MEIL Green Power, Ananthapur, Andhra Pradesh	50 MW, PTC	11.31	Commissioned in November 2014.
4	Lanco Solar, Jaisalmer, Rajasthan	100 MW, PTC	10.49	Under construction (Looking for buyers)
5	KVK Energy Ventures, Jaislamer Rajasthan	100 MW, PTC	11.20	Under construction
6	Corporate Ispat Alloy/Abhijeet, Jaisalmer, Rajasthan	50 MW, PTC	12.24	Under construction (Looking for buyers)
7	Aurum, Porbandar, Gujarat	20 MW, PTC	12.19	Under construction
8	Acme Solar, Bikaner, Rajasthan	2.5 MW, Heliostat	-	Commissioned in 2011, Plant under-performing.
9	National Solar Thermal power testing at simulation facility, Gurgaon, Haryana.	1 MW, PTC and LFR	-	Commissioned in May 2014.
10	Brahmakumari's India One	1 MW, Scheffler	-	Commissioned in 2017

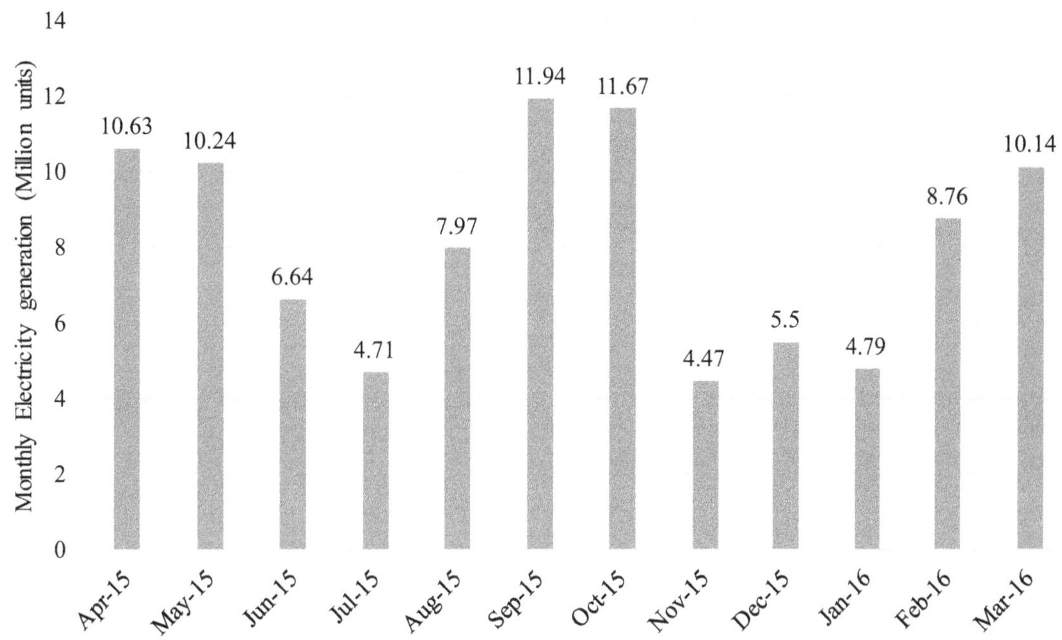

FIG. 4.22: MONTHLY ELECTRICITY GENERATION OF THE GODAWARI CSP PLANT [73].

is located in Bikaner, Rajasthan and is based on the heliostat technology. In October 2013, Indian Institute of Technology Bombay (IIT Bombay) commissioned a 1 MW research and development plant in the campus of NISE, Gurgaon [70,71]. In 2017, Brahmakumari's 'India one CSP plant' was commissioned at Mount Abu, Rajasthan [72]. The plant has 770 Nos. of 60 m^2 parabolic dishes that are equipped with thermal storage and provide on-demand power and heat to 25,000 people on the campus. A detailed list of various project, their size, tariff, and the current status is indicated in Table 4.12. Till date, the cumulative installed capacity of CSP projects in India is 228.5 MW.

Table 4.13 lists the configuration of the CSP plant constructed by Godawari Green Energy Ltd. The plant is based on a parabolic trough collector (PTC) technology and saves 82,228 tons of CO_2 emission annually. The captive power consumption is 9.3% of the annual output.

Figure 4.22 shows the monthly power generation from the plant during the year 2015-16. The plant load factor (PLF) is 22.25%.

It can be seen that when compared with the PV systems, CSP technology has not been successful in India. The possible reasons are as follows:

4.12.1 Unavailability of reliable direct normal irradiation (DNI) data

During the launch of the national solar mission in 2010, MNRE had no data monitoring station at the prospective project sites of the CSP power plants. Therefore, the bidders in Phase I of the solar mission relied on the DNI data available from institutes like NREL, CIEMAT, Solargis, etc. Over the installation period, it was observed that the measured actual DNI at the project site was lower than the estimates provided by these institutes. This had commercial implication as the bidders had to either increase the collector area or expect reduced annual power output. For eg., the annual DNI at the project site of Godawari Green CSP plant, Jaisalmer was 1753 kWh/m^2/year as opposed to the estimated 1825 kWh/m^2/year. This forced the project developer to increase the number of solar loops in the plant from 80 to 120 [56].

4.12.2 Limited experience in CSP technology and failure of plants in Phase I

The 50 kW PTC-based demonstration plant at NISE, Gurgaon in the 1980s was not successful. The 2.5 MW heliostat-based CSP plant, commissioned by ACME solar in 2011, did not yield the desired power output due to low solar radiation and optimization issues [56]. The experience of 2.5 MW plant was not encouraging, and the developers abandoned the idea of expanding to 10 MW under the migration scheme.

As discussed earlier, only 225 MW out of the total allotted capacity of 470 MW under the solar mission is operational. The remaining developers quoted unviable tariffs to win the bids. However, they were not able to execute the projects due to commercial nonviability and are looking for buyers. Thus, the CSP technology failed to generate momentum in Phase I of the solar mission. These factors have demotivated the project developers, and there are no bids for CSP plants in India.

TABLE 4.13

CONFIGURATION OF THE GODAWARI GREEN CSP PLANT [73]

Sr. No.	Parameter	Value
1	Turbine capacity (MW)	50
2	Aperture area (m^2)	3,92,400
3	No. of loops	120
4	Operating temperature (°C)	385
5	Heat rate (kcal/kWh)	2262.3
6	Cooling method	Wet cooling
7	Thermal storage	None

4.12.3 Lack of indigenous manufacturing and skilled workforce

The Indian developers do not have any experience in manufacturing of critical components like low-iron glass, vacuum receiver tubes, heat transfer fluid, tracking systems, and molten salts. These components are imported and have a significant impact on the overall cost of the CSP plant [68]. Similarly, there is a lack of technical know-how and talent pool for the development of CSP technology. There is also limited experience in the manufacturing of components like the structure and foundation of solar collectors etc.

4.12.4 Competition with PV technology

CSP technology has not been able to compete with PV technology in terms of cost. The PV systems have already achieved parity with the grid-based electricity. The tariffs of PV plants have dropped to Rs. 2.44/kWh while the project developers in CSP domain are struggling to construct plants with tariffs of Rs. 10.5-12.2/kWh. Therefore, there has been no interest in pursuing CSP technology in India.

A significant advantage for the PV systems over CSP technology is that they are modular in nature and do not need high-end technology developers for constructing the system. On the other hand, CSP technology is esoteric and needs skilled workforce for designing, construction, commissioning, operation, and maintenance.

The solar plants are typically constructed in areas of high irradiation. In India, these locations are typically arid and have low water availability. The water consumption in the wet-cooled CSP plant is 3.6 m^3/MWh [74] while that of the PV plant is only 0.08 m^3/MWh [75]. This is expected to be a significant factor in the arid or semi-arid areas of Rajasthan and Gujarat that are blessed with good solar radiation. The lower water consumption of PV systems adds to the existing commercial advantage over the CSP systems.

To boost the concentrated solar technology in India, MNRE constituted the 'CSP experts group' in 2016. The group was formed to recommend measures for promoting CSP/CST technologies. The group recommended auxiliary fuel support (hybridization), thermal storage, adequate land arrangement, permitting commissioning period of 3-5 years, using solar heat for fulfilling RPO requirement and providing fiscal and financial incentives for these technologies. However, these recommendations have not yet been adopted, and a final decision is pending.

4.13 SOLAR ENERGY IN NEIGHBOURING COUNTRIES

Figure 4.23 shows a plot of the cumulative installed capacity of solar power in neighbouring countries. Except for Pakistan and Sri Lanka, most of the installed capacity in these countries is for off-grid applications [76]. This section briefly discusses the status of solar power in these countries.

4.13.1 Afghanistan

In 2006, the Ministry of Rural Rehabilitation and Development (MRRD), Afghanistan and Ministry of External Affairs (MEA), India awarded a project to CEL, India for solar electrification of 100 villages in Afghanistan. Within the framework of this project, a total of 101 villages and 100 mosques were electrified by installing 5200 solar home systems (SHS) [77]. In 2017, a 2 MW solar-wind hybrid power plant was commissioned in the western Herat province [78]. The solar capacity in the hybrid plant was 1.7 MW. Recently, the Asian Development Bank (ADB) has provided a $44.76 million grant for setting up 20 MW grid-connected PV plant in Afghanistan [79]. The off-grid solar applications in Afghanistan include street lighting and diesel-hybrid power plants for military applications [80].

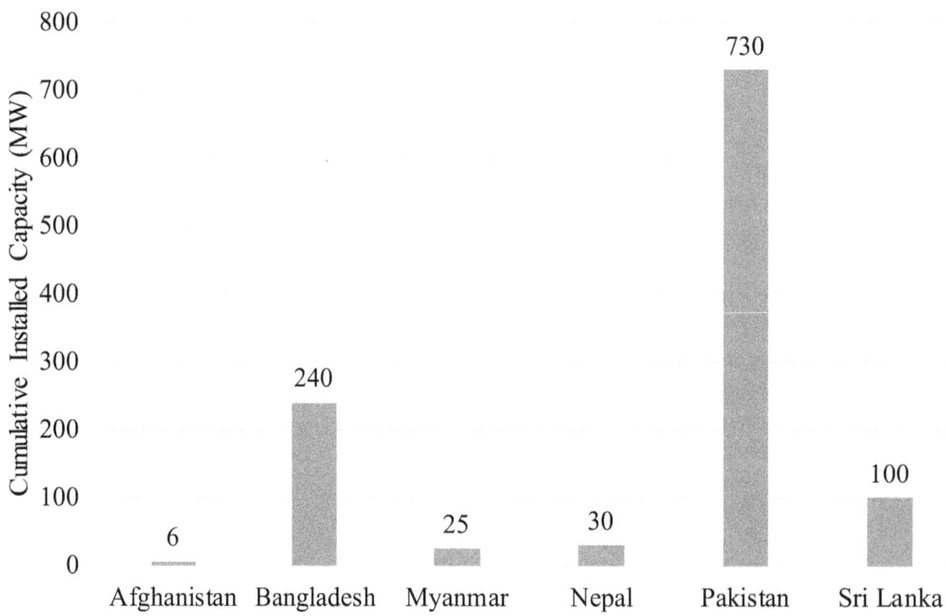

FIG. 4.23 INSTALLED CAPACITY OF SOLAR POWER IN NEIGHBOURING COUNTRIES [76].

4.13.2 Bangladesh

Over 24% of the Bangladesh population has no access to electricity. The Govt. of Bangladesh intends to provide electricity access for all by installing solar home systems [81]. The primary goal is to reduce the dependence on kerosene and diesel generators for home lighting. Therefore, the government is installing more than 65,000 SHS every month that replaces 180,000 tons of kerosene. Apart from SHS, Bangladesh has 80 off-grid solar plants that supply electricity to 1034 villages. There is also a special emphasis on solar systems for irrigation purpose and providing access to pure drinking water. In Bangladesh, grid-connected PV plants are in a very early stage of deployment. Till 2018, the government has only identified non-agricultural land for installing grid-connected power plants. Table 4.14 lists the current status of solar energy in Bangladesh.

4.13.3 Myanmar

Till date, Myanmar has not installed any utility-scale solar plant [83]. Most installations are off-grid SHS and mini-grids [84]. The largest installation is a 120 kW mini-grid system in Myaing [85]. The solar penetration in Myanmar has increased from 9% in 2014 to 25-27% in 2017 due to the installation of off-grid systems. A 130 MW grid-connected PV system is planned to begin operation in 2019. The ministry of electricity and energy has planned to install 1.5 GW solar power across five projects in the future [83].

4.13.4 Nepal

Solar energy is in a very early phase of deployment in Nepal. In 2018, Nepal installed solar PV systems on the rooftop of 21 buildings in the administrative hub with Chinese aid of $ 5.97 million [86]. Other popular solar applications are water pumping, solar lamps, and lanterns. In April 2018, work began for installing a 25 MW grid-connected solar power at Devighar Hydropower station, Nuwakot. The project is financed by the World Bank that is providing $ 130 million to the Nepalese government [87].

4.13.5 Pakistan

The first grid-connected solar plant in Pakistan was inaugurated in 2012 [88]. The plant had a capacity of 178 kW and was installed at the premises of Planning Commission and Pakistan Engineering Council. The plant had a net energy metering system. Subsequently, a 2 MW plant was installed at National Assembly of Pakistan. The success of pilot demonstrations encouraged private developers who obtained a letter of intent for setting up 956 MW of solar power. The government of Pakistan has identified 6500 acres of land in the desert area of Bahawalpur, Punjab to set up 1000 MW Quaid-e-Azam Solar Park. In Phase I of the project, 100 MW capacity was commissioned with a total cost of $ 215 million [89]. The tariffs range from 12.1-15.77 cents/kWh [90]. The government has given a letter of intent for setting up 300 MW and 600 MW in the second and third phase. Apart from PV systems, Pakistan has distributed more than 5000 solar cookers [88]. Till 2013, 14981 solar hot water heating and 1429 irrigations systems were deployed in the country.

TABLE 4.14

SOLAR INSTALLATIONS IN BANGLADESH [76, 81, 82]

Sr. No.	Type	Capacity
1	Solar On-grid installations	17 MW
2	Solar Off-grid installations	223 MW
3	Solar home system	4.5 Million (Nos.)
4	Solar Mini-grid	11 Nos. (2.19 MW)
5	Solar irrigation system	366
6	Solar drinking water system	140

4.13.6 Sri Lanka

Sri Lanka provides 24-hour power supply and meets 100% electricity demand of its citizens [91]. The government of Sri Lanka launched 'Soorya Bala Sangramaya' program on 6th September 2016 to deploy large-scale distributed generation systems [92]. The scheme targets installed capacity of 200 MW by 2020 and 1000 MW by 2025. Till January 31, 2018, Sri Lanka had a cumulative solar installed capacity of 100 MW [93]. For rooftop systems, the government has adopted a net metering policy. The system owners are paid Rs.22.00 per unit for the first seven years and Rs.15.50 for eight to 20 years for the excess energy fed into the grid.

4.14 SOLAR RESEARCH AT IIT BOMBAY

IIT Bombay established the Department of Energy Science and Engineering (DESE) for research in the field of renewable energy. The institute is working on multiple technologies for utilization of solar energy. It has installed and commissioned a 1 MWe solar thermal power plant at NISE, Gurgaon with funding from MNRE [71]. The objectives of the project were:

(i) Establishment of a national research facility on solar thermal power by installing 1 MW(e) grid-connected power plant;
(ii) creation of a test facility for characterization of concentrating collectors, and
(iii) development of a simulation software package for assisting the design of solar thermal plants.

It is expected that the facility would lead to the development of indigenous capabilities and facilitate research and development for cost reduction of CSP plants. The facilities can also be used to provide workforce training. Figure 4.24 and 4.25 show a picture and block diagram respectively of the CSP plant. The plant consists of 3 MW_{th} parabolic trough solar field and 2 MW_{th} LFR system. The parabolic trough field has 30 min of thermal storage. The turbine capacity is 1 MWe.

Figure 4.26 shows the minute-by-minute power output and solar radiation on two sample days at the plant location.

IIT Bombay has also installed a 1 MW PV plant within its campus on the roof-top of various buildings [94]. The institute has set-up a National Centre for Photovoltaic Research and Education (NCPRE) to provide R&D and educational support for meeting India's 100 GW solar target [95]. It has already distributed 1 million solar lamps and has taken up the responsibility to distribute 7 million more such lamps. The institute has also helped in the implementation of a solar PV manufacturing plant owned and operated by tribal women in Dungarpur, Rajasthan [96].

FIG. 4.24: PICTURE OF THE 1 MWE CSP PLANT AT NISE, HARYANA.

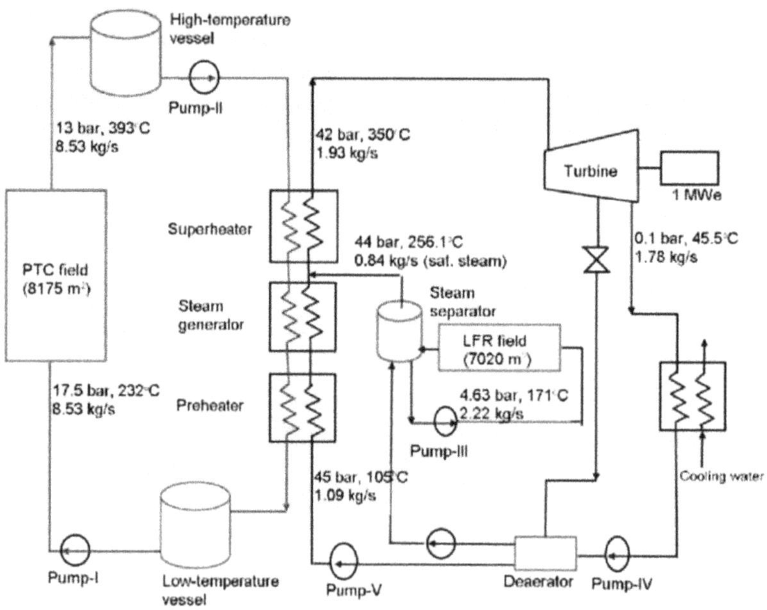

FIG. 4.25: SCHEMATIC OF THE SOLAR THERMAL POWER PLANT INSTALLED BY IIT BOMBAY [71].

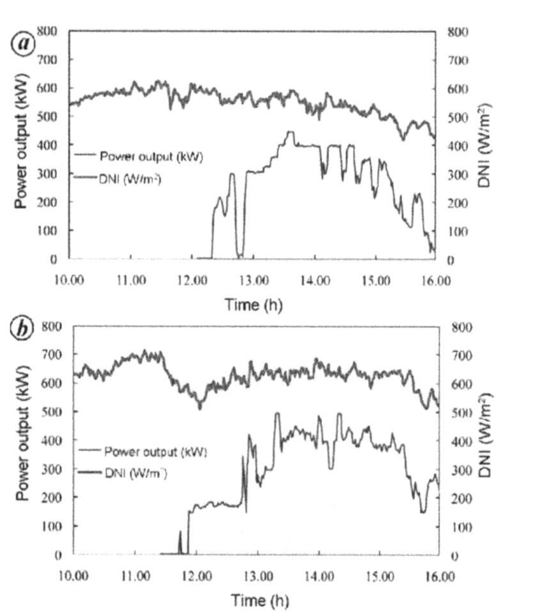

FIG. 4.26: MINUTE-BY-MINUTE POWER OUTPUT DATA FOR A) 28 MAY 2014 B) 4 JUNE 2014 [71].

4.15 SUMMARY AND CONCLUSIONS

India's development strategy is to provide energy access to all and raise the living standard of its citizens. With these objectives in mind, over the last decade, the average electricity consumption in India has increased from 704 kWh/person/year in 2008-09 to 1149 kWh/person/year in 2017-18. The average deficit for energy has also reduced from 8.2% in 2008-09 to 0.8% in 2017-18.

The scarcity of fossil fuels and global warming and climate change problem has resulted in an increased emphasis on renewable energy sources. India intends to have 175 GW of renewable energy installation by 2022. Beyond 2022, NITI (National Institute for Transforming India) Aayog, a government 'Think Tank', shows the possibility of achieving 479 GW of solar power by 2047. Till 2018, the installed capacity of grid-connected renewables is more than 71 GW. The launch of Jawaharlal Nehru National Solar Mission (JNSSM) in January 2010 has provided an impetus to the growth of solar installations in India. The National Solar Mission (NSM) targeted installing 20 GW of solar power by 2022. This target was achieved in 2017-18. In 2015, the government revised the 2022 targets and now intends to install 100 GW solar power by 2022. By 2018, the installed capacity of solar power in India was 21.6 GW. The share of solar power in the generation mix of the country is 2%.

The targetted capacity of 100 GW solar power (by 2022) will be achieved by installing 60 GW of centralized plants. For centralized systems, the government has approved setting-up 45 solar parks that will further boost the solar deployment in the country. Based on the reverse auction mechanism, the tariff in these solar parks has dropped from Rs. 11/kWh in 2011 to Rs. 2.44 /kWh in 2018.

The government intends to install 40 GW (by 2022) of solar power on the rooftop of residential, industrial and commercial sectors. However, it is observed that out of the 21.6 GW of installed capacity, only 2.5 GW are rooftop systems while the rest are centralized power plants. Therefore, based on the current trends, it seems unlikely that the government will achieve solar rooftop targets. The growth in rooftop systems has been primarily driven by the industrial and commercial sectors that pay higher tariffs for grid-based electricity. It is expected that 44% of the rooftop targets will be met by solar installations in the industrial sector alone. The pace of installation in the residential sector has been sluggish due to subsidised electricity for low-end consumers, high capital cost and lack of financing options.

After rapid deployment since 2010, the growth in solar installations has now slowed down in 2018. This is because the electricity distribution companies (DISCOMs) are no longer signing new power purchase agreements (PPAs) with solar power developers. DISCOMs are under financial stress, and the obligation to honour the long-term PPAs signed with the thermal power stations has led to a condition where there are no buyers of solar electricity. The solar rooftop sector has been largely impacted due to the financial issues with the DISCOMs. The state governments are trying to boost the rooftop installations by offering subsidies. For eg., in Madhya Pradesh, the tariff bids for solar rooftop installation has reached Rs. 1.38/kWh because the central and state governments are giving subsidies of 23% and 20% respectively. The central government has taken steps to resolve the standoff between DISCOMS and solar prosumers by bringing them on a common platform. The government has also proposed amendments to the Electricity Act, 2003 that will levy a penalty of Rs. 1-5/kWh on DISCOMs in case of non-compliance with the renewable PPAs. The pace of deployment of solar power in India will now depend on the reforms undertaken in the power sector.

Solar manufacturing in India has not picked up due to the competition with cheap Chinese imports. Since the launch of the solar mission in January 2010, India has spent $11.7 billion on solar imports. Most of these imports are from China. For example, in 2017-18, India imported 89% of its solar requirement from China alone. The module manufacturing capacity in India is 5.5 GW while the solar cell manufacturing capacity is only 1.6 GW. This is because most of the domestic module manufacturers use imported solar cells. Therefore, solar cell manufacturing capability has not reached to the extent of solar installations in the country.

The government is trying to revive the domestic manufacturing industry by imposing anti-dumping duties on solar imports, creating demand by domestic content requirement (DCR) and supporting manufacturers by providing subsidy. This will create employment opportunities in India. However, schemes like DCR cannot be implemented due to WTO (World Trade Organization) rulings. Without government support, it is highly unlikely that the domestic solar cell manufacturing industry will be able to compete with low-cost Chinese imports.

For providing electricity access to rural households, there is a special emphasis on solar home lighting systems, solar pumping and solar lamps. By 2022, the government intends to deploy 10000 renewable-based micro-grids having a capacity of 500 MW. Solar energy is expected to play a significant role in achieving these targets. However, the tariffs for battery integrated 24-hour solar off-grid system (31.25 Rs./kWh) is significantly higher than the grid tariff (3-4 Rs./kWh) . The tariffs in solar-based Microgrids can decrease only with a rapid decline in the cost of batteries.

Solar water heaters for low-grade heating are already viable for residential and commercial applications. These systems are modular and changes in the building codes and a regulatory push have enabled their faster adoption. It is now imperative that all states follow these regulations to accelerate the deployment of solar water heaters. On the technology front, Evacuated tube collectors (ETCs) are more popular than the conventional Flat plate collectors (FPCs). Since 2014, the government has discontinued subsidy on all solar water heaters as they have become commercially viable.

Concentrated solar thermal (CST) systems have a significant potential to replace/reduce oil-fired steam generation in the industry. The total installed area of CST systems in March 2017 was 52,448 m^2. The technologies installed in India are based on Dish, Scheffler, parabolic trough, linear Fresnel and compound parabolic collectors. The feasibility of integrating these technologies into the existing systems has been demonstrated for cooking, cooling, process heat and desalination applications. The manufacturing is indigenous, but components like low-iron glass are still imported that increases the system cost. Therefore, these systems have not commercialized and need subsidies from the government. Apart from subsidies, the deployment of CST systems will need policy intervention from the government. A possible option is creating a demand for CST technologies by awarding extra credits in RPOs (Renewable purchase obligations). Guaranteed annual orders will enable scaling up the technology and ultimately result in lower cost.

The concentrated solar power (CSP) industry has not picked up in India. During Phase I of the national solar mission in 2010, seven power purchase agreements (PPAs) were signed for setting up 470 MW of concentrated solar power (CSP) plants in the country. Based on the reverse bidding mechanism, the tariffs in the range of 10.49-12.24 Rs./kWh were obtained. These tariffs were lower than the government estimate of 15.31 Rs./kWh. The tariffs in the international markets in 2010 were 12.4-18.9 Rs./kWh. It was observed that the bidders quoted unviable tariffs with the only intent of obtaining orders. Due to this reason, only three plants with a cumulative capacity of 225 MW have been commissioned till 2018. The other four bidders are looking to exit the solar thermal business by selling their respective plants. Currently, CSP technology is not commercially viable when compared with the PV system. Therefore, solar thermal power generation is not picking up in India and the government has shown no intent to pursue the technology. Only a market demand for solar power during the peak evening and morning loads might revive the interest in this technology. This will, to a larger extent, depend on the competitiveness of dispatchable CSP power with the battery-integrated PV plants.

Grid-connected solar plants are in a very early stage of deployment in India's neighbouring countries. Pakistan, with 730 MW of solar power, has the highest installed capacity in these countries. Off-grid solar has been popular in Bangladesh and Nepal. A few countries like Sri Lanka and Myanmar have set future targets for grid-connected solar plants.

The national solar mission has achieved its initial objective of 20 GW solar power and now targets to achieve 100 GW by 2022. The large-scale deployment has reduced the cost of solar power from Rs. 11/kWh in 2010 to Rs. 2.44/kWh in 2018. The ability of DISCOMs to pay the renewable energy tariffs will now determine the penetration of solar power in the Indian electric grid. The solar rooftop segment has not grown to its potential and concerns with net-metering systems should be addressed on priority. The integration of renewable energy sources in the electric grid will be a challenge in the future due to the intermittent nature of power generation. The need for grid reliability might need dispatchable solar generation in the future. Therefore, energy storage for integration of solar systems into the existing electricity grid will be the future trend. The mainstreaming of solar technologies will now depend on the innovations and technology solutions that reduce costs of dispatchable power and improve reliability. If the technology developers, researchers, and industry take up the challenge and develop cost-effective solar solutions, India can have an increasing solar share in the future energy mix.

4.16 REFERENCES

1. Central Electricity Authority, Ministry of power (Gov. of India). Executive Summary for the Month of March 2018. New Delhi: 2018.
2. PIB. Ministry of Power, along with Department of Posts, for survey of un- electrified households in Assam, Chhattisgarh, Jharkhand, Madhya Pradesh and Odisha. Press Inf Bur Gov India, Minist Power 2018. http://pib.nic.in/newsite/PrintRelease.aspx?relid=176271 (accessed September 18, 2018).
3. ISA. International Solar Alliance 2015. https://isolaralliance.org/ (accessed October 28, 2018).
4. NITI Aayog. Report of the Expert Group on 175 GW RE by 2022. New Delhi: 2015.
5. Singh R, Banerjee R. Impact of large-scale rooftop solar PV integration : An algorithm for hydrothermal-solar scheduling (HTSS). Sol Energy 2017;157:988–1004. doi:10.1016/j.solener.2017.09.021.
6. MNRE. Welcome to Government of India | Ministry of New and enewable Energy 2019;6:2–3. https://mnre.gov.in/ (accessed May 28, 2019).
7. CEA. Growth of electricity sector in India from 1947-2018. New Delhi: 2018.
8. SOLARGIS. Solar resource maps and GIS data for 180+ countries | Solargis 2018:1. doi:10.1016/j.puhe.2008.10.012.
9. MNRE. Jawaharlal Nehru National Solar Mission - Towards building Solar India 2010:15. https://mnre.gov.in/file-manager/UserFiles/mission_document_JNNSM.pdf (accessed September 10, 2018).
10. MNRE. Scaling of grid-connected solar power projects. Minist New Renew Energy, Gov India 2015:1–15. http://mnre.gov.in/file-manager/grid-solar/100000MW-Grid-Connected-Solar-Power-Projects-by-2021-22.pdf (accessed September 18, 2018).
11. SECI. What's New :: Solar Energy Corporation of India (SECI), A Government of India Enterprise, Under Ministry of New and Renewable Energy 2018. http://www.seci.gov.in/content/news_update/installation-of-500-kwp-rooftop-solar-pv-project-at-kiss--bhubaneswar---largest-in-odisha.php (accessed November 26, 2018).
12. Gupta S. Solar Cell and Module Capacities. Minist New Renew Energy 2017. https://mnre.gov.in/file-manager/UserFiles/information-sought-from-all-Solar-Cell-&-Module-manufacturers-as-on-31052017.pdf (accessed September 24, 2018).
13. Mercom Group. Mercom Exclusive: Can Domestic Manufacturers Capture a Larger Piece of the Growing Indian Solar Market? Merom Cap Gr 2017. http://mercomcapital.com/can-domestic-manufacturers-capture-a-larger-piece-of-the-growing-indian-solar-market (accessed September 24, 2018).
14. Press Trust of India. China accounts for 89% of India's total solar cells imports in 2017-18 _ Business Standard News. Bus Stand 2018. https://www.business-standard.com/article/economy-policy/china-accounts-for-89-of-india-s-total-solar-cells-imports-in-2017-18-118080100849_1.html (accessed September 1, 2018).
15. Verma B. Challenges to solar cell manufacturing in India. Cent Study Sci Technol Policy 2017:3–5. http://www.cstep.in/uploads/default/files/pressroom/stuff/Challenges_to_solar_cell_manufacturing_in_India_-_Infracircle.pdf (accessed September 24, 2018).
16. Rishikumar V. MNRE keen to give big push to solar manufacturing - The Hindu BusinessLine. The Hindu 2018:1. https://www.thehindubusinessline.com/economy/mnre-keen-to-give-big-push-to-solar-manufacturing/article9995618.ece (accessed September 20, 2018).
17. Bloomberg quit. Anti-Dumping Duty_ India Imposes Safeguard Duty On Solar Cells From China, Malaysia. The Quint 2018. https://www.bloombergquint.com/markets/2018/07/30/india-imposes-safeguard-duty-on-solar-cells-imported-from-china-malaysia#gs.6u8zrO4 (accessed September 18, 2018).
18. Azure Power. Azure Power - A Leading Solar Power Company in India. Azur Power 2018. https://www.azurepower.com/ (accessed November 26, 2018).
19. TOI. SECI's auction for 5GW solar manufacturing gets one bidder - Times of India. Times of India 2018:1. https://timesofindia.indiatimes.com/business/india-business/secis-auction-for-5gw-solar-manufacturing-gets-one-bidder/articleshow/65985098.cms (accessed September 30, 2018).
20. Chandrasekaran K, Saluja N. SECI's Manufacture-linked bids put off, for fourth time. Econ Times 2018. https://economictimes.indiatimes.com/industry/energy/power/no-bidders-govt-extends-deadline-for-solar-tenders/articleshow/66174327.cms (accessed October 12, 2018).
21. Chawla K, Waldron M, Dutt A, Aggarwal M, Toril A, Nobuoka Y. Clean Energy Investment Trends 2017- Evolving Landscape for grid-connected renewable energy projects in India. New Delhi: 2018.
22. MNRE. List of approved solar parks. Minist New Renew Energy 2018:4. https://mnre.gov.in/file-manager/UserFiles/List-of-approved-Solar-Parks.pdf (accessed September 24, 2018).
23. Adani. Adani Kamuthi. Adani Gr 2018. https://www.adanigreenenergy.com/kamuthi.html (accessed November 21, 2018).
24. G Jagannath JAGANNATH. TNERC dismisses Adani's plea seeking higher tariff. Deccan Chron 2016. http://www.deccanchronicle.com/nation/current-affairs/181116/tnerc-dismisses-adanis-plea-seeking-higher-tariff.html.
25. Saumy P. Will States Comply with the New Solar RPO Target? Mercom Gr 2018. https://mercomindia.com/states-new-solar-rpo-target/ (accessed May 7, 2019).
26. Chatterjee A. No takers for solar power_ Power minister warns that states are not buying even at lowest rates! Financ Express 2018. https://www.financialexpress.com/industry/no-takers-for-solar-power-power-minister-warns-that-states-are-not-buying-even-at-lowest-rates/1332407/ (accessed November 2, 2018).
27. Business Line. Electricity Act amendments propose stiff penalties for PPA violations. Bus Line 2018. https://www.thehindubusinessline.com/news/electricity-act-amendments-propose-stiff-penalties-for-ppa-violations/article24896715.ece (accessed November 18, 2018).

28. MNRE. State-wise-and-year-wise-target-for-installation-of-40000MWp-GCRT-systems. Minist New Renew Energy, Gov India 2015:3. https://mnre.gov.in/sites/default/files/webform/notices/State-wise-and-year-wise-target-for-installation-of-40000MWp-GCRT-systems_0.pdf (accessed September 18, 2018).
29. Singh R, Banerjee R. Estimation of rooftop solar photovoltaic potential of a city. Sol Energy 2015;115:589–602. doi:10.1016/j.solener.2015.03.016.
30. UN S. Rooftop solar is the fastest-growing segment in India's renewables market. Quartz 2017. https://qz.com/1148332/rooftop-solar-is-the-fastest-growing-segment-in-indias-renewables-market/ (accessed October 28, 2018).
31. UN S. India has quietly downscaled its rooftop solar power target. Quartz 2018. https://qz.com/india/1180435/india-has-quietly-downscaled-its-rooftop-solar-power-target/ (accessed October 21, 2018).
32. Bridge to India. Bridge to India. Bridg to India 2014. https://bridgetoindia.com/reports/ (accessed September 19, 2018).
33. Trivedi S, Ray I, Vulturius G. Scaling up Rooftop Solar Power in India: The Potential of Solar Municipal Bonds. 2018.
34. Bhaskar U. Discom losses plunged to □17,352 crore in FY18, says power minister. Livemint 2018. https://www.livemint.com/Industry/xhM6n0rs45calbkNl8TECN/India-may-install-225-GW-renewable-power-by-2022-Power-mini.html (accessed November 10, 2018).
35. Roy A. Discom plans Rs1.25/unit levy on rooftop solar prosumers. Times of India 2018. https://timesofindia.indiatimes.com/city/nagpur/discom-plans-rs1-25/unit-levy-on-rooftop-solar-prosumers/articleshow/65284257.cms (accessed August 10, 2018).
36. Government of India. Ministry Releases Concept Note on Sustainable Rooftop Implementation for Solar Transfiguration of India (SRISTI); Invites comments/views from stakeholders by 31st December. IMD Releases Long Range Forecast 2014 Southwest Monsoon Seas Rainfall 2014. http://pib.nic.in/newsite/PrintRelease.aspx?relid=104920%0Ahttp://pib.gov.in/newsite/mbErel.aspx?relid=106414 (accessed August 19, 2018).
37. Prateek S. Madhya Pradesh Introduces Innovative Decentralized Solar Policy. Mercom Gr 2018. https://mercomindia.com/madhya-pradesh-introduces-decentralized-solar-policy/ (accessed November 10, 2018).
38. Mercom Group. Lowest Tariff Drops to □1. Mercom Gr 2018. https://mercomindia.com/l1-tariff-1-38-mp-solar-rooftop-auction/ (accessed October 5, 2018).
39. MNRE. Solar Rooftop Installation Online Portal. Minist New Renew Energy, Gov India 2018:1. https://mnre.gov.in/solar-rooftop-grid-connected (accessed October 2, 2018).
40. Kuldeep N, Ganesan K, Gupta V, Ramji A, Chawla K, Aggarwal M. Energy Storage in India: Applications in the Renewable Energy Segment. New Delhi: 2016.
41. Choudhary S. Solar power helps pump up petrol sales amid power shortage. Econ Times 2017. https://economictimes.indiatimes.com/industry/energy/oil-gas/solar-power-helps-pump-up-petrol-sales-amid-power-shortage/articleshow/59343036.cms (accessed September 20, 2018).
42. Su-kam. Su-Kam – Complete Customized Solar Power Solutions for Petrol Pumps. Su-Kam 2018. http://www.su-kam.com/power-solution/solar-solutions/solar-power-for-petrol-pumps (accessed September 20, 2018).
43. Parbat K. Trai to review rules on green energy-driven mobile towers. Econ Times 2016. https://economictimes.indiatimes.com/industry/telecom/trai-to-review-rules-on-green-energy-driven-mobile-towers/articleshow/51335926.cms (accessed September 18, 2018).
44. SBI to go green! 10,000 ATMs to run on solar power in next two years. Bus Today 2018. https://www.businesstoday.in/sectors/banks/sbi-plans-to-install-solar-panels-on-10000-atms-in-next-2-years/story/282564.html (accessed September 20, 2018).
45. Dholakia HH. Solar powered healthcare in developing countries. Nat Energy 2018;3:705–7. doi:10.1038/s41560-018-0205-1.
46. MNRE. MNRE Annual report: 2017-18. New Delhi: 2018.
47. PBI. 63 solar micro grids of 1899 kWp aggregated capacity reported to be installed in the country. Press Inf Bur Gov India, Minist Power 2018. http://pib.nic.in/newsite/PrintRelease.aspx?relid=175300 (accessed September 20, 2018).
48. GNESD. Renewable energy-based rural electrification: The mini-grid experience from India. 2014.
49. Centre for Science and Environment. Mini grids: electricity for all. New Delhi: 2016.
50. Jacob AS, Banerjee R, Ghosh PC. Sizing of hybrid energy storage system for a PV based microgrid through design space approach. Appl Energy 2018;212:640–53. doi:10.1016/j.apenergy.2017.12.040.
51. Jacob AS, Banerjee R, Ghosh PC. Modelling and simulation of a PV battery grid backup system for various climatic zones of India. 2017 IEEE 44th Photovolt Spec Conf PVSC 2017 2018:1–6. doi:10.1109/PVSC.2017.8366394.
52. BHEL. Engineering News Letter. Bharat Heavy Electr Ltd 2006;29.
53. Padigala B, Panwar TS, Mukhopadhyay S, Choudhury P, Palakshappa R, Ghosh A. Renewables beyond electricity - Solar air conditioning & dessalination in India. New Delhi: 2014. doi:10.13140/RG.2.1.5014.5363.
54. Pillai IR, Banerjee R. Methodology for estimation of potential for solar water heating in a target area. Sol Energy 2007;81:162–72. doi:10.1016/j.solener.2006.04.009.
55. Pillai IR. Diffusion of renewable energy technologies. Indian Institute of Technology Bombay, 2008.
56. Kumarankandath A, Goswami N, Bhushan C, Shankar A, Chandan C, Singh S, et al. The State of Concentrated Solar power in India-A roadmap to developing solar thermal technologies in India. New Delhi: 2015.
57. Weiss W, Spork-Dur M. Solar heat worldwide. 2018.
58. MNRE. Ministry of New and Renewable Energy - Solar Cookers & Steam Generating Systems. Minist New Renew Energy 2018. https://mnre.gov.in/concentrating-solar-system-solar-cookers-steam-generating-systems (accessed September 21, 2018).
59. Ministry of New and Renewable Energy. Solar Thermal Cooling Technologies 2014:20.
60. Abhishek K, Udawant R. Highlights of CST field projects supported under UNDP. Sunfocus 2017;4:12–5.
61. Maithel l S, Lalchandani l D, Bhanware l P, Singh l P, Kumar l S, l Ayan, et al. Capturing Sun for heat-Potential, vision and action plant for decentralized solar thermal technologies and applications in India. 2015.
62. Sukhatme SP. Solar Energy: Principle of Thermal Collection and Storage. Mc Graw Hill 2008:62. doi:10.1177/0278364913498122.
63. ATE Enterprises. Solar Thermal Technology. ATE Enterp n.d. http://ategroup.com/energy-efficiency/about-solar-thermal/ (accessed November 21, 2018).
64. Thomas A. Design Methodology For A Small Solar Team Generation System Using The Flash Boiler Concept. Energy Convers Manag 1996;37:1–15. doi:10.1016/0196-8904(95)00022-6.
65. Brahmakumaris. Renewable Energy at Brahma Kumaris centers in India. Brahmakumaris 2016. http://www.brahmakumaris.org/hope/renewable-energy/renewable-energy-at-bk-centers-in-india (accessed November 15, 2018).
66. CliqueSolar. Clique Solar-ARUN Solar Boiler and Concentrated Solar Thermal Company. CliqueSolar 2018. http://www.cliquesolar.com/ (accessed November 14, 2018).
67. KGDS. Solar Thermal Power Plant, Solar Energy Services in India, Solar Systems in India, Solar Desalination Plant Coimbatore. KGDS 2018. http://solar.kgisl.com/ (accessed October 21, 2018).
68. Khanna A, Garg K. Paving the Way for a transformational Future-Lessons from Jawaharlal Nehru National Solar Mission Phase I Pu. Washington, DC: 2013.

69. Acme. ACME Solar. Acme 2018. https://www.acme.in/ (accessed November 27, 2018).
70. Kartheek, N. G. R., Yadav, D., Banerjee, R., Nayak, J. K., Bandyopadhyay, S. and Kedare SB. Experiences in commissioning of a 1 MWe solar thermal power plant in Gurgaon. Int. Conf. Adv. Energy Res., Mumbai: 2013.
71. Nayak JK, Kedare SB, Banerjee R, Bandyopadhyay S, Desai NB, Paul S, et al. A 1 MW national solar thermal research cum demonstration facility at Gwalpahari, Haryana, India. Curr Sci 2015;109:1445–57. doi:10.18520/v109/i8/1445-1457.
72. India One. India one WP test site – India one – WordPress site. Brahmakumari 2018. http://india-one.net/ (accessed October 21, 2018).
73. JS S. Godawari CSP plant: An overview of performance. Sun Focus 2016;4.
74. Burkhardt JJ, Heath GA, Turchi CS. Life cycle assessment of a parabolic trough concentrating solar power plant and the impacts of key design alternatives. Environ Sci Technol 2011;45:2457–64. doi:10.1021/es1033266.
75. IRENA. Water Use in India's Power Sector : Impact of Renewables and Cooling Technologies To 2030. Abu Dhabi, United Arab Emirates: 2016.
76. IRENA. Renewable Capacity Statistics 2018. 2018.
77. Prof. Dr. Abdul Rahman. Energy Sector Afghanistan-Importance of Renewable Energy for Afghanistan 2009. https://carnegieendowment.org/files/Presentation_- Ashraf.pdf (accessed October 9, 2018).
78. Storai Karimi. First-ever solar-wind power plant goes functional in Herat. Pajhwok Afgan News n.d. https://www.pajhwok.com/en/2017/09/17/first-ever-solar-wind-power-plant-goes-functional-herat (accessed October 3, 2018).
79. ADB. ADB Supports First Solar Power Plant to Boost Renewable Energy in Afghanistan. Asian Dev Bank 2018. https://www.adb.org/news/adb-supports-first-solar-power-plant-boost-renewable-energy-afghanistan (accessed October 8, 2018).
80. John Johnston. Solar Just For Greenies_ Nope, US Army Using Solar In Afghanistan. The9billionCom n.d. http://www.the9billion.com/2011/01/19/us-army-using-solar-in-afghanistan/ (accessed October 5, 2018).
81. SREDA. Sustainable & Renewable Energy Development Authority. SREDA 2018. http://www.sreda.gov.bd/ (accessed October 8, 2018).
82. Hil Baky MA, Rahman MM, Islam AKMS. Development of renewable energy sector in Bangladesh: Current status and future potentials. Renew Sustain Energy Rev 2017;73:1184–97. doi:10.1016/j.rser.2017.02.047.
83. Beetz B. 130 MW solar project announced in Myanmar. PV Mag 2018. https://www.pv-magazine.com/2018/08/29/130-mw-solar-project-announced-in-myanmar/ (accessed October 8, 2018).
84. Asian Development Bank. Developing renewable energy mini-grids in Myanmar: A Guidebook 2017. doi:10.22617/TIM178951.
85. Beetz B. The weekend read: Myanmar's solar lifeline. PV Mag 2018. https://www.pv-magazine.com/2018/08/18/the-weekend-read-cells-of-the-future/ (accessed October 9, 2018).
86. Xinhua. Nepali gov't administrative hub to be illuninated by solar power with help of China Aid. Xinhua Net 2018. http://www.xinhuanet.com/english/2018-04/10/c_137100721.htm (accessed October 9, 2018).
87. Kathmandu Post. Work begins on Nepal's largest solar plant 2018. http://kathmandupost.ekantipur.com/news/2018-04-28/work-begins-on-nepals-largest-solar-plant.html (accessed October 9, 2018).
88. Kamran M. Current status and future success of renewable energy in Pakistan. Renew Sustain Energy Rev 2018;82:609–17. doi:10.1016/j.rser.2017.09.049.
89. Woods L. TBEA SunOasis wins Pakistan mega-solar project contracts. PVTech 2014. https://www.pv-tech.org/news/chinese_firm_wins_paksitan_mega_solar_project_contracts (accessed October 9, 2018).
90. AEDB. Solar power: Current status. Altern Energy Dev Board 2018. http://www.aedb.org/ae-technologies/solar-power/solar-current-status (accessed October 8, 2018).
91. Ministry of Power and Energy. Assistance by Finland towards the development of renewable energy sector 2018. http://powermin.gov.lk/english/?p=4960 (accessed October 9, 2018).
92. SLSEA. Sri Lanka Sustainable Energy Authority 2018. http://www.energy.gov.lk/news_1.php (accessed October 9, 2018).
93. Daily News. Solar power electricity in Sri Lanka 2018. http://www.dailynews.lk/2018/01/31/features/141431/solar-power-electricity-sri-lanka (accessed October 9, 2018).
94. IIT Bombay. 1MW Solar PV Power Plant at IIT Bombay 2014. http://www1.iitb.ac.in/SolarPlant/SolarPlant.htm (accessed October 18, 2018).
95. NCPRE. National Centre for Photovoltaic Research and Education (NCPRE). NCPRE 2018. http://www.ncpre.iitb.ac.in/ (accessed November 10, 2018).
96. Bombay I. Off-Grid Solar Power Scenario Of India _ IIT Bombay n.d. http://www.iitb.ac.in/en/story/grid-solar-power-scenario-india (accessed October 18, 2018).

CHAPTER

5

FUTURE SOLAR ENERGY APPLICATIONS

Ram A. Goel

5.1 INTRODUCTION

5.1.1 Harnessing Solar Energy

The Sun is potentially a huge source of renewable, clean energy. Some estimate that sunlight could produce 10,000 times as much power as the Earth used at the turn of the 21st century. There are, however, major technological challenges to be met in harnessing that energy effectively. There are a number of different technologies available, and under development, that use sunlight to provide power.

Sunlight can be used simply to heat water which is then used to provide central heating for homes. Alternatively, it can be used to generate electricity using photovoltaic (PV) cells arranged on solar panels. A third method is to concentrate sunlight on a target to generate heat, which may be used directly for industrial purposes, or to provide electricity.

5.1.2 The Future of Solar Energy

The two widely popular classes of technologies for converting solar energy into electricity — photovoltaics (PV) and concentrated solar power (CSP), sometimes called solar thermal) — in their current and plausible future forms. Because energy supply facilities typically last several decades, technologies in these classes will dominate solar-powered generation between now and 2050. We cannot predict the future of solar energy too far ahead in future because of the following two reasons. First, expanding the solar industry dramatically from its relatively tiny current scale may produce changes we do not pretend to be able to foresee today. Second, we recognize that future solar deployment will depend heavily on uncertain future market conditions and public policies including but not limited to policies aimed at mitigating global climate change.

More than one billion people in the developing world who lack access to a reliable electric grid, the cost of small-scale PV generation is often outweighed by the very high value of access to electricity for lighting and charging mobile telephone and radio batteries. In addition, in some developing nations it may be economic to use solar generation to reduce reliance on imported oil, particularly if that oil must be moved by truck to remote generator sites.

[http://energy.mit.edu/research/future-solar-energy/]

5.2 GLOBAL ENERGY NEEDS AND RESOURCES

Global energy consumption in the last 50 years has increased and is expected to continue to increase with the same pace over the next 50 years but with significant differences. The past increase was stimulated by relatively "cheap" fossil fuels and increased rates of industrialization in North America, Europe, and Japan; yet while energy consumption in these countries continues to increase, additional factors make the situation for the next 50 years more complex. These additional complicating factors include China's and India's rapid increase in energy use as they represent approximately one-third of the world's population, the expected depletion of oil resources in the near future, and the effect of human activities on global climate change. On the positive side, the renewable energy (RE) technologies of wind, biofuels, solar, thermal, and photovoltaics (PV) are finally being developed at a noticeable level and at the ultimate promise of cost competitiveness.

The total primary energy demand in the world increased from 5536 million tons of oil equivalent (MTOE) in 1971 to 10,345 MTOE in 2002, representing an average annual increase of 2% (Figure 5.1). [*Principles of solar energy by D. Yogi Goswami*]

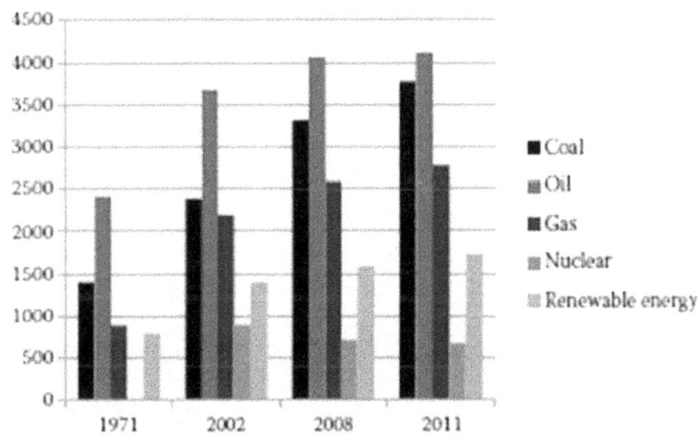

FIG. 5.1: WORLD PRIMARY ENERGY DEMAND (MTOE). (DATA FROM IEA, WORLD ENERGY OUTLOOK 2005, INTERNATIONAL ENERGY AGENCY, PARIS, FRANCE, 2005; IEA, WORLD ENERGY OUTLOOK 2010, INTERNATIONAL ENERGY AGENCY, PARIS, FRANCE, 2010; IEA, WORLD ENERGY OUTLOOK 2013, INTERNATIONAL ENERGY AGENCY, PARIS, FRANCE, 2013.) (*Source:* Principles of solar energy by D. Yogi Goswami)

By 2008, the world energy demand had increased to 12,271 MTOE, with an average annual increase of approximately 3%. The fast-growing energy demand in the Asia Pacific, more specifically China is the main reason for a 50% increase in the annual rate is. The global energy use may continue to grow at this extremely high rate because the per capita energy used in the most populous countries (China and India) is still very small. Because of the deep recession in the United States and Europe where the energy use actually went down, since 2008, the average annual energy growth dropped to approximately 2.1%, mainly.

Even at a 2% growth per year, the energy demand of 12,271 MTOE in 2008 would increase by twice by 2043 and triple by 2063. Of course, the global energy use cannot continue to grow at the same rate forever. The International Energy Agency (IEA 2010) estimated that the global energy use will grow at an average annual rate of 1.2% up to 2035. The global energy use will increase by 38% by 2035, reaching a value of 16,934 MTOE per year even at that optimistic rate. [*Principles of solar energy by D. Yogi Goswami*]

5.2.1 Present Status and Potential of RE

The data in Table 5.1 shows us that in 2011, Renewable Energy was 13.2% of the world's total primary energy. However, approximately 75% of the Renewable Energy supply was from biomass, and in developing countries, it is very inefficient because it is mostly converted by traditional open combustion. Because of its inefficient use, biomass resources presently supply only approximately 20% of that modern, more efficient, available technologies offer.

The total share of all renewables for electricity production in 2011 was approximately 20.1%, a vast majority (78%) of it being from hydroelectric power (Figure 5.1). Even though worldwide solar power capacity represented only 1.4% of the total electricity capacity, it was growing at an average annual rate of approximately 50%. [*Principles of solar energy by D. Yogi Goswami*]

5.2.2 Wind Power

Wind power has been used widely since medieval times. Irrigation pumps were powered by windmills in rural United States. They used to drive small electric generators that were used to charge batteries that provided electricity during the last century. A windmill or wind turbine converts the kinetic energy of moving air into mechanical motion, usually in the form of a rotating shaft. This mechanical

TABLE 5.1

2011 FUEL SHARES IN WORLD TOTAL PRIMARY ENERGY SUPPLY (SOURCE: PRINCIPLES OF SOLAR ENERGY BY D. YOGI GOSWAMI)

2011 Fuel Shares in World Total Primary Energy Supply	
Source	**Share**
Oil	31.4%
Natural gas	21.3%
Coal	28.9%
Nuclear	5.2%
Renewables	13.2%

Source: IEA, *World Energy Outlook 2013,* International Energy Agency, Parts, France, 2013.

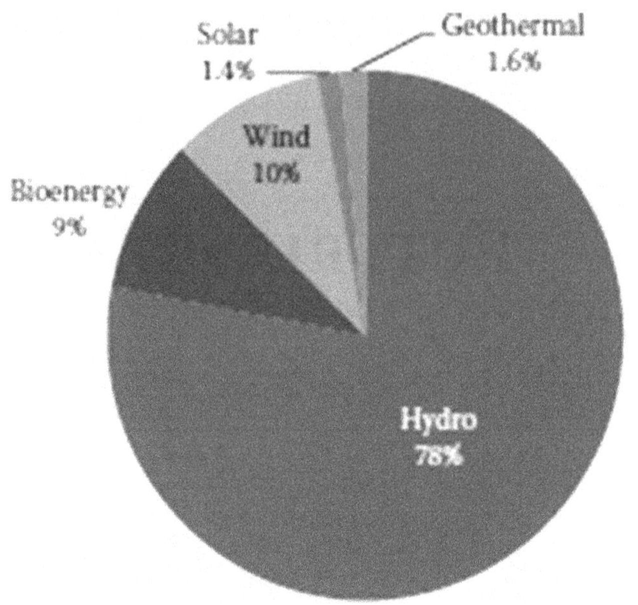

FIG. 5.2: 2011 RENEWABLE RESOURCE SHARES IN WORLD ELECTRICITY CAPACITY (*Source:* Principles of solar energy by D. Yogi Goswami)

motion can be used to drive a pump or to generate electric power. The energy content of the wind increases with the third power of the wind velocity and wind power installations are economical in regions where winds of sufficient strength and regularity occur.

Wind energy technology has progressed significantly over the last two decades. The technology has been vastly improved and capital costs have come down to as low as $1000/kW. At this level of capital costs, wind power is already economical at locations with fairly good wind resources.

Therefore, the average annual growth in worldwide wind energy capacity from 2001 to 2012 was over 25% (Figure 5.1). The average growth in the United States over the same period was 37.7%. The total worldwide installed wind power capacity, which was 24 GW in 2001 (Figure 5.3), reached a level of 282 GW in 2012 (WWEA 2012). The world's total theoretical potential for onshore wind power is around 55 TW with a practical potential of at least 2 TW (UNDP 2004), which is approximately 40% of the entire present worldwide generating capacity. The offshore wind energy potential is even larger. [*Principles of solar energy by D. Yogi Goswami*]

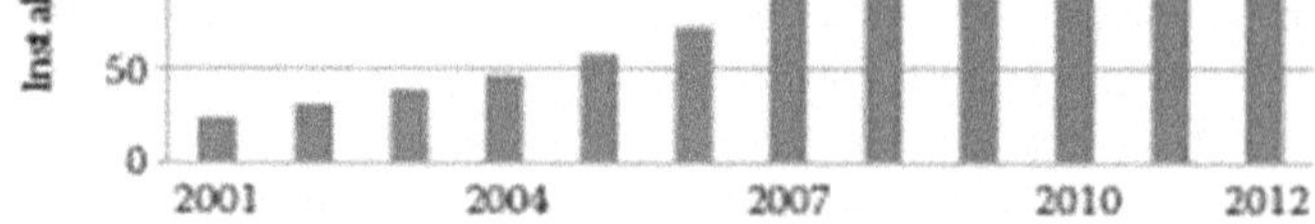

FIG. 5.3: WORLD TOTAL WIND POWER INSTALLED CAPACITY (*Source*: Principles Of Solar Energy By D. Yogi Goswami)

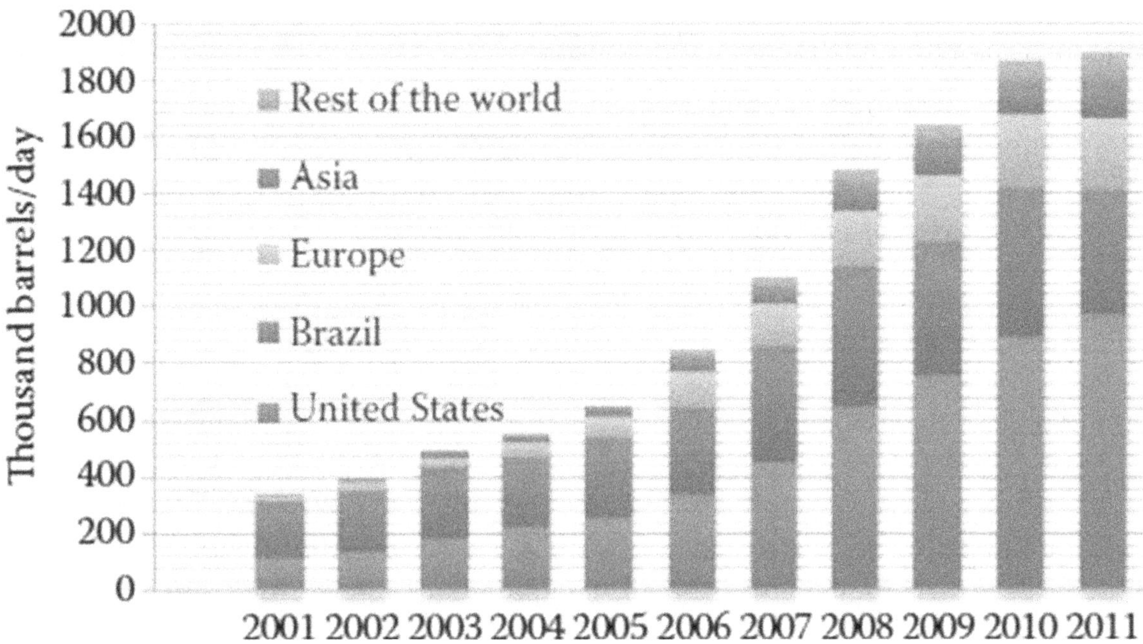

FIG. 5.4: WORLD BIOFUEL PRODUCTION, 2001–2011. (FROM IEA, WORLD ENERGY OUTLOOK 2013, INTERNATIONAL ENERGY AGENCY, PARIS, FRANCE, 2013.) (*Source:* Principles of Solar Energy By D. Yogi Goswami)

5.2.3 Biomass

The total theoretically harvestable biomass energy potential is estimated to be around 90 TW, but the technical and realistic estimate of potential on a sustainable scale is around 8–13 TW or 270–450 exajoules/year. This potential is 3–4 times the present electrical generation capacity of the world. It is estimated that by 2025, even municipal solid waste (MSW) could generate up to 6 exajoules/year.

One of the biggest advantage of biomass as energy is its relatively straightforward transformation into transportation fuels. Biofuels can replace up to 75% of the petroleum fuels in use for transportation in the United States (Worldwatch Institute 2006). This is especially important in view of the declining oil supplies worldwide. Biofuels will not require additional infrastructure development. Therefore, development of biofuels is being favored by governments around the world.

Biofuels, along with other transportation options such as electric vehicles and hydrogen, will help diversify the fuel base for future transportation. Table 5.2, figure 5.4, figure 5.5, figure 5.6 and figure 5.7 show the global production of biofuels from 2001 to 2011. United States, Brazil, and Europe are the top producing countries and region of the world. Biofuel production grew more than 5 times in 10 years, although it started from a much smaller base. In 2005, the world ethanol production had reached approximately 36 billion liters per year while biodiesel production topped 3.5 billion liters during the same year.

The cost of producing ethanol in the present world ranges from approximately 25 Euro cents to approximately 1 Euro per gasoline equivalent liter. While the wholesale price of gasoline is between 40 and 60 Euro cents per liter (Figure 5.5). Biodiesel

TABLE 5.2

UK, 2013, HTTP://WWW.ENERDATA.NET; IEA, *WORLD ENERGY OUTLOOK 2013*, INTERNATIONAL ENERGY AGENCY, PARIS, FRANCE, 2013. *SOURCE:* ENERDATA, *GLOBAL ENERGY STATISTICAL YEARBOOK 2013*, ENERDATA INFORMATION SERVICES, LONDON,

Total Biofuel Production (Thousand Barrels per Day)											
	2001	**2002**	**2003**	**2004**	**2005**	**2006**	**2007**	**2008**	**2009**	**2010**	**2011**
United States	115.7	140.3	183.9	223.3	260.6	335.0	457.3	649.7	747.1	889.8	971.7
Brazil	197.6	216.9	249.4	251.7	276.4	307.3	395.7	486.3	477.5	527.1	438.1
Europe	21.2	29.3	39.3	48.9	76.8	123.9	153.8	198.1	233.2	255.2	250.5
Asia	3.1	8.3	17.2	21.1	28.2	44.9	49.2	75.6	93.8	99.8	118.2
Rest of the world	5.3	8.6	9.6	9.8	14.2	29.6	47.3	67.7	83.8	93.3	118.8
World	342.9	403.5	499.4	554.8	656.3	840.6	1103.3	1477.3	1635.4	1865.4	1897.2

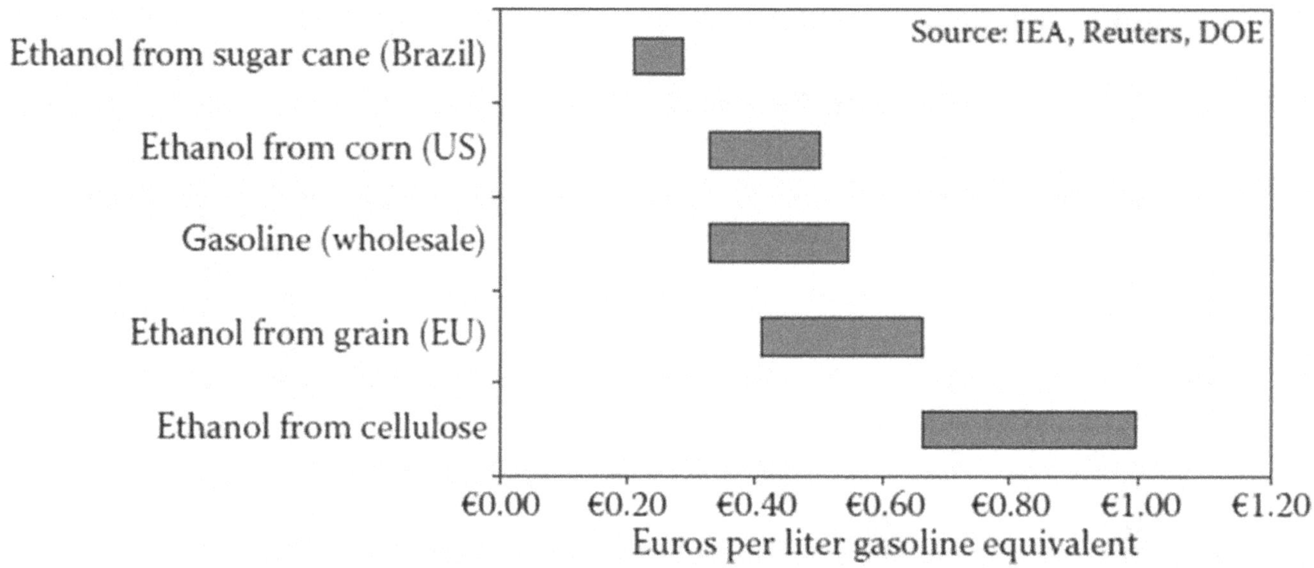

FIG. 5.5: COST RANGES FOR ETHANOL AND GASOLINE PRODUCTION, 2006. (FROM WORLDWATCH INSTITUTE, BIOFUELS FOR TRANSPORTATION—GLOBAL POTENTIAL AND IMPLICATIONS FOR SUSTAINABLE ENERGY IN THE 21ST CENTURY, WORLDWATCH INSTITUTE, WASHINGTON, DC, 2006.) *(Source:* Principles of Solar Energy By D. Yogi Goswami)

costs range between 20 Euro cents to 65 Euro cents per liter of diesel equivalent (Figure 5.6). (Figure 5.7) shows the feedstock used for these biofuels. An important consideration for biofuels is that the fuel should not be produced at the expense of food while there are people going hungry in the world. This would not be of concern if biofuels were produced from MSW or nonfood forest resources.

A city of 1 million people produces approximately 1800 tons of MSW and 1300 tons of organic waste every day according to the Worldwatch Institute report. Using the present-day technology this waste could produce enough fuel to meet the needs of 58,000 persons in the United States, 360,000 in France, and nearly 2.6 million in China at current rates of per capita fuel use (Worldwatch Institute 2006). *[Principles of solar energy by D. Yogi Goswami]*

5.2.4 Ocean Energy Conversion

Almost 3/4 of the world's surface is covered by oceans. Because of the temperature differences produced by the sun as well as the kinetic energy stored in the waves, oceans are the biggest storehouse of solar energy. There are numerous places in the ocean where temperature differences of the order of 20–25 K exist at depths of less than 1000 m, and these temperature differences

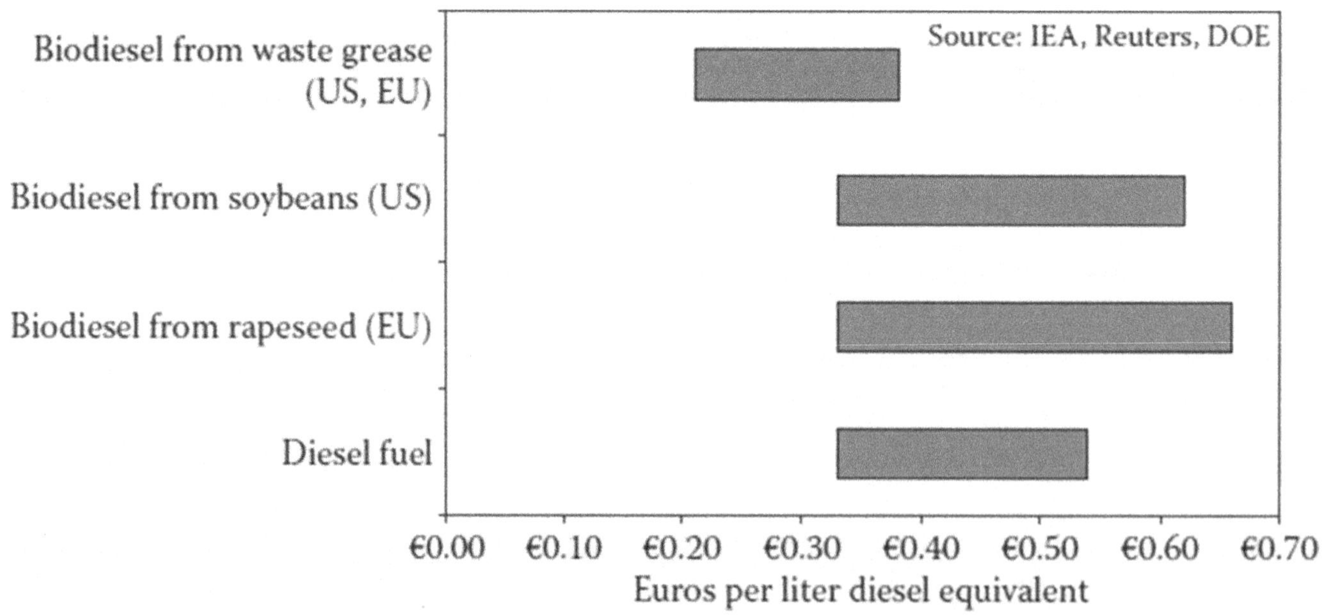

FIG. 5.6: COST RANGES FOR BIODIESEL AND DIESEL PRODUCTION, 2006. (FROM WORLDWATCH INSTITUTE, BIOFUELS FOR TRANSPORTATION—GLOBAL POTENTIAL AND IMPLICATIONS FOR SUSTAINABLE ENERGY IN THE 21ST CENTURY, WORLDWATCH INSTITUTE, WASHINGTON, DC, 2006.) *(Source:* Principles of Solar Energy By D. Yogi Goswami)

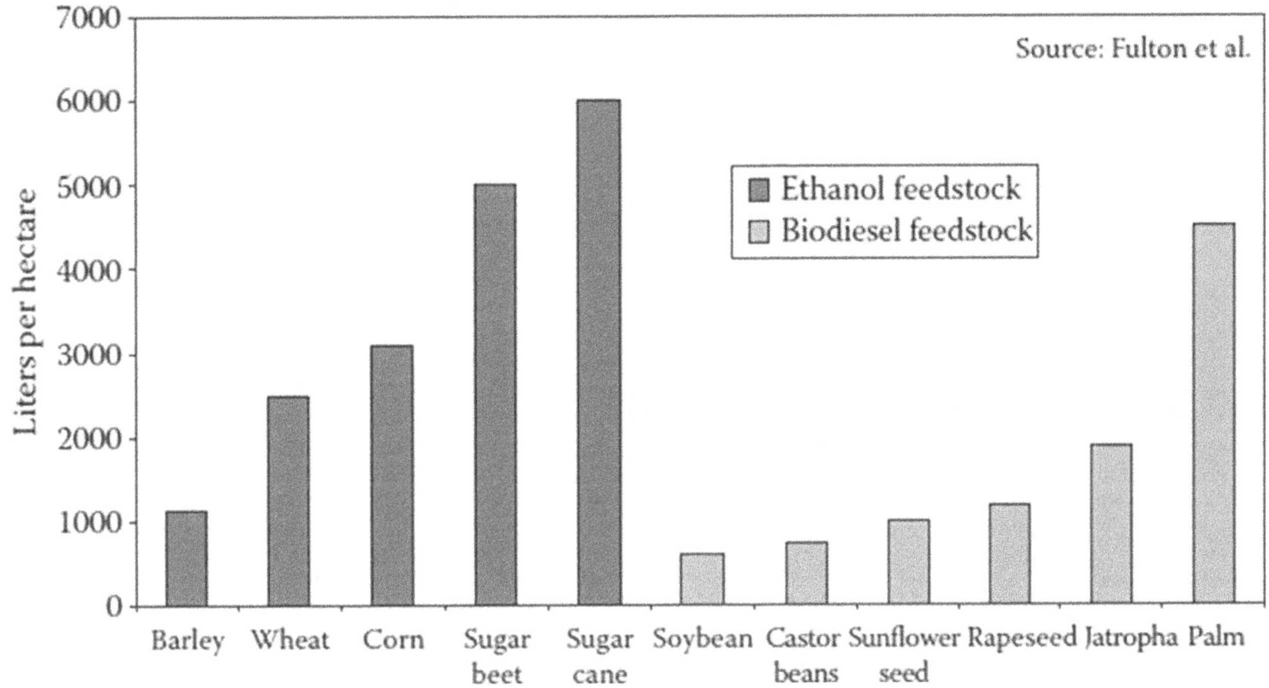

FIG. 5.7: BIOFUEL YIELDS OF SELECTED ETHANOL AND BIODIESEL FEEDSTOCK. (FROM WORLDWATCH INSTITUTE, BIOFUELS FOR TRANSPORTATION—GLOBAL POTENTIAL AND IMPLICATIONS FOR SUSTAINABLE ENERGY IN THE 21ST CENTURY, WORLDWATCH INSTITUTE, WASHINGTON, DC, 2006.) *(Source:* Principles of Solar Energy By D. Yogi Goswami)

could be used to operate low-pressure heat engines. Although the thermodynamic efficiency of a heat engine operating on such a small temperature difference is low, the available amount of thermal energy is very large. However, the difficult part to put this technology into the use of modern world is the development of efficient and cheap heat exchangers that can withstand the rough marine conditions. Since heat-exchange equipment is the most expensive part of any ocean thermal conversion scheme, the cost of using the temperature gradients in the ocean for practical solar energy utilization depends largely on this development. The second method of harvesting energy from the oceans is through ocean waves. Ocean currents and waves possess kinetic energy, which can be converted to power by specially designed turbines. Energy from tides can be harnessed by constructing a dam to hold seawater at a higher elevation when the tide is high and letting it run a turbine when the tide is low.

According to an Electric Power Research Institute study, the total ocean wave and tidal power resource potential for the United States is approximately 2100 TWh/year (EPRI 2007), of which approximately 260 TWh/year could be credibly harnessed. Ocean energy conversion will not be treated in this text. Instead, it will be covered in detail in a companion text on RE conversion. *[Principles of solar energy by D. Yogi Goswami]*

5.2.5 Forecast of Future Energy Mix

It seems very logical to say that the Renewable Energy technologies of solar, wind, and biomass not only will be essential but also will be able to fill the gap and provide a clean and sustainable energy future because of the limited oil, natural gas and nuclear resources, and the serious environmental concerns about coal,. There are many differing views on the future energy mix. However, according to WBGU, which estimates that as much as 50% of the world's primary energy in 2050 will have to come from Renewable Energy (Figure 5.8). However, achieving that level of RE use by 2050 will require a worldwide commitment and an unprecedented global effort. *[Principles of solar energy by D. Yogi Goswami]*

5.3 SOLAR ENERGY

Irradiance is the amount of solar radiant energy falling on a surface per unit area and per unit time. The mean extraterrestrial irradiance normal to the solar beam on the outer fringes of the earth's atmosphere is approximately 1.36 kW/m^2. Since the earth's orbit is elliptical, the sun-to-earth distance varies slightly with time of year, and the actual extraterrestrial irradiance varies by ±3.4% during the year. The angle subtended by the sun when viewed from the earth is only 0.0093 rad (approximately 32 min of arc); the direct beam radiation reaching the earth is therefore almost parallel. Although the brightness of the solar disc decreases from center to edge, for most engineering calculations, the disc can be assumed to be of uniform brightness. *[Principles of solar energy by D. Yogi Goswami]*

The radiant energy from the sun is distributed over a range of wavelengths, and the energy falling on a unit surface area per unit time within a particular spectral band is known as the spectral irradiance; its value is usually expressed in watts per square meter per nanometer of bandwidth. The extraterrestrial spectral irradiance is shown in Figure **5.9**. It can be approximated by the spectrum of a black body at 5800 K. In the upper part of Figure **5.9**, the wave bands typically useable for different solar applications are shown, and the lowest curve shows the spectral direct-beam irradiance

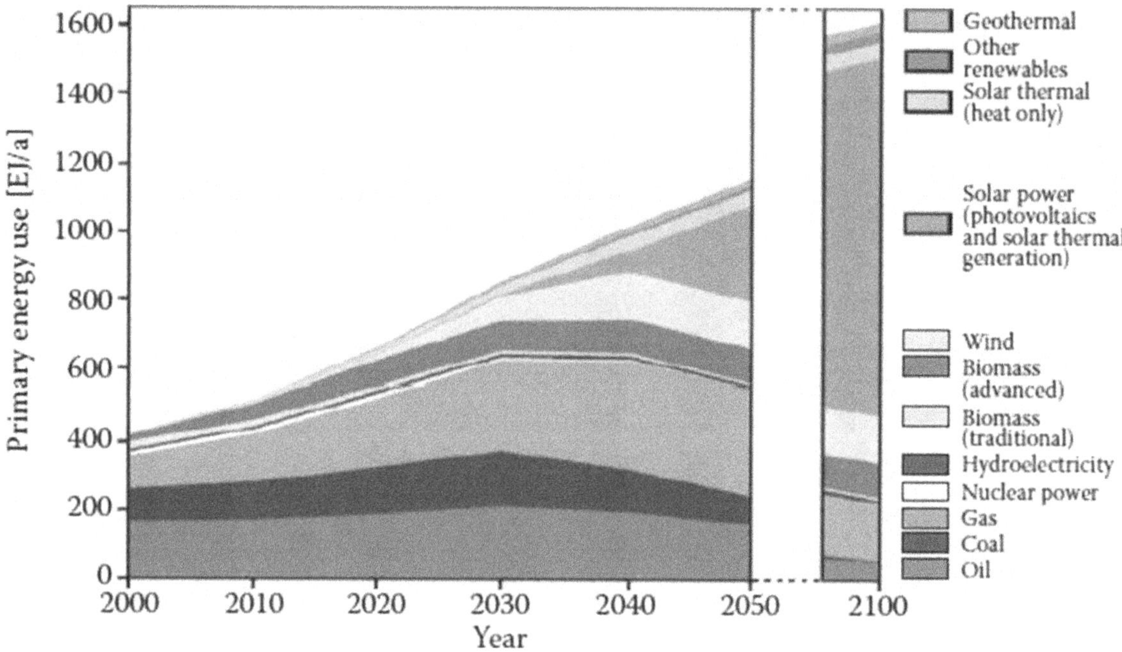

FIG. 5.8: THE GLOBAL ENERGY MIX FOR YEAR 2050 AND 2100 ACCORDING TO WBGU (2003). (*Source:* Principles of Solar Energy By D. Yogi Goswami)

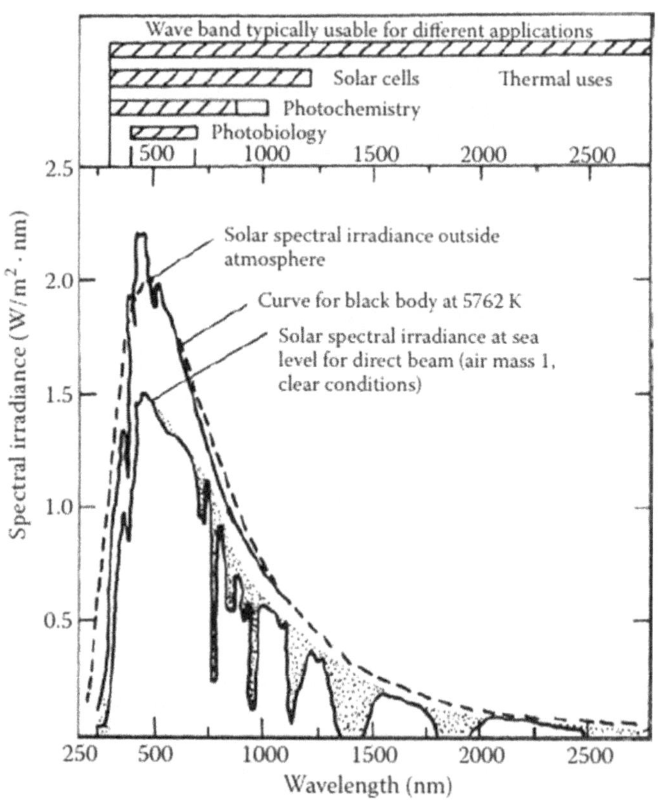

FIG. 5.9: SPECTRAL IRRADIANCE CURVES FOR DIRECT SUNLIGHT ARE EXTRATERRESTRIAL AND AT SEA LEVEL WITH THE SUN DIRECTLY OVERHEAD. SHADED AREAS INDICATE ABSORPTION OWING TO ATMOSPHERIC CONSTITUENTS, MAINLY H2O, CO2, AND O3. WAVELENGTHS POTENTIALLY UTILIZED IN DIFFERENT SOLAR ENERGY APPLICATIONS ARE INDICATED AT THE TOP. (*Source:* Principles of Solar Energy By D. Yogi Goswami)

at sea level on earth under clear sky conditions with the sun overhead. The amount of sunlight striking the earth's atmosphere continuously is 1.75 × 105 TW. Considering a 60% transmittance through the atmospheric cloud cover, 1.05 × 105 TW reaches the earth's surface continuously. If the irradiance on only 1% of the earth's surface could be converted into electric energy with 10% efficiency, it would provide a resource base of 105 TW, while the total global energy needs for 2050 are expected to be approximately 25–30 TW. The present situation of solar energy technologies is such that commercial solar panel efficiencies have reached more than 20%, laboratory multi junction solar cell efficiencies under concentrating sun have exceeded 40%, and solar thermal systems provide efficiencies of 40%–60%. Solar PV panels have come down in cost from approximately \$30/W to approximately \$0.50/W in the last three decades. At \$0.50/W panel cost, the overall system cost is around \$2/W, which is already lower than grid electricity in the Caribbean island communities. Of course, there are many off-grid applications where solar PV is already cost-effective. With net metering and governmental incentives, such as feed-in laws and other policies, grid- connected applications such as Building-Integrated PV have become cost-effective even where grid electricity is cheaper. As a result, the worldwide growth in PV production has averaged more than 43% per year from 2000 to 2012 and 61% from 2007 to 2012 (Figure **5.10**) with Europe showing the maximum growth. *[Principles of solar energy by D. Yogi Goswami]*

The first solar technology that demonstrated its grid power potential was solar thermal power using concentrating solar collectors. A 354 MW concentrating solar thermal power (CSP) plant has been operating continuously in California since 1988. Poor policy and lack of research and development (R&D) impeded the progress of solar thermal power after that time. However, the last 10 years have seen a resurgence of interest in this area and a number of solar thermal power plants around the world are under construction. The largest CSP plant with a capacity of 400 MW came

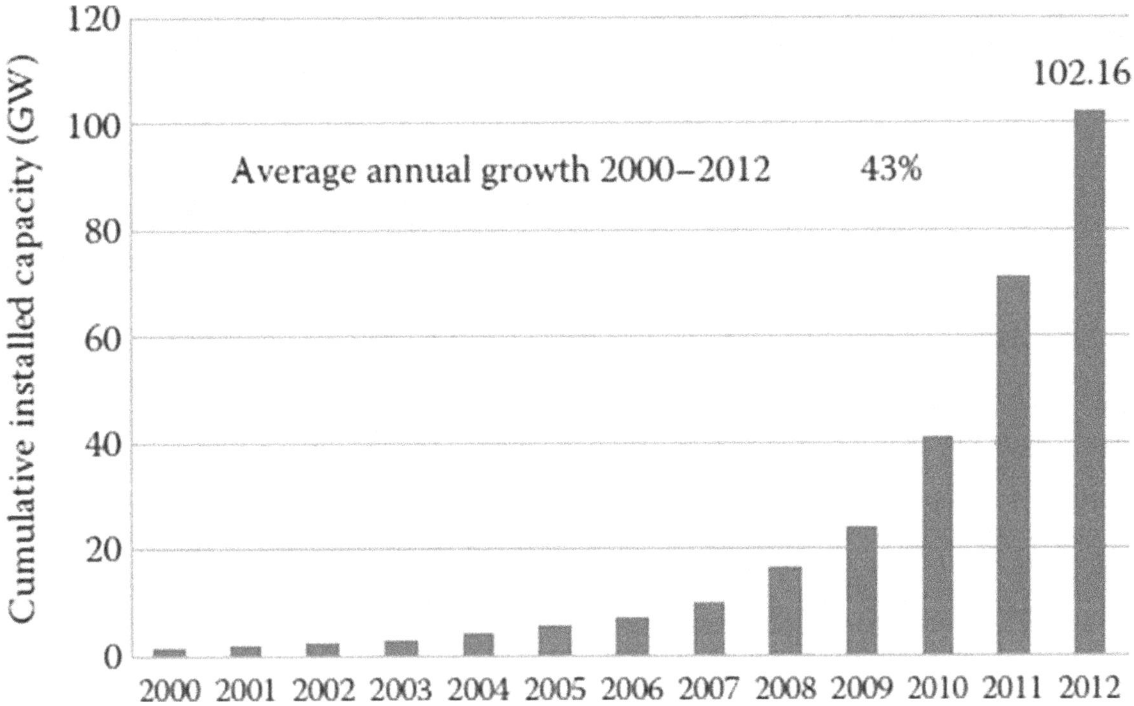

FIG. 5.10: WORLD SOLAR PV PRODUCTION 2000–2012 (GWP). (FROM EPIA, GLOBAL MARKET OUTLOOK FOR PHOTOVOLTAICS, EUROPEAN PHOTOVOLTAIC INDUSTRIES ASSOCIATION, BRUSSELS, BELGIUM, HTTP://WWW .EPIA.ORG.)

on line in Nevada in February 2014. The cost of power from these plants (which is so far in the range of 12 to 16 US cents/kWhe) has the potential to go down to 5 US cents/kWhe with scale-up and creation of a mass market. An advantage of solar thermal power is that thermal energy can be stored efficiently, and fuels such as natural gas or biogas may be used as backup to ensure continuous operation. If this technology is combined with power plants operating on fossil fuels, it has the potential to extend the time frame of the existing fossil fuels. Low-temperature solar thermal systems and applications have been well developed for quite some time. They are being actively installed wherever the policies favor their deployment. Figure **5.11** gives an idea of the rate of growth of solar thermal systems in the world. *[Principles of solar energy by D. Yogi Goswami]*

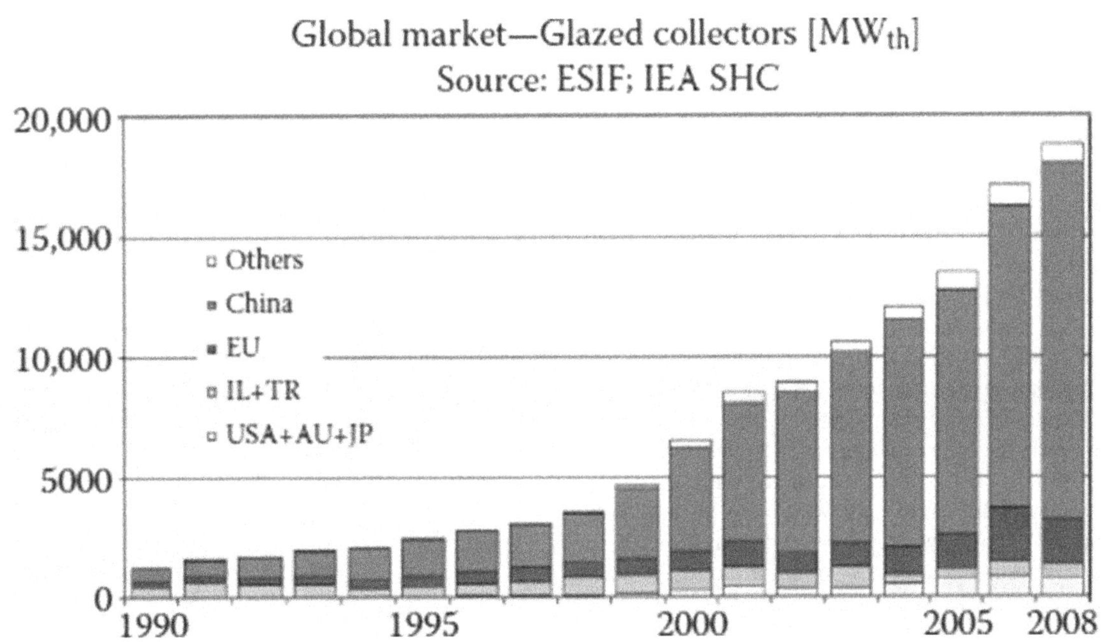

FIG. 5.11: DEPLOYMENT OF SOLAR HEAT (GLAZED) COLLECTORS, MWTH. SOURCE: ESIF; IEA SHC *(Source:* Principles of Solar Energy By D. Yogi Goswami)

5.4 SOLAR THERMAL POWER SYSTEMS USE CONCENTRATED SOLAR ENERGY

Electricity in Solar thermal power (electricity) generation systems is generated by collecting and concentrating sunlight to produce the high temperature heat that produces electricity. All solar thermal power systems have solar energy collectors with two main components: *reflectors* (mirrors) that capture and focus sunlight onto a *receiver*. In most types of systems, a heat-transfer fluid is heated and circulated in the receiver and used to produce steam. The steam is converted into mechanical energy in a turbine, which powers a generator to produce electricity. Solar thermal power systems have tracking systems that keep sunlight focused onto the receiver throughout the day as the sun changes position in the sky.

Solar thermal power systems may also have a thermal energy storage system component that allows the solar collector system to heat an energy storage system during the day, and the heat from the storage system is used to produce electricity in the evening or during cloudy weather. Solar thermal power plants may also be hybrid systems that use other fuels (usually natural gas) to supplement energy from the sun during periods of low solar radiation.

[U.S energy information administration]

[https://www.eia.gov/energyexplained/?page=solar_thermal_power_plants]

5.4.1 Types of concentrating solar thermal power plants

There are three main types of concentrating solar thermal power systems:

- Linear concentrating systems, which include parabolic troughs and linear Fresnel reflectors
- Solar power towers
- Solar dish/engine systems

5.4.2 Concentrated PV Cells

IBM researchers have succeeded in developing a method to make concentrated PV cells that are more efficient in converting the sun's energy into electricity. The researchers have shown that without causing them to melt, it is possible to increase the concentration of light on photovoltaic cells by about 10 times. They say that this makes it possible to boost the amount of usable electrical energy produced by up to five times. The principle behind concentrated PV cells is to use a large lens to focus light onto a relatively small piece of PV semiconductor material. The benefit is the cost reduction because only a fraction of the semiconductor material is used. IBM's solution is to place an ultrathin layer of liquid metal, a compound of gallium and indium, between the two surfaces. The metal has a very high thermal conductivity and because it's a liquid, it is possible to make this layer extremely thin, typically around 10 micrometers. IBM is in talks with solar-cell companies about licensing the technology. Recently, Swiss-based Air light Energy said it has partnered with IBM to bring affordable solar technology to the market by 2017. The system can concentrate the sun's radiation 2,000 times and convert 80% of it into useful energy to generate 12kW of electrical power and 20kW of heat on a sunny day enough to power several average homes.

5.4.2.1 Linear concentrating systems The sun's energy in Linear concentrating systems is collected using long, rectangular, curved (U-shaped) mirrors. The sunlight is focused by mirrors onto receivers (tubes) that run the length of the mirrors. The concentrated sunlight heats a fluid flowing through the tubes. The fluid is sent to a heat exchanger to boil water in a conventional steam-turbine generator to produce electricity. There are two major types of linear concentrator systems: parabolic trough systems, where receiver tubes are positioned along the focal line of each parabolic mirror, and linear Fresnel reflector systems, where one receiver tube is positioned above several mirrors to allow the mirrors greater mobility in tracking the sun. A linear concentrating collector power plant has a large number, or *field,* of collectors in parallel rows that are typically aligned in a north-south orientation to maximize solar energy collection. This configuration enables the mirrors to track the sun from east to west during the day and concentrate sunlight continuously onto the receiver tubes.

[U.S energy information administration]

[https://www.eia.gov/energyexplained/?page=solar_thermal_power_plants]

5.4.2.2 Parabolic troughs A parabolic trough collector has a long parabolic-shaped reflector that focuses the sun's rays on a receiver pipe located at the focus of the parabola. The collector tilts with the sun to keep sunlight focused on the receiver as the sun moves from east to west during the day. Because of its parabolic shape, a trough can focus the sunlight from 30 times to 100 times its normal intensity (concentration ratio) on the receiver pipe, located along the focal line of the trough, achieving operating temperatures higher than 750°F.

Parabolic trough linear concentrating systems are used in the longest operating solar thermal power facility in the world, the Solar Energy Generating System (SEGS). The facility, with nine separate plants, is located in the Mojave Desert in California. The first plant in the system, SEGS I, operated from 1984 to 2015, and the second, SEGS II, operated from 1985 to 2015. The last plant built, SEGS IX, with a electricity generation capacity of 92 megawatts (MW), began operation in 1990. The seven

FIG. 5.12: SOURCE: STOCK PHOTOGRAPHY (COPYRIGHTED) *Source:* https://www.eia.gov/energyexplained/?page=solar_thermal_power_plants]

currently operating SEGS III-IX plants have a combined electricity generation capacity of nearly 357 MW, making them one of the largest solar thermal electric power facilities in the world.

In addition to the SEGS, many other parabolic trough solar power projects operate in the United States and around the world. The three largest projects in the United States after SEGS are

- Mojave Solar Project: a 280 MW project in Barstow, California
- Solana Generating Station: a 280 MW project in Gila Bend, Arizona
- Genesis Solar Energy Project: a 250 MW project in Blythe, California

[U.S energy information administration]

[https://www.eia.gov/energyexplained/?page=solar_thermal_power_plants]

5.4.2.3 Linear Fresnel reflectors In Linear Fresnel reflector (LFR) systems, the mirrors (reflectors) concentrate sunlight onto a receiver located above the mirrors. These reflectors use the Fresnel lens effect, which allows for a concentrating mirror with a large aperture and short focal length. These systems have the capacity of concentrating the sun's energy to approximately 30 times its normal intensity. The only operating linear Fresnel reflector system in the United States is a compact linear Fresnel reflector (CLFR) also referred to as a concentrating linear Fresnel reflector a type of LFR technology that has multiple absorbers within the vicinity of the mirrors. Multiple receivers allow the mirrors to change their inclination to minimize how much they block adjacent reflectors' access to sunlight. This positioning improves system efficiency and reduces material requirements and costs.

[U.S energy information administration]

[https://www.eia.gov/energyexplained/?page=solar_thermal_power_plants]

5.4.3 Solar power towers

A large field of flat, sun-tracking mirrors called heliostats are used solar power tower system to reflect and concentrate sunlight onto a receiver on the top of a tower. Sunlight can be concentrated

FIG. 5.13: SOLAR POWER TOWER; *SOURCE:* NATIONAL RENEWABLE ENERGY LABORATORY (NREL) [U.S ENERGY INFORMATION ADMINISTRATION] [HTTPS://WWW.EIA.GOV/ENERGYEXPLAINED/?PAGE=SOLAR_THERMAL_POWER_PLANTS]

as much as 1,500 times. Some power towers use water as the heat-transfer fluid. Advanced designs are experimenting with molten nitrate salt because of its superior heat transfer and energy storage capabilities. The thermal energy-storage capability allows the system to produce electricity during cloudy weather or at night.

[U.S energy information administration]

[https://www.eia.gov/energyexplained/?page=solar_thermal_power_plants]

5.4.4 Efforts of the U.S. Department of Energy and Electric Utilities

The first demonstration solar power tower was built and operated by The U.S. Department of Energy along with several electric utilities near Barstow, California, during the 1980s and 1990s. Three solar power tower projects now operate in the United States:

1. Ivanpah Solar Power Facility: a 392 MW three-tower project with generation capacities of 126 MW, 133 MW, and 133 MW located in Ivanpah Dry Lake, California
2. Crescent Dunes Solar Energy Project: a 110 MW one-tower project located in Nevada
3. Sierra Sun Tower: a 5 MW two-tower project located in the Mojave Desert in Southern California

Solar dish/engine systems use a mirrored dish similar to a very large satellite dish. The mirrored dish is usually composed of many smaller flat mirrors formed into a dish shape to reduce costs. The dish-shaped surface directs and concentrates sunlight onto a thermal receiver, which absorbs and collects the heat and transfers it to an engine generator. The most common type of heat engine used in dish/engine systems is the Stirling engine. This system uses the

FIG. 5.14: SOLAR DISH/ENGINES. SOURCE: STOCK PHOTOGRAPHY (COPYRIGHTED) [U.S ENERGY INFORMATION ADMINISTRATION] [HTTPS://WWW.EIA.GOV/ENERGYEXPLAINED/?PAGE=SOLAR_THERMAL_POWER_PLANTS]

fluid heated by the receiver to move pistons and create mechanical power. The mechanical power runs a generator or alternator to produce electricity. Solar dish/engine systems always point straight at the sun and concentrate the solar energy at the focal point of the dish. A solar dish's concentration ratio is much higher than linear concentrating systems, and it has a working fluid temperature higher than 1,380°F. The power-generating equipment used with a solar dish can be mounted at the focal point of the dish, making it well suited for remote locations, or the energy may be collected from a number of installations and converted into electricity at a central point. The U.S. Army is developing a 1.5 MW system at the Tooele Army Depot in Utah with 429 Stirling engine solar dishes. The system is scheduled to be fully operational in 2017.

[U.S energy information administration]

[https://www.eia.gov/energyexplained/?page=solar_thermal_power_plants]

5.4.5 Economics of Solar energy

U.S. is the leading consumer of renewable energy even though RE is only a fraction of total energy consumed. The solar still only accounts for 0.4% of total energy used in the U.S despite the increase of available solar energy over the past 10 years. Solar power also trails hydropower, biomass and wind in terms of preferred sources of renewable energy, making up 4% of total U.S. renewable consumption. Currently, only two types of solar technology exist which are capable of converting the sun's energy into a source of power: solar thermal and photovoltaic. Solar thermal collectors absorb the sun's radiation in order to heat a home or water. Photovoltaic devices use sunlight to replace or supplement the electricity provided on the utility grid.

https://www.investopedia.com/articles/investing/061115/economics-solar-power.asp

5.4.5.1 Solar Power Adoption Solar energy systems that were only afforded by the wealthy or fanatical before, are now affordable and accessible universally for everyone mainly due to sharply declining costs. In the early 2000s, the average U.S. solar system cost $10 per watt; in 2013, the price per watt was just below $4. As a result, the number of photovoltaic systems installed in the U.S. has drastically increased among residential and commercial spaces. Over the past decade, it is estimated that the global output from photovoltaics has increased 40% each year.

Solar energy has experienced a global growth in consumption as more countries become aware of the harmful effects of burning fossil fuels. Increased competition within the solar power industry has resulted in sharp declines in installation costs. Many of the largest economies, including the U.S., China, India and several European nations, have begun to implement solar energy. In an effort to combat pollution, China has made the biggest push into renewable energy and installed the most photovoltaics in 2014. Similarly, India, which is also plagued by pollution, is making a $160 billion plan for solar energy expansion.

Big businesses are also investing in reusable solar systems. Walmart (WMT), Verizon (VZ) and Apple (AAPL) are switching stores, offices and facilities to solar energy. In the largest ever solar procurement deal, Apple purchased 130 megawatts for $850 million from First Solar (FSLR) in February.

Although solar power continues to account for a small share of overall energy supply, the residential and commercial sectors are slowly embracing renewable energy. As prices continue to decline, it is expected that solar energy systems become more prevalent. In Europe, the price per kilowatt hour is expected to decline to between 4 and 6 cents in 2025 and further decrease to as low as 2 cents in 2050.

Solar photovoltaics will be amongst the cheapest sources of energy according to the current forecasts. With declining prices, the IEA conservatively estimates solar systems to supply 5 percent of global electricity consumption in 2030, rising to 16 percent by 2050. Achieving this vision would require increasing the global capacity of solar energy from 150 gigawatts in 2014 to 4600 gigawatts by 2050. As a result, this would avoid the emission of 4 Gt of carbon dioxide annually. In conjunction with increased production of renewable energy, there is an increasing commitment to declining greenhouse gas emissions from burning fossil fuels. Many cities and countries around the world have committed to cutting greenhouse gas emissions 80 percent by 2050, including New York City. Besides cutting emissions, California has committed to producing 33% of total energy by renewable resources by 2020.

[https://www.investopedia.com/articles/investing/061115/economics-solar-power.asp]

5.4.5.2 Tax Credits Even though solar energy systems are more cost-effective today, residential and commercial usage still receive government subsidies. In the U.S., the Renewable Energy Tax Credit decreases the tax liability of solar energy users. A taxpayer can claim a credit of 30% of qualified expenditures for systems that serve an occupied space. The U.S. government applies the same credit to wind and geothermal systems. Many European countries impose a Feed-In-Tariff scheme to increase the appeal of renewable energy systems. Under a feed-in-tariff scheme, renewable energy system owners can collect money from the government. Costs are based on per kilowatt-hour (kWh), with prices varying between countries.

[https://www.investopedia.com/articles/investing/061115/economics-solar-power.asp]

5.4.5.3 The Bottom Line For the most part, the commitment to renewable resources has come from individuals, big businesses and countries. Besides solar energy, companies such as Google and Amazon have committed to using wind to power company facilities. With big businesses, individuals and countries continuing to transition to renewable energy sources, adverse environmental effects from burning fossil fuels can hopefully be moderated.

[https://www.investopedia.com/articles/investing/061115/economics-solar-power.asp]

5.5 10 MAJOR APPLICATION OF SOLAR ENERGY

The evidence of increasing interest in solar power is not only on roofs, but also in innovative ideas to power up almost every part of our lives. Some of the major applications of solar energy are:

- Solar electric power generation
- Solar thermal power production
- Solar heating of buildings
- Solar Green Houses
- Solar water heating
- Solar pumping
- Solar furnaces

- Solar drying of agricultural and animal products
- Solar cooking
- Solar distillation.

Photovoltaic cells have been used to operate rail road crossing warnings, navigational signals, highway emergency call systems, automatic meteorological stations etc. in areas where it is difficult to lay power lines. They are also used for weather monitoring and as portable power sources for televisions, calculators, watches, computer card readers, battery charging and in satellites etc. Besides these, photovoltaic cells are used for the energization of pump sets for irrigation, drinking water supply and for providing electricity in rural areas i.e. street lights etc.

[http://www.yourarticlelibrary.com/energy/solar-energy-10-major-application-of-solar-energy-explained/28197]

5.5.1 Solar Electric Power Generation

Photovoltaic cells are used to produce electric energy or electricity directly from solar energy by means of. The photovoltaic cell is an energy conversion device which is used to convert photons of sunlight directly into electricity. It is made of semi- conductors which absorb the photons received from the sun, creating free electrons with high energies. High energy free electrons are induced by an electric field, to flow out of the semiconductor to do useful work. This electric field in photovoltaic cells is usually provided by a p-n junction of materials which have different electrical properties. There are different fabrication techniques to enable these cells to achieve maximum efficiency. These cells are arranged in parallel or series combination to form cell modules. Some of the special features of these modules are high reliability, no expenditure on fuel, minimum cost of maintenance, long life, portability, modularity, pollution free working etc.

[http://www.yourarticlelibrary.com/energy/solar-energy-10-major-application-of-solar-energy-explained/28197]

5.5.2 Solar Thermal Power Production

The conversion of solar energy into electricity through thermal energy is called solar thermal power production. In this method, solar energy is first used to heat up a working fluid, gas, water or any other volatile liquid. This heat energy is then converted into mechanical energy from a turbine. Finally a conventional generator coupled to a turbine converts this mechanical energy into electrical energy.

[http://www.yourarticlelibrary.com/energy/solar-energy-10-major-application-of-solar-energy-explained/28197]

5.5.3 Solar Heating of Buildings

Solar energy can be used for space heating of buildings in many ways namely:

a) Collecting the solar radiation by some element of the building itself i.e. solar energy is admitted directly into the building through large South-facing windows.
b) Using separate solar collectors which may heat either water or air or storage devices which can accumulate the collected solar energy for use at night and during inclement days. When the building requires heat then from these collectors or storage devices, the heat is transferred by conventional equipment such as fan, ducts, air outlets, radiators and hot air registers etc. to warm up the living spaces of a building. When the building does not require heat, the heated air or water from the collector can be moved to the heat storage device such as well insulated water tank or other heat holding material. For inclement days, an auxiliary heating system using gas, oil or electricity is required as a backup system.

[http://www.yourarticlelibrary.com/energy/solar-energy-10-major-application-of-solar-energy-explained/28197]

5.5.3.1 Solar Windows New solar electric window technologies collect the invisible rays contained in sunlight and transforming them into electricity while simultaneously allowing visible light to shine through glass panels. These applications are still exploratory, but one company is working with Sun Shot to make the technology practical across large sheets of glass, so the windows of commercial buildings can serve dual purposes. [Office of energy efficiency and renewable energy] [https://www.energy.gov/eere/articles/solar-power-does-what-4-unusual-uses-photovoltaic-technology]

Solar windows are made of solar panels that harvest the energy from the sun, converting it into electricity. These panels retrofit all types of windows, helping building owners achieve partial independence from the power grid. [##] Over 54 percent of the electricity consumed globally comes from burning fossil fuels, contributing to 99 percent of greenhouse gas emissions. The installation of solar windows can significantly reduce our dependence on fossil fuels, ensuring a sustainable future for all. Organizations have used organic photovoltaic (OPV) technology to develop electricity-generating see-through windows, turning commercial buildings, towers, and skyscrapers into vertical power generators. Scientists working on this project claim that installing solar windows on a 50-story tower can reduce carbon emissions equivalent to that emitted by 2.2 million vehicles per year. The National Renewable Encrgy Laboratory (NREL) is working on introducing switchable solar windows with glasses that change from clear to colored in the presence of sunlight. Though this project has a long way to go before it enters the commercialization phase, it is expected to drastically improve our dependence on renewable energy resources. The solar window can convert every window into a source of electricity, paving the way for a zero-emission and sustainable future.

[https://www.impero.co.in/whatsnew/2018/07/19/future-innovation-of-solar-energy/]

FIG. 5.15: SOLAR WINDOWS (CREDIT: PAUL WELLMAN / NEXT ENERGY TECHNOLOGIES) [HTTPS://WWW.ENERGY.GOV/EERE/ARTICLES/SOLAR-POWER-DOES-WHAT-4-UNUSUAL-USES-PHOTOVOLTAIC-TECHNOLOGY]

Solar windows may seem like a fantasy, but they could be on the market soon. A solar window is a transparent glass window that uses sunlight to create electricity, just like a solar panel. Solar windows are still in the development stage in the US. The purpose of a window is to let light through, but solar window technology requires striking the right balance of captured light for energy and remaining light that can brighten a room. The good news is that companies in the US and abroad have produced solar windows that work, suggesting that these technologies may be available to consumers sooner than we think. In 2017, Solar-Window Technologies announced a partnership with Tri-view Glass Industries to create Solar-Window™ products that can generate electricity. The key to this new invention is in Solar-Window Technologies' liquid transparent coatings that make the glass act as a photovoltaic device. Once it reaches the commercialization stage, the Maryland-based company plans to use these windows to turn skyscrapers into vertical power generators. "Our company is essentially writing the chapter on large-area commercialization of organic photovoltaic devices and panels," says CEO John A. Conklin. Buildings consume 40 percent of electricity in the US, and the company anticipates that Solar-Window coatings could cut the annual $150 billion dollars it costs to pay for their electricity in half, according to company models, which feature Solar-Window on all four sides of a commercial building. In addition to its commercial glass products, the company is developing window glass for residential homes. Meanwhile, Physee, a power company in the Netherlands, has patented similar solar windows made with transparent double-paned glass. Physee has already successfully installed these Power-Windows in commercial office buildings in Amsterdam's Zuidas business district and plans to install 1,850 square meters of them in Amsterdam's BOLD apartment high-rise.

[https://www.greenamerica.org/new-green-tech-promise-and-pitfalls/bright-future-solar-power]

5.5.3.2 Solar Walls and Exposed Surfaces A thin-film perovskite solar cells able to be printed directly onto silicon solar cells is being developed at Oxford Photovoltaics Ltd., a spin-off from the University of Oxford. In December, Oxford PV said it got 8.1 million pounds ($10 million) of additional funding from investors including Statoil ASA. "We expect to have a product that meets industry requirements by the end of 2017," Frank Averdung, chief executive officer at Oxford PV said by email. "Adding some time for qualification, certification and production, our first product could be commercially available towards the end of 2018."

[https://www.renewableenergyworld.com/articles/2017/03/the-wonder-material-that-may-make-spray-on-solar-pv-reality.html]

The challenge is to come up with a way to ensure the material remains stable outdoors for long periods of time. Methods for painting the material on large surfaces must also be improved, said Masanori Iida, an official at the technology and design sector at Panasonic. "It is difficult to continuously make the coating even," Iida said by email. "Perovskites aren't ready for prime time," said Robert Armstrong, director of the Massachusetts Institute of Technology's Energy Initiative. "They degrade too quickly," he said, but added that progress is being made to make them more stable.

Others are also cautious about the commercial viability of this new solar technology. "It certainly is going to be more than five years and it could be never," Green, the professor at the University of New South Wales, said, referring to how soon perovskite solar cells will be commercially available. "For any technology to be marketable, it has to be certified and the testing process could take time. You need someone to bankroll an operation to scale up to go into manufacturing."

[https://www.renewableenergyworld.com/articles/2017/03/the-wonder-material-that-may-make-spray-on-solar-pv-reality.html]

5.5.3.3 Solar Panels A photoelectric effect is when certain substances produce a small electric current when exposed to light. Photovoltaic use this effect to produce electricity. Solar panels consist of large arrays of these devices positioned to receive as much sunlight as possible. Together, they can generate significant amounts of electricity. Although, they are still relatively inefficient, they normally have very low running and maintenance costs, and can be very effective in providing power for homes. A great deal of research is taking place into improving efficiency and constructing cells from cheaper materials.

FIG. 5.16: SOLAR WINDOW CAN CONVERT EVERY WINDOW INTO A SOURCE OF ELECTRICITY (SOURCE:HTTPS://WWW.GREENAMERICA.ORG/NEW-GREEN-TECH-PROMISE-AND-PITFALLS/BRIGHT-FUTURE-SOLAR-POWER)

FIG. 5.17: SOLAR WINDOWS ARE LIKE INVISIBLE SOLAR (*SOURCE:*HTTPS://WWW.GREENAMERICA.ORG/NEW-GREEN-TECH-PROMISE-AND-PITFALLS/BRIGHT-FUTURE-SOLAR-POWER)

Most solar panels consist of crystalline silicon PV cells, which are 14-16% efficient in converting sunlight into electricity. A number of alternatives, however, are under investigation. Thin film cells can be made from a variety of materials. Although these are currently less efficient than standard PV cells, they are lightweight, flexible, and cheaper to manufacture. Multi junction cells can achieve an efficiency of more than 43%. They are structured so that different parts of the cell are tuned to capture sunlight at specific wavelength ranges, rather than having a single receptor that misses a portion of the available energy.

Another promising area is the dye-sensitized solar cell (DSSC), sometimes called the Gratzel cell, after Michael Gratzel, who first developed it in the 1990s. These use a dye to capture solar energy and produce a flow of electrons, which are replenished via a liquid electrolyte layer below. Although potentially cheap to produce, they are only about 12% efficient, and there are problems with durability that may affect commercial use. For example, the liquid can freeze in cold conditions, or expand when it is warm, and may leak. Researchers have developed a version of the cell in which the troublesome liquid electrolyte is replaced with a solid material, opening the way for cheap and durable solar panels.

An important part of the future of solar energy lies in the construction of new buildings and the retrofitting of many older ones. Some experts predict that most, if not all, new buildings will have solar panels installed on roofs. Since these are also easily fitted, many older buildings may receive upgrades to run on power from the Sun. Experts and environmentalists hope that green-energy building will be encouraged by governments through generous tax incentives, exemptions and grants for alternative energy use.

Solar panels on a roof can, in some areas, provide all or most of a home's energy needs. In cases where people live in multi-story accommodation, however, the amount of roof space is very small compared to the number of homes. While small, individual, applications can take some of the strain off the electrical grid, if the Sun is to provide the power needs of cities and industries, its future must lie in large solar powered electricity generating stations.

The space required to build the power plants is the biggest problem facing the harnessing of solar energy using PV cells. A plant is comprised of thousands of solar panels, not unlike those currently installed on alternative energy homes. Because of this, they require a consistently sunny area and a considerable amount of space. Currently, the one of the largest power stations in the world covers more than 10 square miles (16.9 km^2) and creates enough electricity to run about 200,000 homes. Some experts suggest that to provide power for the entire United States, an area approximately 100 miles (160.9 km) per side would be required, probably somewhere in the desert climate of the American Southwest directly for industrial purposes, or to provide electricity for residential and commercial usage.

[http://www.dokiosolar.com/news/show.asp?i=5]

5.5.3.4 Alternatives to Solar Panels A lot of alternatives exist for the large-scale harnessing of the Sun's power. One example is concentrating solar power (CSP) technology. Rather than generating electricity directly, these concentrate sunlight to heat water, providing steam to drive a turbine that produces electricity just like a conventional power plant. They may consist of arrays of parabolic mirrors that focus sunlight on a linear tube filled with liquid. Alternatively, heat from the Sun may be focused by a parabolic mirror to heat a fluid that drives a Stirling engine, which provides the mechanical energy for electricity generation.

Another proven system is the "power tower," in which an array of sun-tracking flat mirrors focuses heat from the Sun on a container of liquid that is used to provide steam for a generator. A number of plants are in operation, producing 10-20 megawatts of electricity. Future plants may deliver up to 200 megawatts. [http://www.dokiosolar.com/news/show.asp?i=5]

5.5.3.5 Light Weight Solar Panels for Home and Business Sunflare, a Los Angeles-based startup, is looking to become the next solar heavyweight with its lightweight panels. Compared to conventional solar panels, Sunflare's panel products have the same metallic blue look, but not much else in common. The company's panels are thin, flexible, and lightweight. Instead of silicon and glass, Sunflare uses a stainless steel substrate with copper, indium, gallium, and selenide to make a semiconductor that's only a few micro meters thick.

[https://www.greenamerica.org/new-green-tech-promise-and-pitfalls/bright-future-solar-power]

Rectangular Sunflare panels come in both standard and custom sizes. With super-thin panels come more possibilities. Rather than relying on heavy aluminum frames and skilled installers, Sunflare panels will stick almost anywhere with a special adhesive, and because they're flexible, they can follow a curve instead of being limited to flat installations.

If a branch shades a traditional solar panel on a home's roof, it will trigger the shutoff of that panel or a series of panels, explains Sunflare's chief marketing officer Elizabeth Sanderson. She notes that Sunflare panels have more diodes that help bypass just the shaded area. In addition, these unique panels can be used with off-grid homes that can't support regular panels, like trailer homes, tiny homes, or RVs.

One of Sunflare's first customers stuck the panels onto her tiny home in Spain, allowing her to live off the grid. And Vestibule, a camping-trailer manufacturer, has worked with Sunflare to incorporate panels into the roof of its teardrop-shaped trailers.

FIG. 5.18: LIGHT WEIGHT SOLAR PANELS FOR HOME AND BUSINESS (*Source:*HTTPS://WWW.GREENAMERICA.ORG/NEW-GREEN-TECH-PROMISE-AND-PITFALLS/BRIGHT-FUTURE-SOLAR-POWER)

Sanderson says she's excited about the products' commercial applications: "Silicon panels are four times heavier than the Sunflare panels. Warehouses and distribution center roofs often can't handle that kind of weight. Also, commercial building owners aren't very happy when they have to puncture their roofs to install racking. But they are happy with Sunflare installation; its peel and stick. And the panels are durable and can provide more insulation for the building. It's a win-win-win." Because Sunflare is in the initial phases of scale-up and in the beginning of its technology lifecycle, its prices are about 50-100 percent higher than traditional panels. However, that gap narrows as you take its much cheaper installation costs into account.

[https://www.greenamerica.org/new-green-tech-promise-and-pitfalls/bright-future-solar-power]

5.5.3.6 Solar Skin Design A high percentage of homeowners do not consider solar panels to be pleasing to eyes. But, Sistine Solar, a Boston-based design firm, is making major strides with the concept of aesthetic enhancement that allow solar panels to have a customized look. The MIT startup has created a "solar skin" product that makes it possible for solar panels to match the appearance of a roof without interfering with panel efficiency or production.

[https://news.energysage.com/solar-panel-technology-advances-solar-energy/]

5.5.3.7 Perovskites Due to the declining cost of silicon solar panels and a global push for clean energy sources to address climate change, the future of solar power has looked increasingly bright in recent years. In the most optimistic scenarios for the power source, according to data from the International Energy Agency, solar generation will increase 16-fold between 2016 and 2040. But now new research into a class of materials known as perovskites is underway in U.S. government labs and other facilities around the world which promises to accelerate rapidly even the most optimistic projections for solar deployment. If they make it from the lab to commercial use of perovskites could dramatically change the market for the renewable energy source offering potential for panels to capture energy virtually any surface facing the sun.

"The future of perovskites is breathtaking," says Varun Sivaram, a fellow at the Council on Foreign Relations and author of "Taming the Sun," a book on the future of solar power. "For solar to really anchor a clean energy transition we're going to need technological innovation and perovskite is the frontrunner in my mind." You don't need a physics or engineering degree to understand the primary reasons why perovskites potentially serve as a superior alternative to traditional solar cells. For one, perovskite solar cells are flexible and easy to produce. They can be painted or sprayed on a surface from an ink solution or churned out of a printer like a newspaper. That flexibility means they can be attached to virtually anywhere.

During a visit to the National Renewable Energy Lab in Golden, Colo., researchers Joseph Berry and David Moore excitedly point out the various research projects underway to transform perovskites from a lab phenomenon to a commercial reality. Moore lifts a small surface with visible perovskite solar cells he painted on himself with a brush purchased at a local craft store, showing how easily the cells can be applied. Berry points to an experiment where perovskite cells are being held under bright light to test their durability. Moore holds up a light to a solar cell that dims in response to the light, mimicking potential protection against sunlight. "The materials have the right properties that make them good at this," says Berry.

Researchers also think they are a more efficient material for capturing the sun's energy. Silicon solar panels, the primary type used today on rooftops and in utility-scale power plants alike, tend to achieve about 20% efficiency, meaning they capture a fifth of the energy that passes through and turn it into electricity. A perovskite solar cell reached a record 22.7% efficiency last summer and researchers say that number is likely to continue to rise with continued to research. The theoretical maximum efficiency of perovskites tops 40%. And because perovskites are so thin, they could be applied to silicon modules that have already been built, boosting electricity production.

[http://time.com/5297011/solar-energy-perovskite-national-lab/]

In the past, various solar technologies have offered either flexibility or efficiency, but researchers have struggled to combine the two. "Perovskites shatter that trade off and promise to do better on literally every performance characteristic," says Sivaram.

[http://time.com/5297011/solar-energy-perovskite-national-lab/]

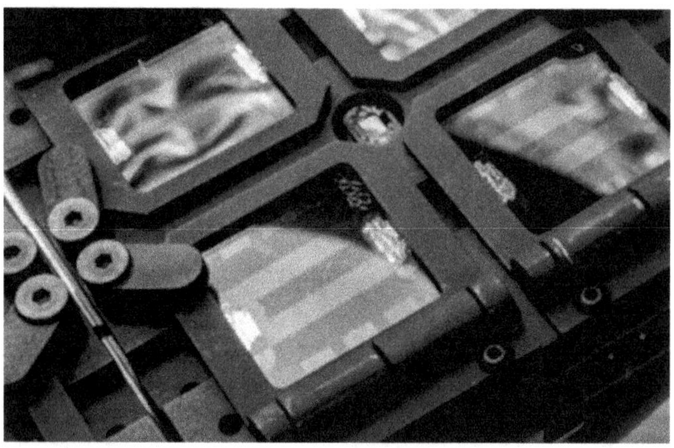

FIG. 5.19: PICTURE OF PEROVSKITES Source: Perovskite Solar Cells lab at the National Renewable Energy Laboratory –Dennis Schroeder --NREL [http://time.com/5297011/solar-energy-perovskite-national-lab/]

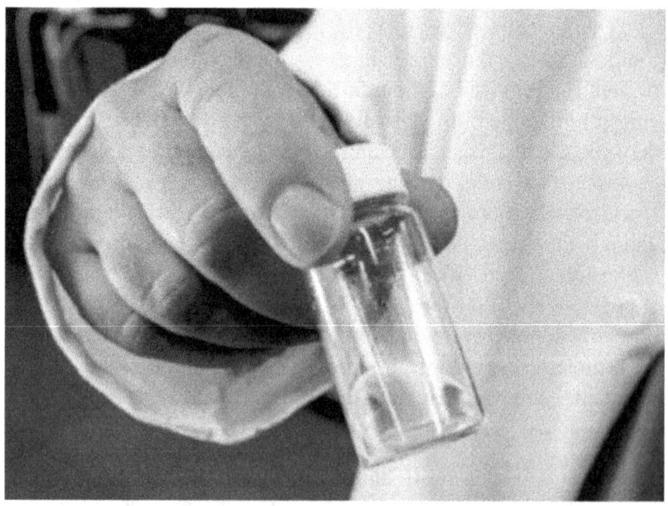

FIG. 5.20: NREL SCIENTIST DAVID MOORE HOLDS A CONTAINER HOLDING A PEROVSKITE INK FORMULATION WHICH CAN BE PAINTED ONTO A SURFACE TO CREATE A SOLAR CELL. (WERNER SLOCUM—NREL) [HTTP://TIME.COM/5297011/SOLAR-ENERGY-PEROVSKITE-NATIONAL-LAB/]

5.5.3.8 Solar Roof Tiles Solar Roof tiles look similar to regular roof tiles, whether you have basic gray shingles, slate tiles, or clay Spanish tiles. Tesla began taking orders for its Solar Roof in May, 2018. Tesla tiles generate solar power, invisibly. The Solar Roof currently comes in four styles and can be installed on the entire roof or just on the sunny side. Consumers thought it would be years or decades before the Solar Roof would hit the market. But in January 2018, Tesla confirmed that the roofs had been installed on homes of company executives, and installation has begun at the homes of non-employee customers. Tesla has an online calculator to estimate how much you'll spend on a Solar Roof per month based on the size of your roof, minus the value of the energy your Solar Roof would generate and any available tax incentives. Tesla guarantees the tiles will last through the life of your house (or infinity, whichever comes first) even if the solar capacity stops working after 30 years. Tesla recently announced it would put kiosks in 800 Home Depot stores across the country to sell its solar panels, Powerwall batteries, and Solar Roofs, and Lowe's is in talks to sell Tesla solar products as well.

[https://www.greenamerica.org/new-green-tech-promise-and-pitfalls/bright-future-solar-power]

5.5.4 Solar Green Houses

A green house is a structure covered with transparent material (glass or plastic) that acts as a solar collector and utilizes solar radiant energy to grow plants. It has heating, cooling and ventilating devices for controlling the temperature inside the green house. Solar radiations can pass through the green house glazing but the thermal radiations emitted by the objects within the green house cannot escape through the glazed surface. As a result, the radiations get trapped within the green house and result in an increase in temperature. As the green house structure has a closed boundary, the air inside the greenhouse gets enriched with CO_2 as there is no mixing of the greenhouse air with the ambient air. Further, there is reduced moisture loss due to restricted transpiration. All these features help to sustain plant growth throughout the day as well as during the night and all year round.

[http://www.yourarticlelibrary.com/energy/solar-energy-10-major-application-of-solar-energy-explained/28197]

FIG. 5.21: SOLAR ROOF TILES (*Source*: Https://Www.greenamerica.org/New-Green-Tech-Promise-And-Pitfalls/Bright-Future-Solar-Power)

5.5.5 Solar Water Heating

A solar water heating unit consists of a blackened flat plate metal collector with an associated metal tubing facing the general direction of the sun. The plate collector has a transparent glass cover above and a layer of thermal insulation beneath it. The metal tubing of the collector is connected by a pipe to an insulated tank that stores hot water during cloudy days. The collector absorbs solar radiations and transfers the heat to the water circulating through the tubing either by gravity or by a pump. This hot water is supplied to the storage tank via the associated metal tubing. This system of water heating is commonly used in hotels, guest houses, tourist bungalows, hospitals, canteens as well as domestic and industrial units.

[A Text Book of Energy Engineering By AVINASH KAMBLE]

5.5.6 Solar-Pumping

In solar pumping, the power generated by solar-energy is utilized for pumping water for irrigation purposes. The requirement for water pumping is greatest in the hot summer months which coincide with the increased solar radiations during this period and so this method is most appropriate for irrigation purpose. During periods of inclement weather when solar radiations are low then the requirement for water pumping is also relatively less as the transpiration losses from the crops are also low.

[A Text Book of Energy Engineering By AVINASH KAMBLE]

5.5.7 Solar Furnaces

In a Solar furnace, high temperature is obtained by concentrating the solar radiations onto a specimen using a number of heliostats (turn-able mirrors) arranged on a sloping surface. The solar furnace is used for studying the properties of ceramics at extremely high temperatures above the range measurable in laboratories with flames and electric currents. Heating can be accomplished without any contamination and temperature can be easily controlled by changing the position of the material in focus. This is especially useful for metallurgical and chemical operations. Various property measurements are possible on an open specimen. An important future application of solar furnaces is the production of nitric acid and fertilizers from air.

[https://www.linkedin.com/pulse/solar-energy-10-major-application-explained-marcelo-cotroneo/]

[A Text Book of Energy Engineering By AVINASH KAMBLE]

5.5.8 Solar Drying of Agricultural and Animal Products

Agricultural products are dried in a simple cabinet dryer that consists of a box insulated at the base, painted black on the inner side and covered with an inclined transparent sheet of glass. At the base and top of the sides ventilation holes are provided to facilitate the flow of air over the drying material which is placed on perforated trays inside the cabinet. These perforated trays or racks are carefully designed to provide controlled exposure to solar radiations. Solar drying, especially of fruits improves fruit quality as the sugar concentration increases on drying. Normally soft fruits are particularly vulnerable to insect attack as the sugar content increases on drying but in a fruit dryer considerable time is saved by quicker drying minimizing gap the chances of insect attack.

The present practice of drying chilies by spreading them on the floor not only requires a lot of open space and manual labor for material handling but it becomes difficult to maintain its quality and taste unless drying is done in a controlled atmosphere. Moreover, the products being sun dried very often get spoiled due to sudden rains, dust storms or by birds. Besides, reports reveal that it is not possible to attain very low moisture content in the sun-dried chilies. As a result, the chilies become prone to attack by fungi and bacteria. In sun-drying sometimes, the produce is over dried and its quality is lost. Solar energy operated dryer helps to overcome most of these disadvantages.

Other agricultural products commonly solar-dried are potato-chips, berseem, grains of maize and paddy, ginger, peas, pepper, cashew-nuts, timber and veneer drying and tobacco curing. Spray drying of milk and fish drying are examples of solar dried animal products.

[https://www.linkedin.com/pulse/solar-energy-10-major-application-explained-marcelo-cotroneo/]

[A Text Book of Energy Engineering By AVINASH KAMBLE]

5.5.9 Solar Cooking

A variety of fuel like coal, kerosene, cooking gas, firewood, dung cakes and agricultural wastes are used for cooking purposes. Due to the energy crisis, supply of these fuels are either deteriorating (wood, coal, kerosene, cooking gas) or are too precious to be wasted for cooking purposes (cow dung can be better used as manure for improving soil fertility). This necessitated the use of solar energy for cooking purposes and the development of solar cookers. A simple solar cooker is the flat plate box type solar cooker. It consists of a well-insulated metal or wooden box which is blackened from the inner side. The solar radiations entering the box are of short wavelength. As higher wave length radiations are unable to pass through the glass covers, the re-radiation from the blackened interior to outside the box through the two glass covers is minimized, thereby minimizing the heat loss. The heat loss due to convection is minimized by making the box airtight. This is achieved by providing a rubber strip between the upper lid and the box for minimizing the heat loss due to conduction, the space between the blackened tray and outer cover of the box is filled with an insulting material like glass wool, saw-dust, paddy husk etc. When placed in sunlight, the solar rays penetrate the glass covers and are absorbed by the blackened surface thereby resulting in an increase in temperature inside the box. Cooking pots blackened from outside are placed in the solar box. The uncooked food gets cooked with the heat energy produced due to increased temperature of the solar box. Collector area of such a solar cooker can be increased by providing a plane reflector mirror. When this reflector is adjusted to reflect the sun rays into the box, then a 15°C to 25°C rise in temperature is achieved inside the cooker box.

The solar cooker requires neither fuel nor attention while cooking food and there is no pollution, no charring or overflowing of food and the most important advantage is that nutritional value of the cooked food is very high as the vitamins and natural tastes of the food are not destroyed. Maintenance cost of the solar cooker is negligible. The main disadvantage of the solar cooker is that the food cannot be cooked at night, during cloudy days or at short notice. Cooking takes comparatively more time and chapattis cannot be cooked in a solar cooker.

[https://www.linkedin.com/pulse/solar-energy-10-major-application-explained-marcelo-cotroneo/]

[A Text Book of Energy Engineering By AVINASH KAMBLE]

5.5.10 Solar Distillation

Potable water is scarce in arid semi and or coastal areas. But the abundance of sunlight in these areas can be used for converting saline water into potable distilled water. This is done by the method of solar distillation. In this method, solar radiation is admitted through a transparent air tight glass cover into a shallow blackened basin containing saline water. Solar radiation passes through the covers and is absorbed and converted into heat in the blackened surface causing the water to evaporate from the brine (impure saline water). The vapors produced get condensed to form purified water in the cool interior of the roof. The condensed water flows down the sloping roof and is collected in the troughs placed at the bottom and from there into a water storage tank to supply potable distilled water in areas of scarcity, in science laboratories of colleges and schools, defense labs, petrol pumps, hospitals and pharmaceutical industries. Per liter distilled water cost obtained by this system is cheaper than distilled water obtained by other electrical energy-based processes.

[A Text Book of Energy Engineering By AVINASH KAMBLE]

5.6 FURTHER APPLICATIONS OF SOLAR ENERGY

Some applications of solar energy we can see in the near future are (i) Photo-biological Cells; (ii) Solar Thermal Fuel (STF); (iii) Solar Can Power Places That Otherwise May Not Have Access to the Grid; (iv) Production of Power through Solar Ponds; (v) Solar Transportation; (vi) Solar Desalination; (vii) Floating Panels, Floating Solar Farms; (viii) Solar Power Harvesting Trees; (ix) Solar Charging Stations; (x) Solar Powered Trash; (xi) Stored up Solar Energy; and (xii) Solar Fashion.

5.6.1 Photo-biological Cells

The BioSolar Cells project aims for a sustainable production of energy, biomass, and food using photosynthesis. For the use of BioSolar Cells, an extensive research project involving ten knowledge institutions and 45 private industries, is exploring the role of photobiological cells in producing green energy. These cells improve the efficiency of the process by which plants, algae, and bacteria capture energy from the sun. Furthermore, researchers at the University of Cambridge have developed novel fuel cells, namely biological photovoltaics (BPV) that use a photosynthetic material in algae to capture solar energy, directly converting it into electricity. Once these 'living solar panels' are ready for use, they will emerge as environmentally-friendly and low-cost approach to harness solar power and generate green energy. Several winning innovations are helping people and organizations harness the unlimited power of the sun. The above-mentioned innovations will help appreciate the technological advancements that are shaping the future of solar power.

[https://blueandgreentomorrow.com/energy/6-innovations-that-shaping-future-solar-energy/]

5.6.1.1 Bio-Solar Cells For the first time, nine biological-solar (bio-solar) cells were connected into a bio-solar panel and continuously produced electricity from the panel and generated the most wattage of any existing small-scale bio-solar cells. Recently, the group took steps towards building a better bio-solar cell by changing the materials used in anodes and

FIG. 5.22: BIOLOGICAL-SOLAR (BIO-SOLAR) CELLS INTO A BIO-SOLAR PANEL (*Source*: http://onlineresize.club/pictures-club.html)

cathodes (positive and negative terminals) of the cell and also created a miniature microfluidic-based single-chambered device to house the bacteria instead of the conventional, dual-chambered bio-solar cells. However, this time, the group connected nine identical bio-solar cells in a 3x3 pattern to make a scalable and stackable bio-solar panel. The panel continuously generated electricity from photosynthesis and respiratory activities of the bacteria in 12-hour day-night cycles over 60 total hours.

The current research is the latest step in using cyanobacteria, which can be found in almost every terrestrial and aquatic habitat on earth—as a source of clean and sustainable energy. Even with the breakthrough, a typical "traditional" solar panel on the roof of a residential house, made up of 60 cells in a 6x10 configuration, generates roughly 200 watts of electrical power at a given moment. The cells from this study, in a similar configuration, would generate about 0.00003726 watts. So, it isn't efficient just yet, but the findings open the door to future research of the bacteria itself. "Once a functional bio-solar panel becomes available, it could become a permanent power source for supplying long-term power for small, wireless telemetry systems as well as wireless sensors used at remote sites where frequent battery replacement is impractical," said Seokheun 'Sean' Choi, an assistant professor of electrical and computer engineering in Binghamton University's Thomas J. Watson School of Engineering and Applied Science, and co-author of the paper, in a 11 April press statement. The findings are currently available online and will be published in the June edition of the journal Sensors and Actuators B: Chemical.

[https://www.sciencedaily.com/releases/2016/04/160411152653.htm]

5.6.2 Solar Thermal Fuel (STF)

The biggest drawback of solar power as an energy source is storage. Despite of the incredible growth of the PV industry in recent years, the future of solar involves an affordable storage solution that will make solar a truly sustainable energy source 24 hours a day. Though solar batteries are a storage option, they are still not economically viable for the mainstream. Luckily, MIT Professor Jeffrey Grossman and his team of researchers have spent much of the past few years developing alternative storage solutions for solar, the best one appears to be solar thermal fuels (STFs). The technology and process behind STFs is comparable to a typical battery. The STF can harness sunlight energy, store it as a charge and then release it when prompted. The issue with storing solar as heat, according to the team's findings, is that heat will always dissipate over time, which is why it is crucial that solar storage tech can charge energy rather than capture heat. For Grossman's team, the latest STF prototype is simply an improvement of a prior design that allowed solar power to be stored as a liquid substance. 2016 saw the invention of a solid state STF application that could be implemented in windows, windshields, car tops, and other surfaces exposed to sunlight.

[https://news.energysage.com/solar-panel-technology-advances-solar-energy/]

5.6.3 Solar Can Power Places

Places that may not have access to the grid can be provided electricity with solar power. Solar-powered lanterns are lighting up rural parts of India. Solar-powered shipping container that can host medical facilities in Haiti can also power other areas in need of such services.

[https://www.solarpowerworldonline.com/2013/05/7-cool-solar-energy-applications/4/]

5.6.4 Production of Power through Solar Ponds

A solar pond is a natural or artificial body of water used for collecting and absorbing solar radiation and storing it as heat. It is very shallow (5-10 cm deep) and has a radiation absorbing (black plastic) bottom. It has a curved fiber glass cover over it to permit the entry of solar radiation but reduces losses by radiation and convection (air movement). Loss of heat to the ground is minimized by providing a bed of insulating material under the pond. Solar ponds utilize water for collecting and storing the solar energy which is used for many applications such as space heating, industrial process heating and to generate electricity by driving a turbine powered by evaporating an organic fluid with a low boiling point.

[https://www.solarpowerworldonline.com/2013/05/7-cool-solar-energy-applications/7/]

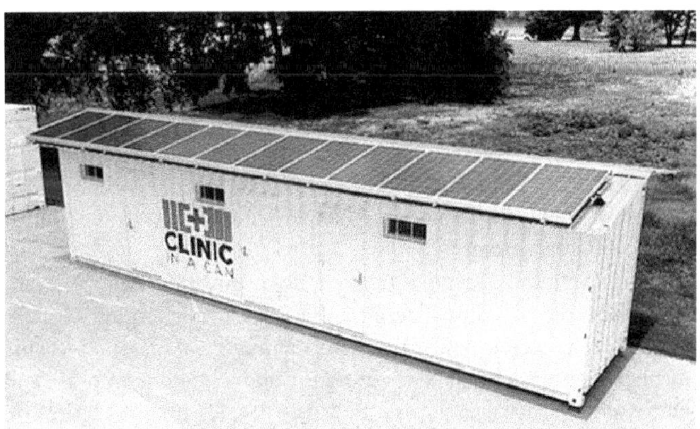

FIG. 5.23: SOLAR POWERED PLACES THAT OTHERWISE MAY NOT HAVE ACCESS TO THE GRID (*Source*: https://www.solarpowerworldonline.com/2013/05/7-cool-solar-energy-applications/4/)

FIG. 5.24: SOLAR PV CAN HEAT-UP A SWIMMING POOL. (*Source*: https://www.solarpowerworldonline.com/2013/05/7-cool-solar-energy-applications/7/)

5.6.5 Solar Transportation

In 2013, Australia introduced the first solar-powered bus that generates zero emissions. After that, solar transportation has come a long way. The airline industry is also undergoing a revolutionary change with new prototypes, like the Solar Impulse 2; it was the first solar-powered airplane and it made a historic trip around the globe. The aircraft carries more than 17 thousand photovoltaic cells that charge the solar batteries, enabling the jet to stay aloft for days on end.

[https://www.impero.co.in/whatsnew/2018/07/19/future-innovation-of-solar-energy/],

Several countries are likely to harness the power of solar energy by introducing numerous solar airline projects, solar-powered bus stops, solar boats, and other vehicles, ensuring a clean and fossil fuel-free future in transportation.

[https://www.impero.co.in/whatsnew/2018/07/19/future-innovation-of-solar-energy/]

Research and development in solar-powered vehicles are increasing day by day. There're solar carports charging electric cars, solar-powered boats, rickshaws and the Solar Impulse. This airplane has already made it across Europe without a drop of fuel, and is making its way across the United States this spring before attempting to fly around the entire globe on solar.

5.6.5.1 Solar-power use in military The U.S. military is a leader to take note of the military's initiatives to reduce its dependence on fossil fuels. The military will play a key role in bringing

FIG. 5.25: SOLAR POWERED AIRCRAFT (*Source*: https://www.impero.co.in/whatsnew/2018/07/19/future-innovation-of-solar-energy/)

solar to the mass market, with the slogan "If it's good enough for the military, it's good enough for the American public".

[https://www.solarpowerworldonline.com/2013/05/7-cool-solar-energy-applications/5/]

5.6.6 Solar Desalination

Solar desalination is one of the most applications of solar energy. Desalination is a method of extracting salt and minerals from saline water and has been in use for generations. In fact, an estimated 300 million people get their freshwater supply from nearly 19 thousand desalination plants around the globe. With solar energy touching every aspect of human life, solar-powered machines can now help solve the looming water crisis. Scientists have come up with a revolutionary solution to make freshwater available in drought-plagued areas. The solar desalination system uses a combination of membrane distillation technology and sunlight-harvesting nanophotonic cells to convert salty or brackish water into fresh drinking water. With the planet running perilously low on accessible freshwater supplies, this novel technology is all set to solve the water shortage in parched and arid regions.

[https://www.impero.co.in/whatsnew/2018/07/19/future-innovation-of-solar-energy/]

FIG. 5.26: SOLAR POWERED IMPULSE AIRPLANE (*Source:* https://www.impero.co.in/whatsnew/2018/07/19/future-innovation-of-solar-energy/)

FIG. 527: SOLAR-POWERED SHIPPING CONTAINER THAT CAN HOST MEDICAL FACILITIES CAN POWER OTHER AREAS AS WELL. (*Source*: https://www.solarpowerworldonline.com/2013/05/7-cool-solar-energy-applications/5/)

5.6.7 Floating Panels, Floating Solar Farms

In many countries, there is a lack of space to install large-scale ground-mount solar systems. As authorities wish to avoid taking away large farmlands for ground-mount solar systems, companies are introducing ecological alternative solutions. One such firm is French company Ciel & Terre International, which has been developing large-scale floating solar solutions since 2011. Its Hydrelio Floating PV system allows standard PV panels to be installed on large bodies of water such as drinking water reservoirs, quarry lakes, irrigation canals, remediation and tailing ponds, and hydroelectric dam reservoirs. This simple and affordable alternative to ground-mounted systems is particularly suitable for water-intensive industries that cannot afford to waste either land or water. This is how it works. The main float is constructed of high-density thermoplastic (HDPE) and is set at a 12-degree angle to support a standard 60-cell PV solar module. A secondary non-slip HDPE float is then used to link the main floats together and provide a platform for maintenance and added buoyancy. According to Ciel and Terre, the system is easy to install and dismantle, can be adapted to any electrical configuration, is scalable from low- to high-power generation and requires no tools or heavy equipment. It is also eco-friendly, fully recyclable, has low environmental impact and is cost-effective. To date, the system has been installed in the UK. The company has also set up projects for floating solar farms in countries such as India, France and Japan.

[https://www.livemint.com/Industry/lGM9f97bTxoboIZebYtq9N/10-technologies-shaping-the-future-of-solar-power.html]

5.6.7.1 WaterNest 100 EcoFloLife has developed the WaterNest 100 eco-friendly floating house, exclusively designed by renowned Italian architect Giancarlo Zema. It is an over 100 sq. m residential unit, 12 meters in diameter and 4 meters high, made entirely of recycled laminated timber and a recycled aluminum hull. Balconies are conveniently located on the sides and thanks to the large windows, permit enjoyment of fascinating views over the water. Bathroom and kitchen skylights are located on the wooden roof, as well as 60 sq. m of amorphous photovoltaic panels capable of generating 4kWp, which are used for the internal needs of the floating house. The interior of WaterNest 100 floating house can include a living room, dining area, bedroom, kitchen and bathroom or have other configurations according to the different housing or working needs. The floating house can be positioned along river courses, lakes, bays, atolls and sea areas with calm waters. The use of materials and sustainable production systems make this unit recyclable up to as much as 98%. It has a hull that is made entirely of aluminum, a light alloy, highly resistant to impact, corrosion and 100% recyclable. Its photovoltaic panels installed on the wooden roof differ from conventional ones due to the low-energy consumption required for their production. From an aesthetic point of view, they can be curved to fit almost any type of roofing. [https://www.livemint.com/Industry/lGM9f97bTxoboIZebYtq9N/10-technologies-shaping-the-future-of-solar-power.html]

5.6.8 Solar Power Harvesting Trees

Scientists at the VTT Technical Center of Finland have developed prototype plants that harvest solar energy and store it to power devices namely mobile phones, laptops, LED street lights, and electric vehicles.

Solar trees will be a crucial architectural component of the exhibition pavilion for upcoming Expo 2020 to be hosted in Dubai. These powerhouses can be placed in deserts, office car parks, golf courses, business parks, and malls; adding aesthetic value to

FIG. 5.28: VTT TECHNICAL CENTER OF FINLAND HAVE DEVELOPED PROTOTYPE PLANTS. *Source:* https://www.solarpowerworldonline.com/2013/05/7-cool-solar-energy-applications/)

the area whilst harnessing renewable energy. Researchers at the VTT Technical Research Centre in Finland have used solar and 3D printing technologies to develop prototypes of what they have christened as "energy harvesting trees". The tiny leaves generate and store solar energy and can be used to power small appliances and mobile devices. They flourish indoors and outdoors and can also harvest kinetic energy from wind and temperature changes in the surrounding environment. The tree's leaves are actually flexible organic solar cells, printed using well-established mass-production techniques. Each leaf has a separate power converter, creating a multi-converter system that makes it possible to collect energy from a variety of sources such as solar, wind and heat temperature. The more solar panels there are in a tree, the more energy it can harvest. The tree trunk is made with 3D technology by exploiting wood-based biomaterials VTT has developed. They are mass producible and can be infinitely replicated.

5.6.9 Solar Charging Stations

Innovative companies are developing options for charging personal devices for free, using the sun's energy. Just a few years ago we didn't have much to charge on our bedside table. But now we often run out of outlets to charge our tablet, MP3 player, cell phone, e-reader, laptop and other portable devices. Not only is the growing popularity of portable personal devices a strain on electric bills, but it also puts pressure on the power grid.

[https://www.solarpowerworldonline.com/2013/05/7-cool-solar-energy-applications/]

5.6.10 Stored-up Solar Energy

Hurricane Sandy caused huge power outages at New York, New Jersey and other areas along the East Coast, and even at Cleveland, With solar power we can have back-up power but is still not an option until it's available in every Home Depot across the country, which with time will make Solar a viable option.

[https://www.solarpowerworldonline.com/2013/05/7-cool-solar-energy-applications/6/]

5.6.11 Solar Fashion

The concept of solar-reliant attire has been in development for over a decade, with prominent fashion labels longing to lure shoppers with innovative apparel and fashion accessories that can charge mobile phones and wearable devices. With the world moving towards a sustainable future, researchers are working on embedding flexible solar panels into the fabric, enabling every person to harness renewable solar energy. Pauline van Dongen, a well-known Dutch fashion designer, has created an entire collection of shirts, parkas, and high-end fashion wear that can produce up to one watt of electricity – enough to charge phones, MP3 players, and other hand-held gadgets for a few hours.

Several other firms are introducing solar-powered sunglasses, jewelry, watches, and backpacks to revolutionize the fashion industry while moving towards a renewable future.

[https://www.pluginsolar.co.uk/?p=5138]

FIG. 5.29: BED SIDE BATTERY SOLAR CHARGING STATIONS (*Source*: https://www.solarpowerworldonline.com/2013/05/7-cool-solar-energy-applications/**)**

FIG. 5.30: WITH STORED UP ENERGY SOLAR CAN BE A VIABLE ENERGY OPTION. (*Source*: Inhabitat.com) (Source:https://www.solarpowerworldonline.com/2013/05/7-cool-solar-energy-applications/)

5.7 FUTURE APPLICATION OF SOLAR ENERGY

Because of population growth at unprecedented rates and technological advancements, world's energy demand is growing fast. It is therefore important to go for reliable, cost effective and everlasting renewable energy source for energy demand arising in future. Solar energy, among other renewable sources of energy, is one of the most promising and freely available energy source for managing long term issues in energy crisis. Solar industry is developing steadily all over the world because of the high demand for energy while major energy source, fossil fuel, is limited and other sources are expensive. It has become a tool to develop economic status of developing countries and to sustain the lives of many underprivileged people as it is now cost effective to expedite its development. The solar industry would definitely be a best option for future energy demand since it is superior in terms of availability, cost effectiveness, accessibility, capacity and efficiency compared to other renewable energy sources. Constant researches are underway to upgrade solar industry, its potential applications and barriers for better solar industry in future in order to resolve energy crisis.

[https://ideas.repec.org/a/eee/rensus/v62y2016icp1092-1105.html]

Our planet receives 173 thousand terawatts of solar energy each year. This makes solar energy the largest source of renewable and clean energy for mankind. This energy source offers ten-thousand times the energy required by people worldwide. Because of its real potential for replacing fossil fuels and fostering a green future, solar energy is making it into the mainstream. Technology is empowering people to salvage the sun's energy in a low-maintenance and cost-effective manner. Over the past decade, the solar power industry has seen several technological innovations, making a serious impact on the world's energy systems. If you desire to live an eco-friendly lifestyle, here are six intriguing solar innovations that you need to understand.

[https://blueandgreentomorrow.com/energy/6-innovations-that-shaping-future-solar-energy/]

5.7.1 Reshaping Solar Spectrum to Turn Light Into Electricity

A team of chemists at the University of California has succeeded in finding a way to produce more energy form solar cells. In a paper that was published in Nano Letters, an American Chemical Society publication, the researchers said that by combining inorganic semiconductor nanocrystals with organic molecules, they succeeded in "up-converting" (two low-energy photons into one high-energy photon) photons in the visible and near-infrared regions of the solar spectrum. The infrared region of the solar spectrum passes right through the photovoltaic materials that make up today's solar cells, explained Christopher Bardeen, a professor of chemistry in a press release on 27 July 2015. This up-converted photon is readily absorbed by photovoltaic cells, generating electricity from light that normally would be wasted, according to Bardeen. He added that these materials are essentially "reshaping the solar spectrum" so that it better matches the photovoltaic materials used today in solar cells. The ability to use the infrared portion of the solar spectrum could boost solar photovoltaic efficiencies by 30% or more. Besides solar energy, the ability to up-convert two low-energy photons into one high-energy photon has potential applications in biological imaging, data storage and organic light-emitting diodes, says Bardeen.

[https://www.livemint.com/Industry/lGM9f97bTxoboIZebYtq9N/10-technologies-shaping-the-future-of-solar-power.html]

5.7.2 Transmitting Solar Power Wirelessly From Space

The Japanese Space Agency (JAXA)'s Space Solar Power Systems (SSPS) is trying to transmit energy from an orbiting solar panels by 2030. Mitsubishi Heavy Industries Ltd (MHI) was able to conduct a ground demonstration test of "wireless power transmission", a technology that will serve as the basis for the SSPS. In the test, 10kW of electricity was successfully transmitted via a microwave unit. Power reception was confirmed at a receiver located 500 meters away. LED lights on the receiver confirmed the transmission. This marks a new milestone in transmission distance and power load (enough to power a set of conventional kitchen appliances).

Potentially, a solar battery in orbit (36,000km above earth) could generate power that would then be transmitted to earth via microwave/laser, without relying on cables. JAXA anticipates that this new technology could become a mainstay energy source that will simultaneously solve both environmental and energy issues on earth. The estimated lifecycle carbon dioxide emission for the operational SSPS indicates that it is almost the same as from nuclear power system and much less than fossil fuel power system, JAXA claims on its website. Countries such as India, China and Japan are investing heavily in these technologies right now.

[https://www.livemint.com/Industry/lGM9f97bTxoboIZebYtq9N/10-technologies-shaping-the-future-of-solar-power.html]

5.7.3 Ways to Boost Solar Power

Scientists at the University of Toronto have discovered a new type of light-sensitive nanoparticles called colloidal quantum dots. This technology will provide a less expensive and more flexible material for solar cells. Specifically, the new materials use n-type and p-type semiconductors, but ones that can actually function outdoors. This is possible because the new colloidal quantum dots don't bind to air (unlike traditional n-type materials that bind to oxygen), a quality that also helps increase radiant light absorption besides offering stability outdoors. The researchers claim that panels using this new technology were found to be up to 8% more efficient at converting sunlight.

Researchers at Imperial College University in London said they discovered a new material, gallium arsenide that could make solar PV systems nearly three times more efficient than existing products on the market. The solar cells are called "triple junction cells" and they are much more efficient because they can be chemically altered in a manner that optimizes sunlight capture. The model uses a sensor-driven window blind that can track sun light along with "light-pipes" that guide the light into the system.

Novatec Solar recently commissioned an energy storage solution for solar PV systems using a molten salt storage technology. The process uses inorganic salts to transfer energy generated by solar PV systems into solar thermal using heat transfer fluid rather than oils as some storage systems have. The result is that solar plants can operate at temperatures over 500^0 Celsius, which would result in a much higher power output. This means that costs to store solar power would be lowered significantly and utility companies could finally use solar power plants as base load plants rather than to meet peak demand during prime daylight hours.

In a project funded by the US Department of Energy, Ohio State University researchers recently announced they created a battery that is 20% more efficient and 25% cheaper than anything on the market today. The secret to the design is that the rechargeable battery is built into the solar panel itself, rather than operating as two stand-alone systems. By conjoining the two into one system, scientists said they could lower costs by 25% compared to existing products. Scientists are exploring ways to actually line up highways and roads with solar panels, which would then be used to deploy large amounts of electricity to the grid. Called solar roadways, they have already popped up in countries such as the Netherlands and promise to save on land space.

[https://www.livemint.com/Industry/lGM9f97bTxoboIZe-bYtq9N/10-technologies-shaping-the-future-of-solar-power.html]

5.7.4 Solving the Storage Problem

In 2015, Tesla launched its Powerwall, a giant solar-storing lithium ion battery that solves that issue. Each Powerwall battery can hold ten Kilowatt hours (kWh) of electricity. By comparison, a car battery holds about 1.2 kWh. The Powerwall is the best-known of a few products that let people store solar energy. LG, Panasonic, BMW, and Nissan, as well as several start-up companies, all sell "home batteries" similar to the Powerwall. (Nissan uses recycled batteries from its electric cars to power its x-Storage battery, which is about $1,000 cheaper than Tesla's.)

[https://www.greenamerica.org/show-ga-blog?nid=8736]

5.7.5 Solar Paint

In June 2017, researchers at Australia's Royal Melbourne Institute of Technology (RMIT) have developed a solar paint that could eventually generate solar power on its own. Led by Dr. Torben Daeneke, the team invented a paint made with titanium oxide, a common ingredient in paint and sunscreen, and synthetic molybdenum-sulphide, a compound that absorbs moisture in the air. Both sunlight and moisture are used by the chemical combination to convert water into hydrogen fuel, as well as oxygen. Dr. Daeneke explained to the Sydney Morning Herald that the final product of the paint would need to include membranes that can harvest hydrogen selectively for energy use. In the past, hydrogen energy has depended on unclean processes like the burning of fossil fuels. RMIT's paint demonstrates that hydrogen can be a clean energy source when hydrogen is generated with renewables.

[https://www.greenamerica.org/show-ga-blog?nid=8736]

5.7.6 Solar in the City

Danny Kennedy has been in the solar business a long time, as a clean energy advocate at Greenpeace in the '90s who went on to establish rooftop solar company Sungevity in 2007. He's part of Green America's Solar Circle, which has brought together industry leaders to push the solar industry into quicker progress since 2002. Now, Kennedy works with green entrepreneurs at the California Clean Energy Fund, an early stage fund for clean-energy projects, and Powerhouse, an incubator for clean-energy startups. He's excited about anything that makes solar easier to get to customers. But what he really gets a kick out of is seeing people starting to put solar panels in unexpected places like solar-powered trash cans from Big Belly. "Who knew you needed that? Because solar is cheap enough for people to put an engine in a trash can, municipalities can buy trash cans that compact the trash and can call to be emptied when they're full," Kennedy explains. "The result is less litter on the streets and more efficient litter collection, which saves cities tens of thousands of dollars."

Kennedy names bike sharing systems as another model that wouldn't be the same without solar especially dockless systems like that of the company Mobike. Mobike bikes have a solar panel in the bike basket, connecting the bike's lock and GPS to a phone app, so customers can easily locate, reserve, and unlock the bikes they want to use. Without solar, the whole automated process would require a very expensive battery, Kennedy says. He adds that providing more bike share options means fewer trips by car, so it decreases traffic pollution and congestion.

[https://www.greenamerica.org/show-ga-blog?nid=8736]

5.7.6.1 Solar Powered Trash We are seeing more and more cities with solar-powered trash and recycling compactors. Not only are they cool, but they're saving cities millions by reducing litter and cutting down collections (they hold about five times more than the average trash bin). Cities such as Philadelphia, New York and Chicago have already installed them and more are catching on (including my hometown of Cleveland).

[https://www.solarpowerworldonline.com/2013/05/7-cool-solar-energy-applications/2/]

5.8 SOLAR TECHNOLOGY: WHAT'S NEW?

So far we have discussed the future of Solar Energy based upon what is already either in vogue globally or is in exploratory stages either in University or Department of Energy (DOE) or professional organizations Laboratories. However given the potential of the ubiquitous Solar Energy it is worth addressing what the potential holds in terms of (a) A Bright Future; (b) Squeezing More Out Of the Sun; (c) Solar Powered Roads; (d) Wearable Solar; and (e) Solar Batteries: Innovation in Solar Storage. These are discussed in the following paragraphs.

FIG. 5.31: SOLAR POWERED TRASH RECEPTACLE (*Source_https*://www.solarpowerworldonline.com/2013/05/7-cool-solar-energy-applications/2/)

5.8.1 A Bright Future

In 2016, solar became the fastest growing source of energy in the US, and the biggest employer of all energy industries, with around 370,000 employees, compared to 187,000 in all fossil fuels combined, according to the US Department of Labor. The price of solar fell 165 percent from 1975 to 2015, as the megawatts installed globally surged from two to 65,000 in the same period, according to a report by Bloomberg and the Earth Policy Institute.

Due to the harsh realities of the climate crisis, the world's impending approach to peak coal and peak oil, and the rapid adoption of solar worldwide, the future looks very bright indeed for solar power, and we can expect more solar innovations to keep coming. Photovoltaics, invented by Bell Labs in 1954, already have and are going to continue to completely transform civilization until we become a completely solar-powered society.

[https://www.greenamerica.org/new-green-tech-promise-and-pitfalls/bright-future-solar-power]

5.8.2 Squeezing More Out Of the Sun

Even though the sun is undoubtedly the greatest sustainable energy source for earth, the problem is the low efficiency: 80% of installed PV panels worldwide have a performance of 15% or lower; if the panels are not tracked with the sun, the average of annual tilt losses add up to minus 70%.

German architect Andre Broessel believes his company Rawlemon can "squeeze more juice out of the sun", even during the night hours and in low-light regions. He has created a spherical sun power generator prototype called the beta ray. His technology will combine spherical geometry principles with a dual axis tracking system, allowing twice the yield of a conventional solar panel in a much smaller surface area. The futuristic design is fully rotational and is suitable for inclined surfaces, walls of buildings, and anywhere with access to the sky. It can even be used as an electric car charging station. By using a high-efficiency multi-junction cell, Broessel's company claims to have reduced the cell surface down to 1% compared to the same power output as a conventional silicon cell in optimal conditions. "In combination with dual axis tracking, our system generates twice the yield of a conventional panel. In addition, our smaller cell area has a lower carbon footprint because its production requires fewer precious semiconductor or other building materials," he says on the company website. Rawlemon has also introduced a USB spherical sun charger called beta.ey. [https://www.livemint.com/Industry/lGM9f97bTxoboIZebYtq9N/10-technologies-shaping-the-future-of-solar-power.html]

5.8.3 Solar Powered Roads

Recently a new PV technology – solar powered roads was tested is US. The sidewalks along Route 66, America's historic interstate highway, were chosen as the testing location for solar-powered pavement tech. These roadways are heralded for their ability to generate clean energy, but they also include LED bulbs that can light roads at night and have the thermal heating capacity to melt snow during winter weather. The next stop following sidewalk tests is to install these roadways on designated segments of Route 66.

[https://news.energysage.com/solar-panel-technology-advances-solar-energy/]

5.8.4 Wearable Solar

Though wearable solar devices are nothing new (solar-powered watches and other gadgets have been on the market for several years), 2017 saw an innovation in solar textiles: tiny solar panels can now be stitched into the fabric of clothing. The wearable solar products of the past, like solar-powered watches, have typically been made with hard plastic material. This new textile concept makes it possible for solar to expand into home products like window curtains and dynamic consumer clean tech like heated car seats. This emerging solar technology is credited to textile designer Marianne Fairbanks and chemist Trisha Andrew.

[https://news.energysage.com/solar-panel-technology-advances-solar-energy/]

5.8.5 Solar Batteries: Innovation in Solar Storage

The popularity of off-grid solar and solar plus storage has increased a lot in U.S. markets. The industry-famous Tesla Powerwall, a rechargeable lithium ion battery product launched in 2015, continues to lead the pack with regard to market share and brand recognition for solar batteries. Tesla offers two storage products, the Powerwall for residential use and the Power pack for commercial use and the clean auto behemoth is expected to launch its Powerwall 2.0. Solar storage is still a fairly expensive product in 2016, but a surge in demand from solar shoppers is expected to bring significantly more efficient and affordable batteries to market very soon.

[https://news.energysage.com/solar-panel-technology-advances-solar-energy/]

5.9 ADVANCES IN SOLAR ENERGY: THE LATEST SOLAR TECHNOLOGY BREAKTHROUGHS

In spite of being repetitive it is worth chronicling the latest advances in solar energy that will pave way for further advances, few of which are: (1) The Wonder Material That May Make Spray-On Solar PV Reality; (2) Solar Tracking Mounts; (3) Advances in Solar Panel Efficiency; (4) Solar Water Purifiers; (5) What New Solar Technology Means For Homeowners In 2018; (6) Solar-Powered Car Surfaces; (7) Vaccine Refrigerators (8) Smart Solar Cities;.(9) Turning Sunshine into Liquid Gold; (10) Momentum Builds, and (11) Efficiency Gains.

5.9.1 The Wonder Material That May Make "Spray On Solar" a PV Reality

Solar researchers think there's a good chance the economics of the $42 billion industry will soon be disrupted by something called perovskites, a range of materials that can be used to harvest light when turned into a crystalline structure. The hope is that perovskites, which can be mixed into liquid solutions and deposited on a range of surfaces, could play a crucial role in the expansion of solar energy applications with cells as efficient as those currently made with silicon. One British company aims to have a thin-film perovskite solar cell commercially available by the end of 2018. "This is the front-runner of low-cost solar cell technologies," said Hiroshi Segawa, a professor at the University of Tokyo who's leading a five-year project funded by the Japanese government that groups together universities and companies such as Panasonic Corp. and Fujifilm Corp. to develop perovskite technology. Not everyone is sold on perovskite as a game-changer from the industry's heavy reliance on silicon photovoltaic cells. That said, recent research pointing to the material's potential continues to grip the solar energy research community.

The World Economic Forum picked the material as one of its top 10 emerging technologies of 2016. Meanwhile, solar panel makers and top universities in Europe, the U.S. and Asia are racing to commercialize the technology, with researchers churning out as many as 1,500 papers a year on the material.

[https://www.renewableenergyworld.com/articles/2017/03/the-wonder-material-that-may-make-spray-on-solar-pv-reality.html]

5.9.2 Solar Tracking Mounts

Trackers allow solar panels to maximize electricity production by following the sun as it moves across the sky. PV tracking systems tilt and shift the angle of a solar array as the day goes by to best match the location of the sun. Though this panel add-on has been available for some time, solar manufacturers are truly embracing the technology. GTM Research recently unveiled a recent report that shows a major upward trend in the popularity of tracking systems. GTM projects a 254 percent year-over-year increase for the PV tracking market this year. The report stated that by 2021, almost half of all ground mount arrays will include solar tracking capability.

[https://news.energysage.com/solar-panel-technology-advances-solar-energy/]

5.9.3 Advances in Solar Panel Efficiency

There has been a race in the solar industry to the top in terms of solar cell efficiency. The biggest achievements is held by premium panel manufacturer "Sun Power" with a current record for solar panel efficiency of 23.5 percent.

The solar cell types used in mainstream markets could also undergo improvements in cost per watt, a metric that compares relative affordability of solar panels. Thanks to Swiss and American researchers, Perovskite solar cells (as compared to the silicon cells that are used predominantly today) have seen some major breakthroughs in the past two years. The result will be a solar panel that can generate 20+ percent efficiency, which is one of the lowest cost options on the market. Of course, the work doesn't stop there, as MIT researchers reminded when they announced new technology that could *double* the efficiency of solar cells overall. The MIT lab team revealed a new tech concept that captures and utilizes the waste heat that is usually emitted by solar panels. This typically released and non-harnessed thermal energy is a setback and opportunity for improvement for solar technology which means this innovation could help the cost of solar to plummet even further.

[https://news.energysage.com/solar-panel-technology-advances-solar-energy/]

5.9.4 Solar Water Purifiers

Stanford University researchers collaborated with the Department of Energy this year to develop a new solar device that can purify water when exposed to sunlight. The minuscule tablet (roughly half the size of a postage stamp) is not the first solar device to filter water, but it has made major strides in efficiency compared to past inventions. Prior purifier designs needed to harness UV rays and required hours of sun exposure to fully purify water. By contrast, Stanford's new product can access *visible* light and only requires a few minutes to produce reliable drinking water. As the technology behind solar purifiers continues to improve, expect these chiclet-sized devices to come to market with hikers and campers in mind as an ideal consumer audience.

[https://news.energysage.com/solar-panel-technology-advances-solar-energy/]

5.9.5 Solar-Powered Car Surfaces

Even though there are solar-powered charging stations for electric vehicles in numerous U.S. cities, solar technology is making its way directly into the body of some cars. The 2010 Toyota Prius featured a solar panel on its roof, but the generated electricity only helped to power the car's climate control system. An updated version of the same car, now being manufactured in Japan, connects the solar panel directly to the car's battery, increasing the efficiency and reliability of the electrical system. This is a far leap from the first semi-transparent solar sunroof technology developed in the 1980s that augmented the car's ventilation.

[https://www.energy.gov/eere/articles/solar-power-does-what-4-unusual-uses-photovoltaic-technology]

5.9.6 Solar Vaccine Refrigerators

In developing countries, 24-hour electricity isn't guaranteed, and in many cases, there is no electrical grid. In places with a grid, the infrastructure is often so poor that chronic power outages occur daily. Private companies have been manufacturing solar-powered vaccine refrigerators so healthcare workers in remote areas can administer critical medication to those who need it and camels are used for transporting Solar Vaccine Refrigerators. This tech-

FIG. 5.32: SOLAR POWER CALCULATOR (CREDIT: ARCO SOLAR [HTTPS://WWW.ENERGY.GOV/EERE/ARTICLES/SOLAR-POWER-DOES-WHAT-4-UNUSUAL-USES-PHOTOVOLTAIC-TECHNOLOGY])

nology solution has been saving lives for several years. [https://www.energy.gov/eere/articles/solar-power-does-what-4-unusual-uses-photovoltaic-technology]

5.9.7 Smart Solar Cities

Solar-powered streetlights are currently used all across urban areas. The sun charges a battery during the day so streetlights, now primarily utilizing light-emitting diodes (LEDs), can shine at night. Some cities, like San Diego, California, are using streetlights to optimize infrastructure. The Smart City San Diego Initiative is incorporating smart sensors into streetlights that have the ability to direct drivers to open parking spaces and help first responders during emergencies. Combining internet-linked sensors with solar powered streetlights saves both time and money. [https://www.energy.gov/eere/articles/solar-power-does-what-4-unusual-uses-photovoltaic-technology])

5.9.8 Turning Sunshine into Liquid Gold

Tiny metallic-gold particles are being used to convert sunlight into fuel. The technology is being developed in South Australia to store solar energy as an alternative to battery storage. Researchers from Flinders University and the University of Adelaide, in collaboration with a number of international institutions, have converted solar energy directly into chemical energy in the form of methane and methanol. The process uses dynamic nano-clusters, which consist of a specific number of metallic-gold atoms that interact with the molecules in UV light. The team built a lab-scaled device in South Australia's capital Adelaide and testing its effectiveness using artificial ultraviolet light. While it is still being scaled up, researchers say it has potential for industrial, commercial and domestic applications. The U.S. Army is so interested in using the process as a potential mobile generator, which would allow troops in the field to store energy that it is now funding the research. Lead researcher Gunther Andersson said as the world moved toward implementing more renewable energy initiatives there was a growing need for more reliable storage.

"This is not completely a new idea, chemical energy storage, but the thing that is special in our work is that we use specific nano-clusters, which make the conversion far more efficient," he said. "Using a gold-based catalyst we get about 10 times more product out of it than what a contemporary catalyst would give us." "Our lab product is working quite efficiently, at least in our research terms, and we will start to scale up the project to further test the process in the coming months."

The technology has the potential for large-scale application that could be used to help power whole cities or just your regular home. The dynamic shape of the nano-clusters makes them catalytic

FIG. 5.33: Camels used for Transporting Solar Vaccine Refrigerators (*Credit:* NESTE Advanced Power Systems [https://www.energy.gov/eere/articles/solar-power-does-what-4-unusual-uses-photovoltaic-technology])

FIG. 5.34: SOLAR-POWERED STREETLIGHTS (*Source:* Credit: Robert Ashworth [https://www.energy.gov/eere/articles/solar-power-does-what-4-unusual-uses-photovoltaic-technology])

active, creating a more efficient production of chemical energy in the form of methanol or methane. Andersson said these materials were ideal substances because they were already in frequent use and easily stored. "Obviously, there are large-scale batteries, really powerful ones, but the problem is then availability. The batteries are also not quite developed for all these things yet," Andersson said. "To get long-lasting batteries for a reasonable price is a major challenge at the moment, plus you have the problem that the chemicals you use in batteries are not very environmentally friendly."

According to a recent report by the World Economic Forum, the cost of solar energy is now the same price or cheaper than fossil fuel in more than 30 countries and has declined at a 20 percent compounded annual rate. South Australia is cementing itself as a world leader in green energy, as companies begin cracking the code to make renewables commercially viable. Nearly a quarter of houses in South Australia have installed rooftop solar panels, making it one of the highest penetration rates in the world. It is also the largest producer of wind energy in Australia the state's 1.5 GW of wind energy represents almost half of the country's capability.

South Australia made headlines around the world when it was announced that this state had been powered by 100 percent renewable energy for an entire working day in 2015.

Andersson's solar-storage device is an international collaboration involving researchers from Flinders University, University of Adelaide, Canterbury University, Victoria University and the University of Utah.

South Australia's capital Adelaide has three long-standing public universities, Flinders University, University of South Australia and the University of Adelaide, each of which is consistently rated highly in the international higher education rankings

https://www.renewableenergyworld.com/articles/2017/02/turning-sunshine-into-liquid-gold.html

5.9.9 Momentum Builds

Perovskite's usefulness was first hinted in 2006 when Tsutomu Miyasaka, a professor at Toin University of Yokohama, was approached by a graduate student interested in testing how well the material could convert sunlight to electricity. Though he'd been testing a number of different materials for solar panels, the Japanese academic had never heard of the synthesized crystal, Miyasaka said in an interview,

The idea to use perovskite, which is based on the same structure as a mineral named after Russian mineralogist Lev Perovski, initially went nowhere. Its structure was poorly understood and the industry had already latched on to silicon as the best material to convert sunlight into electricity. Silicon solar cells have been around since the early days of the space program and now dominate the industry, with global shipments of solar products expected to have totaled $41.9 billion in 2016, according to market researcher IHS Markit. But they have limitations tremendous amounts of energy are needed to produce the silicon in solar cells. Things began to change for perovskite with the first publication of research on the material by Miyasaka's group in the Journal of the American Chemical Society in 2009. "We had been turned down by magazines like Nature and Science and I suspect that's because it was low in efficiency and also the material was unheard of," Miyasaka said. "We talked about perovskite on many occasions but there was no feedback. Ninety-nine percent of people didn't understand the structure of perovskite and they decided to ignore it."

[https://www.stltoday.com/business/local/new-material-may-make-spray-on-solar-power-a-reality/article_83403f28-a4a4-5364-bbe3-abec9b3f880d.html]

5.9.10 Efficiency Gains

Since then, the buzz has grown, thanks to research showing perovskite can convert sunlight more efficiently than initially thought. The big breakthrough came in 2012 when the material's conversion efficiency -- the portion of sunlight that can be converted into electricity -- rose above 10 percent for the first time. Passing that threshold attracted the attention of researchers toiling away on different types of solar cells that were then yielding lower efficiency, according to Martin Green, a professor at the University of New South Wales who also studies perovskite.

The efficiency of perovskite cells has improved further -- exceeding 20 percent in the lab -- to reach a level that took silicon cells years to achieve. Though conventional solar cells are still more efficient at about 25 percent, they've been stuck at that level for about 15 years, according to the World Economic Forum. In December, engineers at Green's University of New South Wales announced they achieved a record 12.1 percent efficiency rating on a cell measuring 16 square centimeters. That's the highest efficiency on a large-size perovskite solar cell to date, according to the university's website. Higher efficiency ratings have been reported on smaller surfaces. In September, Ecole Politechnique Federale de Lausanne (EPFL) in Switzerland said their scientists achieved an efficiency rating of 21.6 percent by adding rubidium to improve stability. Panasonic, a Japanese maker of silicon-based solar panels, took part in the Swiss project, Researchers at Stanford University and the University of Oxford wrote in October their technology obtained a 20.3 percent efficiency, according to the Science website. The gain was achieved by stacking two perovskite cells to capture low-energy and high-energy light waves.

[https://www.stltoday.com/business/local/new-material-may-make-spray-on-solar-power-a-reality/article_83403f28-a4a4-5364-bbe3-abec9b3f880d.html]

5.11 TECHNOLOGIES SHAPING THE FUTURE OF SOLAR POWER

5.11.1 The Future

One encouraging trend relating to the future of solar energy is that many of the world's greatest innovators are choosing to focus their talents and funds on improving alternative energy technology. Many award schemes funded by various governments around the world focus on providing solar power economically, and on a large scale. In many countries, citizens are given financial incentives for switching to "green" energy sources, and installing their own solar panels.

Although, as of 2013, there are many encouraging signs that governments are acknowledging the need for alternative energy sources, and promoting research into solar energy, the answer lies in part in the hands of the world's citizens. What ordinary citizens choose to buy and support will influence the trends of the future. By installing solar panels, donating to research organizations involved in alternative energies, taking a university degree in a related subject, and voting for measures promoting alternative energy development, anyone can have a say the future of solar energy.

[https://www.wisegeek.com/what-is-the-future-of-solar-energy.htm].

5.11.2 A New Way for Converting Solar Energy into Electricity: Photo-Bio-Electrochemical Cells

Researchers from The Hebrew University of Jerusalem in Israel, and the University of Bochum in Germany, reported a new paradigm for the development of photo-bioelectrochemical cells in Nature Energy this January, providing a means for the conversion of solar energy into electricity.

While photosynthesis is the process by which plants and other organisms make their own food using carbon dioxide, water and sunlight, bioelectrochemical systems take advantage of biological capacities (microbes, enzymes, plants) for the catalysis of electrochemical reactions.

Researchers pointed out that although significant progress has been achieved in the integration of native photosystems with electrodes for light-to-electrical energy conversion uniting photosystems with enzymes to yield photo-bioelectrocatalytic solar cells remains a challenge. Researchers constructed photo-bioelectrochemical cells using the native photosynthetic reaction and the enzymes glucose oxidase, or glucose dehydrogenase. The system comprises modified integrated electrodes that include the natural photosynthetic reaction center, known as photosystem I, along with the enzymes. The native proteins are electrically wired by means of chemical electron transfer mediators. Photo-irradiation of the electrodes leads to the generation of electrical power, while oxidizing the glucose substrate acts as a fuel. The system provides a model to harness the native photosynthetic apparatus for the conversion of solar light energy into electrical power, using biomass substrates as fuels. Itamar Willner, a professor at the Hebrew University's Institute of Chemistry, said in a statement: "The study results provide a general approach to assemble photo-bioelectrochemical solar cells with wide implications for solar energy conversion, bioelectrocatalysis and sensing."

[https://www.livemint.com/Industry/lGM9f97bTxoboIZebYtq9N/10-technologies-shaping-the-future-of-solar-power.html]

5.12 REFERENCES

1. MIT Energy Initiative (http://energy.mit.edu/research/future-solar-energy/)
2. Principles of solar energy by D. Yogi Goswami, CRC Press, Boca Raton, FL 2015
3. U.S energy information administration (https://www.eia.gov/energyexplained/?page=solar_thermal_power_plants)
4. Investopedia (https://www.investopedia.com/articles/investing/061115/economics-solar-power.asp)
5. Youth Article Library (http://www.yourarticlelibrary.com/energy/solar-energy-10-major-application-of- solar-energy-explained/28197)
6. Office of energy efficiency and renewable energy (https://www.energy.gov/eere/articles/solar-power-does-what-4-unusual-uses-photovoltaic-technology)
7. Impero power project (https://www.impero.co.in/whats-new/2018/07/19/future-innovation-of-solar-energy/)
8. Green America (https://www.greenamerica.org/new-green-tech-promise-and-pitfalls/bright-future-solar- power)
9. Time (http://time.com/5297011/solar-energy-perovskite-national-lab/)
10. Linkedin/Marcelo Cotroneo (https://www.linkedin.com/pulse/solar-energy-10-major-application-explained-marcelo-cotroneo/)
11. A Text Book of Energy Engineering By AVINASH KAMBLE
12. Blue and green tomorrow (https://blueandgreentomorrow.com/energy/6-innovations-that-shaping-future-solar-energy/)
13. Online resize (http://onlineresize.club/pictures-club.html)
14. Science Daily (https://www.sciencedaily.com/releases/2016/04/160411152653.htm)
15. Energy's Age (https://news.energysage.com/solar-panel-technology-advances-solar-energy)
16. Solarpowerworldonline (https://www.solarpowerworldonline.com/2013/05/7-cool-solar-energy-applications/4/)
17. LiveMint (https://www.livemint.com/Industry/lGM9f97bTxoboIZebYtq9N/10-technologies-shaping-the-future-of-solar-power.html)
18. Renewable Energy World (https://www.renewableenergyworld.com/articles/2017/03/the-wonder-material-that-may-make-spray-on-solar-pv-reality.html)
19. St. Louis Post Dispatch (https://www.stltoday.com/business/local/new-material-may-make-spray-on-solar-power-a-reality/article_83403f28-a4a4-5364-bbe3-abec9b3f880d.html)
20. Wise Geek (https://www.wisegeek.com/what-is-the-future-of-solar-energy.htm)

CHAPTER

6

NASA GLENN'S HISTORICAL ROLE IN PHOTOVOLTAIC SOLAR ENERGY CONVERSION

Jeremiah S. McNatt and Sheila G. Bailey[1]

6.1 INTRODUCTION

Since the beginning of the National Aeronautics and Space Administration (NASA) over 60 years ago there has been a strong link between advances in photovoltaic power generation and exploring the solar system. NASA Glenn Research Center (GRC), formerly NASA Lewis Research Center, has been involved in developing solar cell and array technologies for decades and continues as the Agency lead for photovoltaics. In this time, solar cells have matured from converting only a small fraction of the Sun's energy to over 30% conversion efficiency and have allowed humans to better observe the Moon, other planets, and the stars. They have powered countless satellites, rovers on the surface of Mars, and the International Space Station. This chapter provides a brief snapshot of the historical role photovoltaics and NASA GRC's role in powering past, current, and future missions.

6.2 FOUNDATIONS OF PHOTOVOLTAICS FOR SPACE

Photovoltaics have been used to power spacecraft for over 60 years starting with Vanguard 1, the world's first solar power satellite launched on March 17, 1958 [1]. The photoelectric effect had been observed and reported over 100 years prior but it was not until 1954 that modern day photovoltaics began with the work of Chapin, Fuller, and Pearson where they reported a conversion efficiency of 6% with a single-junction silicon device [2]. At around the same time, Reynolds at the U.S. Air Force Laboratory in Dayton, Ohio was developing the first solar cells based on cadmium sulfide. While these cells had much lower performance, approximately 1.5% conversion efficiency, they represented the beginning of thin film photovoltaics [3]. In the following years, the first III-V (GaAs, InP) single-junction photovoltaics were developed with reported efficiencies of around 6% for a GaAs cell [4]. These advances were the motivation to push development efforts to raise the conversion efficiency of photovoltaics. By 1958, research level silicon devices were reported to have a conversion efficiency of 14% under terrestrial illumination.

Solar cell efficiency had grown significantly by the time of the National Aeronautics and Space Act of 1958. This federal statute which created the National Aeronautics and Space Administration (NASA) had multiple objectives for the new agency including "The improvement of the usefulness, performance, speed, safety, and efficiency of aeronautical and space vehicles." [5] Photovoltaics were at a stage to help meet this objective. Vanguard 1, which began as a program at the Naval Research Laboratory in Washington was transferred to NASA shortly after the agency's founding along with personnel to continue the mission. Vanguard 1 was designed primarily to test launch capabilities and to monitor the effect of the space environment on satellites in Earth orbit. Vanguard 1 along with its radio transmitters was powered by a mercury battery and a solar array of 6 silicon solar cells mounted around the body of the spacecraft. The placement of these cells allowed for illumination while the satellite rotated (Figure 6.1). The solar cell powered transmitter operated until 1964 whereas the battery powered transmitter ceased operation after 20 days. It is not certain, but it was suspected that the failure of the solar powered transmitter was due to circuitry failure, not the solar cells. The cells were fabricated by Hoffman Electronics for the U.S. Army Signal Research and Development Laboratory at Fort Monmouth. The successful demonstration of these silicon solar cells moved photovoltaics from novelty to primary power generation for spacecraft for years to come.

1 Sheila G. Bailey was the initial author of this chapter that appeared as chapter 6 entitled "Chapter 6 Role of NASA in Photovoltaic and Wind Energy" by Sheila G. Bailey and Larry A. Viterna in the publication "Energy and Power Generation Handbook - Established and Emerging Technologies" edited by K. R. Rao, published by ASME Press, New York, NY 2011, which has been completely revised by Jeremiah S. McNatt for this publication. - Editor in Charge, Renewable Energy Series, ASME Press.

FIG. 6.1: IMAGE OF VANGUARD 1. THE SOLAR CELLS CAN BE SEEN IN THE SQUARE PANELS ON THE SURFACE OF THE SPHERICAL SATELLITE. (*Credit*: NASA)

The adoption of solar cells was not without its challenges. While silicon cells continued to provide power for multiple NASA and military missions as well as for the commercial satellite industry, their use on the surface of the Earth also grew and differences in the functionality and performance of the cells on Earth and in orbit became evident. The impacts of thermal cycling and the radiation like that trapped in the Van Allen belts degraded the lifetime of the solar cells significantly faster than their terrestrial counterparts. Research was needed now to not just develop new cell technologies and increase efficiency but to ensure mission operations and success.

Missions continued to use silicon solar cells and research facilities were formed to better understand the technology and improve performance. In 1961, many of the staff of the Fort Monmouth silicon research team were transferred to NASA Glenn Research Center (formerly NASA Lewis Research Center) in Cleveland, Ohio to establish a photovoltaics branch for the agency. From that time to the present, the GRC branch has served as the research and development base for NASA's solar power requirements. Their activities spanned from fundamental cell research to performance of large solar arrays. Over time, other NASA centers and many other government facilities have established photovoltaic teams to test device performance and to understand the impact of the space environment for their own specific mission needs.

Throughout the 1960s and 70s, NASA reached further beyond Earth orbit by sending humans to the Moon and robotic spacecraft to Venus and Mars and beyond. Many of these robotic missions used solar cells for primary power. Research during that time led to improvements in radiation resistance, an increase in blue light response and current output, and the use of back-surface fields for increased voltage which raised the efficiency of the silicon solar cell and improved its mission lifetime. Cells went from being mounted to the body of the spacecraft to being attached to deployed fold out arrays once in orbit.

Solar arrays continued to grow in size to meet the demands of higher power missions. Comparing the needs of early satellite missions at less than a few hundred watts to Skylab at over 10 kW, it was clear that research was needed to reduce the overall mass and stowed volume of the array. Specific power (watts per kilogram) became a driving metric since launch costs were closely tied to spacecraft mass. Research teams looked back at thin film technologies such as cadmium sulfide and cadmium telluride for potentially lighter weight and lower cost alternatives to the conventional silicon cell but both technologies lacked comparable conversion efficiencies and reliability.

Space photovoltaics were now powering many spacecraft, providing higher power for longer missions. But, this came at a cost. In the early 70s, silicon cells were in the range of $100 per watt. This is a significant decrease (300 percent drop) [6] from 1965 but still too high for most terrestrial applications. Efforts to reduce the cost of solar cell technology further would take support from both the space community and the terrestrial energy community.

6.3 BRIDGING TERRESTRIAL AND SPACE PHOTOVOLTAIC TECHNOLOGIES

NASA and the ERDA (Energy Research and Development Administration, predecessor of the Department of Energy), recognizing the need for cooperation and collaboration jointly sponsored the first Terrestrial Photovoltaic Measurement Workshop in March 1975. This meeting, held at NASA Lewis Research Center, had nearly 100 attendees from all segments of the photovoltaic community. The workshop focused on the quality and fidelity of solar cell performance measurements with sessions related to solar intensity and spectral conditions, sunlight simulation, and methodology for measurement and calibration. While the emphasis was on terrestrial cells and measurement methods, the testing procedures developed at the workshop were equally important to the space community.

Reliability and lifetime testing was performed throughout the 70s to demonstrate that low-cost solar arrays could achieve 20 year operational lifetimes. These efforts were supported through the National Photovoltaic Program, established by the ERDA, which included an outdoor installation at NASA Lewis (Figure 6.2). These array modules demonstrated not only different encapsulation and manufacturing techniques to extend performance lifetime but also included panels configured for high voltage output. These panels were likely the first of their kind to demonstrate voltages of approximately 200 V dc in real environmental conditions. The data provided from these panels would later be used to better understand high voltage operation of solar panels in space applications.

Throughout the '70s, NASA GRC managed multiple photovoltaic projects for the Department of Energy. One of these projects, *Tests and Applications*, was intended to introduce photovoltaic power systems to a wide range of potential users with near-term applications in order to accelerate their entry into the commercial market. [7] Cost-effectiveness, technical feasibility, promotional value, and use for multiple applications were key considerations in these efforts. Additionally, NASA supported the U.S. Agency for International Development to help implement photovoltaic installations for remote applications. NASA employees worked closely with the power users to design arrays that would meet the power loading and environmental needs of the site. The systems included power for remote automated weather stations, forest lookout towers, highway dust storm warning signs, insect survey traps, and water chillers among others ranging in size from 25 watts to 3.5 kW. These systems led to the development of many advances to the

FIG. 6.2: THE STF SITE AT NASA LEWIS RESEARCH CENTER AS SEEN IN 1977 DURING A TELEVISION INTERVIEW. SOLAR CELLS OF DIFFERENT SIZES, SHAPES, AND PACKAGING ARE TESTED FOR PERFORMANCE AND DURABILITY. (*Credit*: NASA)

silicon solar cell which were eventually transitioned to space solar arrays. Larger area cells, new blanket and laminating technologies, welding methods, and large area encapsulants were amount these developments that would improve performance and extend lifetimes for space missions.

Higher efficiency, reduced mass, lower costs, and increased radiation tolerance for space applications continued as the primary focus of the photovoltaic program at NASA through the '70s and early '80s. [8] One of these accomplishments was the development of wrap around contacts for high efficiency silicon cells to enable automated array assembly and reduced cost. This technology would later be implemented for the International Space Station (ISS).

6.4 DEVELOPING THE POWER SYSTEM FOR THE INTERNATIONAL SPACE STATION

The idea of developing and operating a space station has been part of NASA since its formation in 1958. In the early years, there was a higher priority for the creation of an orbital science and observation platform than even landing on the moon. That is until President Kennedy's announcement in 1961 to land a man on the moon within a decade which postponed the space station efforts. In 1969 following Apollo 11's successful trip to lunar surface, the Space Task Group issued recommendations on the future direction of the Agency. As part of the group's report was the goal to explore and use space and the "development of a new space transportation capability and space station modules." [9]

In 1973, Skylab, NASA's first space station, was launched. Skylab was designed to be powered by a set of deployed solar panels which would provide up to 10 kW in orbit [10]. The solar arrays can be seen in the artist concept shown in Figure 6.3. In operation, the station's main power was generated from only 1 primary power panel (Figure 6.3) as one wing was torn off during ascent when an external shield ripped off. Sklyab also had 4 windmill-like solar arrays attached to the Apollo Telescope Mount which consisted of a total of eighteen independent panels (each wing had 4 full panels and one half panel). Each panel contained 20 solar cell modules. These arrays powered instruments for solar astronomy. The cells used were silicon devices of an n-on-p construction. Diode isolation and cell interconnections were optimized to reduce shadowing effects. At that time, it was the largest solar array to be flown. Many important lessons were learned related to solar cells and arrays through Skylab for future missions that would be in low Earth orbit. The extreme thermal cycling between illumination and eclipse periods shows that there needed to be a closer matching between solar cells and interconnect materials as well as the to eliminate soldered interfaces between the cells.

In 1984 President Reagan directed NASA to develop a permanently manned space station within a decade. From this directive, NASA embarked on a definition phase to completely define cost, capabilities, and components needed to develop such a space station [11]. The original solar array design for this station was to provide 75 kW of continuous electrical power (60 kW for science payloads and 15 kW for station needs) that could grow to 150 kW if needed. Because of the extremely large arrays and emerging technologies, the power system was designed to incorporate technology that evolve or be replaced by next generation systems. This work

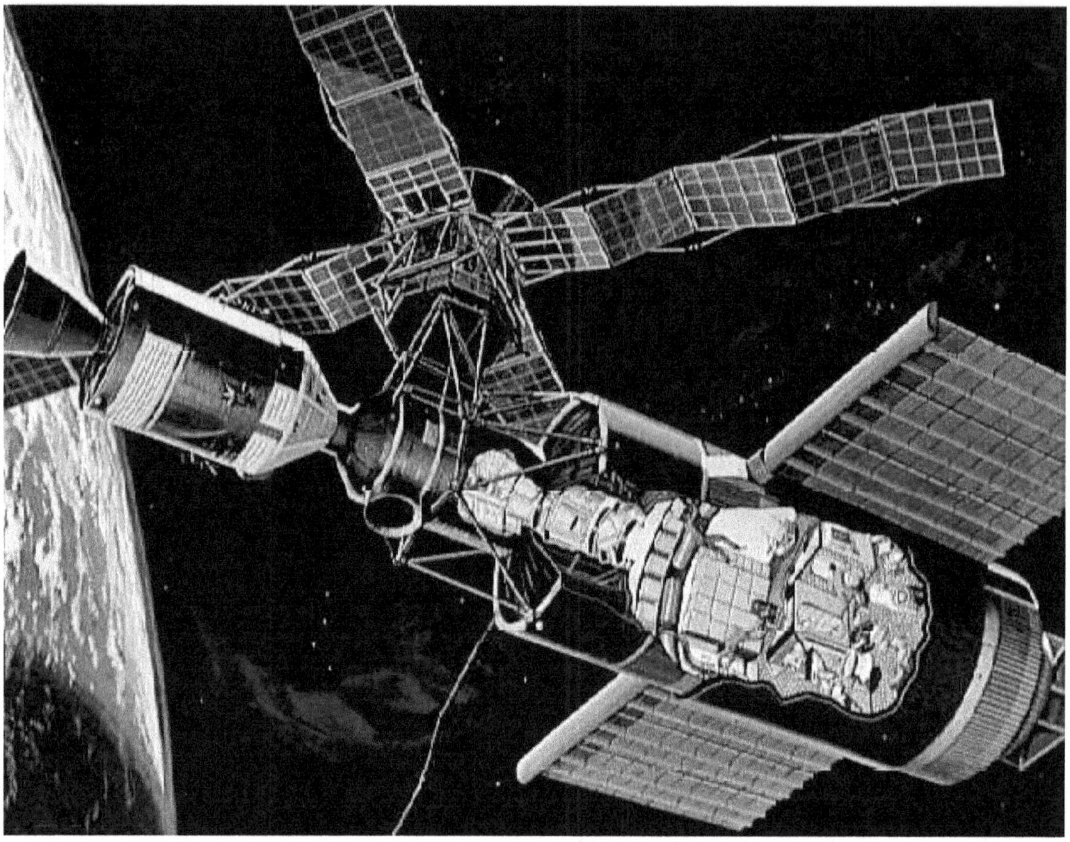

FIG. 6.3: ARTIST'S RENDERING OF SKYLAB WITH A CUT-AWAY TO SHOW THE INTERIOR. BOTH PRIMARY POWER SOLAR ARRAYS CAN BE SEEN ON THE HABITATION MODULE. THE TELESCOPE SOLAR ARRAYS CAN BE SEEN IN THE UPPER PORTION OF THE IMAGE. (*Credit*: NASA)

FIG. 6.4: IMAGE OF SKYLAB IN ORBIT AROUND EARTH. ONLY ONE PRIMARY POWER SOLAR ARRAY IS DEPLOYED. (*Credit*: NASA)

grew into the later development of the solar array power system for the International Space Station (ISS). The ISS was, and still is to date, the largest solar power system ever flown in space.

The ISS array uses silicon cells, demonstrating one of the first uses of large active area (8cm by 8cm) devices, with wrap-through contacts allowing all the electrical connections to be made from the backside. The front surface of this cell is shown in Figure 6.5. The ISS cells were assembled into 40 cell modules with 5 modules per panel. The panels were assembled into accordion-folded wings 107 ft by 38 ft (34 m by 12 m) (Figure 6.6), producing ~32 kW of power per wing. Over the course of nearly 9 years and four space shuttle missions, the final array was delivered to the Space Station in March, 2009. The eight solar arrays allow the ISS electrical power system to provide a total of 75 kW of continuous orbital power, during both the sunlit and eclipse portion of the orbit, with nickel-hydrogen batteries providing the storage capacity for the eclipse operations (the batteries are currently being replaced with lithium-ion based technology). The system operates at about 110 kW of average power, that after battery charging, life support, and distribution, supplies ~46 kW of continuous power for research experiments. The array assembly has a total of 262,400 solar cells that could produce up to nearly 250 kW of photovoltaic array power (8 wings at 32 kW each) at deployment. The arrays have been providing continuous power for over 10 years and have degraded over this time (granted, at a lower rate than originally expected) [12]. Since the ISS is expected to have an operational lifetime extending well into the 2020s, augmentations to the solar arrays is planned in the near future. The solar array system for ISS was managed by NASA GRC and built by Lockheed Martin. Today, GRC still provides monitoring and problem solving for the power system of ISS.

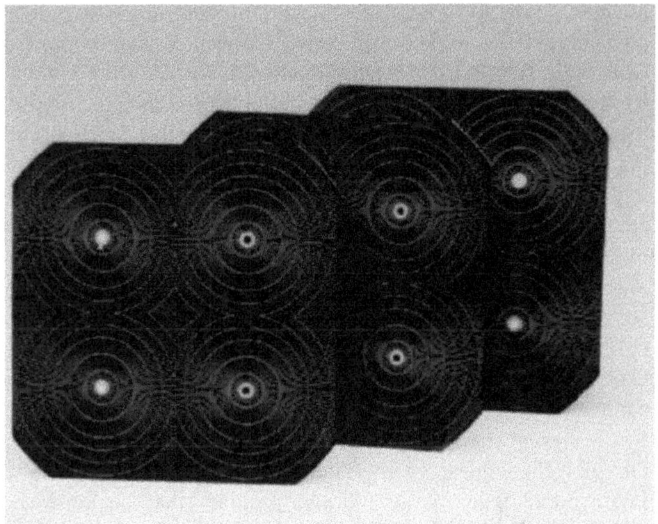

FIG. 6.5: SILICON SOLAR CELLS LIKE THOSE USED ON THE INTERNATIONAL SPACE STATION. (*Credit*: NASA)

FIG. 6.6: INTERNATIONAL SPACE STATION SOLAR ARRAY WING PANELS PHOTOGRAPHED BY A CREWMEMBER OF SPACE SHUTTLE ENDEAVOUR (STS-118) WHILE DOCKED WITH THE STATION. (*Credit*: NASA)

6.5 ADVANCES IN SEMICONDUCTOR GROWTH LEAD TO NEW PHOTOVOLTAIC TECHNOLOGIES

In the '90s the gap between theoretical efficiencies and experimental efficiencies for silicon, gallium arsenide and indium phosphide started to converge. Thin film cells of amorphous silicon, $CuInSe_2$ and CdTe renewed the enthusiasm for the promise of lower costs for the terrestrial community and the potential for increasing the thin film efficiency leading to flexible substrates for the space community. Satellites grew in both size and power, leading to new requirements and structures for deploying large solar arrays. However, the mass and fuel penalty for attitude control in flight dictated the move to more efficient cells that could provide enough power in a smaller areal footprint. The first GaAs solar cells reached flight status in 1983 on the Living Plum Shield (LIPS)-II satellite [13]. In 1997 the first dual-junction GaInP/GaAs cells were launched into space, a technology that quickly supplanted single-junction GaAs, and then was itself supplanted by the current state of the art triple-junction (III-V) GaInP/GaAs/Ge cell technologies. NASA JPL completed in 1991 the Advanced Photovoltaic Solar Array (APSA) program that demonstrated a lightweight deployable flexible array wing using 55 micron thick silicon cells and 100 micron thick GaAs/Ge tandem solar cells. Much progress was made in space photovoltaics and arrays through efforts funded by the Department of Defense (DOD) agencies including the Air Force Research Laboratory and the Naval Research Laboratory. The DOD has regularly had sufficient funds to develop cell production and array assembly capabilities concurrently while as NASA's cell and array development has generally followed a very different path, governed by limited funds and lack of a specific mission need [14]. Program needs have followed from the demonstration of a space-ready components or devices for specific mission requirements.

Thin film cells were generally considered easier to manufacture due to the growth processes utilized. The semiconductors used in these cells are usually amorphous or polycrystalline in stoichiometry allowing for physical vapor deposition and chemical bath deposition growth. The growth controls and purity of the materials needed to be monitored and well understood but defects in the crystal structure were acceptable and still allowed the cell to produce power. NASA GRC has investigated thin film photovoltaics through multiple programs in the past and continue to today [15]. Efforts to reduce defects and improve device structures to develop a high efficiency, low cost, and light weight device continue to this day but these cells remain limited in efficiency due to their single junction nature.

In an effort to raise efficiency, tandem and multijunction devices were developed. These required greater complexity in growth because of the need to match the crystal lattice of the semiconductor to a substrate and multiple other semiconductor layers. The III-V family of semiconductors offered a path to achieve this lattice matched growth. The III-V semiconductor materials required more precise growth techniques to ensure higher purity as defects had greater impact to their single crystal structure. Metal-organic vapor phase epitaxy (MOVPE) offered a path to grow these materials by chemical vapor deposition at high purity and relatively fast rates. NASA GRC has operated at least one MOVPE system continuously since the late 80s, leading to many advances in III-V materials for photovoltaics.

During the 2000s, triple junction cells became the standard for most space missions. Today's space cells offer conversion efficiencies of ~30% in Earth orbit along with improved radiation tolerance that was not available with silicon (the prior standard). These triple junction cells comprise three semiconductor junctions, grown using MOVPE in a lattice-matched monolithic stack on a germanium substrate. The three junctions include the III-V materials $GaInP_2$, GaInAs, and an activated region of the germanium substrate. Each subcell is optimized to convert a different portion of the solar spectrum to electrical current. In the United States the majority of III-V space solar cells are produced by two companies, SolAero Technologies and Spectrolab, Inc. Each manufacturer offer similar cells based on the $GaInP_2$, GaInAs, Ge structure in a variety of sizes to meet spacecraft power needs and have been used on countless NASA, DOD, and commercial satellite provider spacecraft.

6.6 UNIQUE MISSIONS REQUIRE UNIQUE PHOTOVOLTAIC SOLUTIONS

Solar cells have been used to power spacecraft beyond Earth's orbit since Pioneer 5's mission to investigate the interplanetary space between Earth and Venus in 1960 and Mariner 2's encounter with Venus in 1962. The first solar powered spacecraft had only been launched a few years earlier but NASA knew the capabilities of photovoltaics to enable and extend missions to new locations. Within the next decade, NASA sent solar powered spacecraft towards Mars and the Moon to photograph their surfaces and launched satellites in orbit around Earth to observe the stars.

In 1966, solar arrays were used to power the Surveyor spacecraft, the first US missions to land on the moon. Solar arrays also powered the Apollo Lunar Surface Experiments Package, ALSEP, scientific package left on the moon by the Apollo-11 mission. It is noted that later ALSEP packages used radioisotope power.

NASA was uncertain of the amount of light available on the Martian surface due the planet orbiting at a greater distance from the Sun and observations which indicated the planet's cloudy and dusty atmosphere. The first lander on Mars, the Viking mission in 1976, was powered by a radioisotope source to ensure operation on the surface. Advancements in photovoltaic efficiency and greater understanding of the amount of light hitting the surface of the red planet convinced NASA to launch the solar powered Mars Pathfinder Lander and Sojourner rover (Figure 6.7) in 1997. Pathfinder and Sojourner demonstrated the first use of solar power on the surface of another planet [16]. The cells that powered Pathfinder were single-junction GaAs on germanium devices [17]. Dust settling and accumulating on the solar array surface was a concern for mission life which was estimated to last from a week to a month. However the rover exceeded this goal and operated successfully for almost three months (Figure 6.8). Since dust accumulation was thought to be the limiting problem with using solar arrays on Mars, the Materials Adherence Experiment (MAE) instrument monitored the amount of dust that built up on the solar array during mission. The MAE measurements indicated a steady dust accumulation at a rate of ~0.29% per day [18]. This allowed NASA to design and plan for further solar powered missions to Mars.

In 2003 the Mars Exploration Rovers (MER) (Figure 6.9) were launched. The MER rovers, Spirit and Opportunity, have now

FIG. 6.7: ARTIST'S RENDERING OF PATHFINDER AND SOJOURNER ON THE SURFACE OF MARS. (*Credit*: NASA)

FIG. 6.8: PHOTOGRAPH OF SOJOURNER ROVER TAKEN FROM THE PATHFINDER LANDER. (*Credit*: NASA)

been on the surface of Mars for over 15 years. Among their many accomplishments, they demonstrated the first use of triple-junction GaInP/GaAs/Ge solar cells on Mars [19]. A significant finding of the MER mission was that in the Martian spring, wind events remove dust from the solar arrays, and hence dust accumulation was not an insuperable limit to solar array lifetime [20]. In between wind events, dust did build up on the array surface. Photos of the solar array taken by the rover show significant dust covering (Figure 6.10). Each rover had a planned mission life of 90 Martian solar days but due to dust clearing events the missions continued much longer. Spirit was active until March 22, 2010, while Opportunity was active until June 10, 2018. Opportunity currently holds the record for the longest distance driven by any off-Earth wheeled vehicle.

The success of a photovoltaic power systems on Mars has been remarkably demonstrated. As opposed to the more common mode of spacecraft operation the MER power generation is so minimal and variable that the mission power profile must be adjusted to the expected power generated. The impact of unpredictable factors such as atmospheric conditions and dust accumulation and removal on the solar panels limits the accurate prediction of array power. NASA engineers monitored the solar array performance of the rovers throughout their whole mission lifetime and used the data to help predict performance of future systems.

FIG. 6.9: DIGITAL RENDER OF A MARS EXPLORATION ROVER ON THE SURFACE OF THE RED PLANET. THE SOLAR PANELS ARE FOLDED UP UNTIL LANDING ON THE SURFACE TO PROTECT THEM DURING DESCENT. (*Credit*: NASA)

FIG. 6.10: THE SOLAR ARRAYS ARE DIFFICULT TO SEE UNDER A COATING DUST ON THEIR SURFACE. AFTER A SMALL WIND STORM, MUCH OF THE DUST IS REMOVED, ALLOWING THE CELLS TO PRODUCE THE EXPECTED AMOUNT OF POWER. (*Credit*: NASA)

In 2007 the Mars Phoenix mission was launched, arriving in May of 2008. The Phoenix mission was the first chosen in NASA's Scout program, sent to the polar region of Mars. It used a lander that was intended for the 2001 Mars Surveyor project that was cancelled. The lander was powered by two 1.8m diameter solar arrays and lasted a little more than five months in the Martian northern plains. A picture of the spacecraft during assembly with extended arrays can be seen in Figure 6.11 and deployed on Mars in Figure 6.12. The Phoenix lander uses the UltraFlex array similar to that currently used on the Cygnus spacecraft which provides resupply to the ISS. The mission was not designed to survive the polar winter, during which it was expected to be covered with several meters of dry ice, and as expected, it fell silent due to low power fault in early Martian autumn, on November 2, 2008, and was not subsequently recovered. Following the winter, a flyover by Mars Reconnaissance Orbiter suggested there was severe ice damage to the lander's solar panels.

Launched in 1999, the Stardust mission was a probe to gather dust from the comet Wild-2 and return the material to Earth. Because the mission would operate at low temperatures, lab testing indicated that 10-Ohm-cm silicon cells would outperformed the current state of the art GaAs solar cells [21]. At that time, Stardust broke the record for the farthest spacecraft from the sun powered by solar energy, at a farthest distance of 2.72 Astronomical Units (AU) from the sun (Figure 6.13). This record was later broken by the Dawn mission to the asteroids Vesta and Ceres [22], launched in launched in July 2007, with a power of 10.5 kW at near Earth illumination, and a power at 3 AU estimated at 1.4 kW. The distance record was broken again by the Juno mission to Jupiter [23].

NASA has also been interested in missions operating at distances closer to the sun. Starting in 1962, NASA sent a number of probes inward toward Venus, and one probe, Mariner 10, to fly past Venus on the way to a series of three encounters past Mercury. This was the first NASA mission that came closer than half an astronomical unit to the sun, encountering Mercury on March 29, 1974 [24]. To avoid overheating the solar panels at solar intensity approaching ten times the intensity at Earth orbit, the solar panels were rotated up to 76° away from normal to the incident sunlight. By this technique, the solar cells were kept at a temperature under about 115°C (239°F) at the closest approach of the spacecraft to the Sun.

Two other spacecraft also flew at the high-intensity region, Helios A, launched on 10 December 1974, which reached 0.31 AU; and Helios B, launched on 15 January 1976, which reached 0.29 AU. Both of these spacecraft used silicon cells that were slightly modified for high-intensity use in conjunction with surface mirrors to reduce the intensity on the array.

Following Mariner and Helios, there were no further missions inward of Venus until the ambitious MESSENGER mission to orbit Mercury, launched August 3, 2004. The spacecraft, built by Johns Hopkins Applied Physics Laboratory for the NASA mission, was designed to operate under solar power at a distance as close as 0.3 astronomical units from the sun, where the solar intensity is 11 times that seen in Earth orbit. To minimize temperature, the same technique as used on the Helios mission was

FIG. 6.11: PHOENIX WITH BOTH SOLAR ARRAYS DEPLOYED DURING FINAL ASSEMBLY. (*Credit*: NASA)

FIG. 6.12: IMAGE SHOWS A SELF-PORTRAIT OF PHOENIX LANDER TAKEN IN VERTICAL PROJECTION. (*Credit*: NASA)

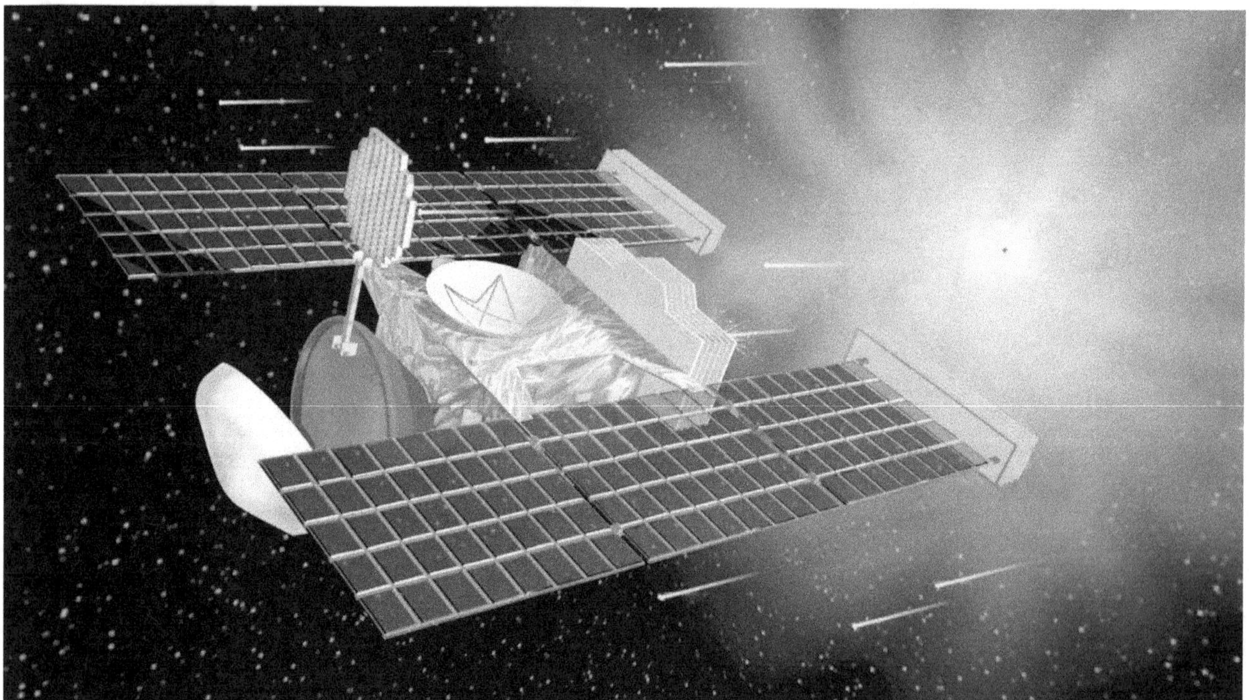

FIG. 6.13: ARTIST'S CONCEPT OF STARDUST SPACECRAFT. (*Credit*: NASA)

adapted to the MESSENGER array, and much of the surface of the solar cell array was populated with Optical Solar Reflector (OSR) mirrors (Figure 6.14), with a cell to OSR ratio of 1:2, to limit the solar absorbance and avoid overheating the panels. The arrays are also operated at a tilt angle to the sun, to reduce the projected area illuminated, limiting the array temperature below the maximum of 150°C (actual maximum temperatures seen on the mission are closer to 100°C). The solar cells were 0.14-mm thick, 3 cm by 4 cm Advanced Triple Junction cells with a minimum efficiency of 28% at 1-sun conditions. This mission operated successfully at distances close to the sun, reached an orbit around Mercury on March 18, 2011 and was deorbited 4 years later [25].

Currently, NASA and the Johns Hopkins Applied Physics Laboratory are monitoring the solar array performance of the Parker Solar Probe (Figure 6.15) as it makes it journey to the outer corona of the Sun, within 0.05 AU from the center of our nearest star. The primary power for the probe is a dual system of solar panels. A primary array is used for the portion of the mission outside 0.25 AU. This array is retracted behind a shadow shield during the close approach to the Sun. A much smaller secondary array powers the spacecraft through closest approach. This secondary array uses pumped-fluid cooling to maintain operating temperature of the solar panels and prevent damage from overheating. The Parker Solar Probe has already broken records becoming the closest ever

FIG. 6.14: IMAGE OF THE MESSENGER SPACECRAFT DURING FINAL ASSEMBLY. THE SOLAR PANELS CONTAIN MANY MIRRORS TO REFLECT HEAT. (*Credit*: NASA)

FIG. 6.15: DIGITAL RENDER OF THE PARKER SOLAR PROBE AS IT APPROACHES THE SUN. (*Credit*: NASA)

man-made object to the Sun. The previous record, 26.55 million miles from the Sun's surface, was set by the Helios 2 spacecraft in April 1976. The Parker Solar Probe.

6.7 CALLS FOR HIGHER EFFICIENCY, IMPROVED RADIATION TOLERANCE, AND LOWER MASS GUIDE RESEARCH AND DEVELOPMENT EFFORTS

NASA has interest in continuing to explore space with mission locations close to our Sun, on the surfaces of other planets and moons, and to the outer regions of our solar system and beyond. In many cases, these mission locations are exposed to sunlight and are potential candidates for photovoltaics. To meet mission requirements of power, lifetime, and mass, current state of the art solar cell technologies are being modified and new technologies are being developed. Increasing efficiency and enabling operation in these specific environments are driving photovoltaic research and development efforts at NASA.

The primary approach to increase efficiency is to optimize the junctions in the multijunction solar cells to have similar photon absorption and power conversion to every other junction in the cell. Each semiconductor should be current-matched to the others to maximize efficiency. Current state of the art cells are actually not current matched as well would desired but have been limited to lattice matched materials. In principle, higher efficiencies are possible using similar chemistries if the absorption and current can be better matched. NASA and the space photovoltaic industry have been developing new methods to manufacture solar cell materials that better match the current of each junction. These methods include inverted structures and metamorphic growth layers to allow for materials that do not have the exact lattice parameter to be grown on each other.

The term metamorphic in regards to the lattice parameter and crystal structure refers to a mismatch between the atomic spacing of the materials. Typically graded multilayers of material are grown alternatively on top of each other slowly changing the lattice spacing until the necessary lattice constant is reached. This buffer layer requires very precise growth to control the formation of defects into the active junctions. Upright metamorphic multi-junction and inverted metamorphic multi-junction (IMM) solar cells use these methods to add additional junctions and increase conversion efficiency. These cell technologies are still in development and qualification programs for use on future missions.

The IMM cells are grown in an inverted structure as their name indicates. To allow for sunlight to properly pass through the junctions, the structure is removed from the growth substrate, inverted, and attached to a carrier. Depending on the thickness of the carrier, the IMM cell is much thinner than that of traditional III-V devices grown on germanium substrates. Advances with the carrier and encapsulation are expected to lead to lighter weight, flexible devices. Such devices would be enabling for very large solar arrays and solar electric propulsion missions to Mars and the outer solar system. Additionally, concepts for exploring the atmosphere of Venus and Mars would benefit from thin, flexible cells as minimal mass is needed for the overall vehicle.

High radiation environments exist in the Van Allen belts around Earth and in the magnetosphere of Jupiter. Jupiter and its inner moons are of great interest to NASA and the many missions pass through the Van Allen belts on their way to other planets. The high energy electrons and protons in these environments degrade cell performance by creating defects in the crystal lattice of the semiconductor material. These defects reduce the overall performance of the solar cells and limit available mission power production. To overcome this issue, state of the art solar cells are shielded with thick cover-glass that aids in blocking the amount of radiation that reaches the cell. The cover-glass does cause a mass penalty on the overall solar array system but is needed to allow for long term mission lifetimes in these environments. The solar cells on the Juno mission (Figure 6.16) which is in orbit near Jupiter use cover-glasses three times the thickness of typical Earth orbit cells (300 microns versus 100 microns).

FIG. 6.16: DIGITAL RENDER OF JUNO SPACECRAFT NEAR JUPITER. (*Credit*: NASA)

Another concern for missions to Jupiter (and potentially as far as Saturn) is the low irradiance, low temperature (LILT) environment due to reduced sunlight. At Jupiter, there is approximately only 4% intensity of sunlight compared to near Earth. The low irradiation reduces the current generated within the solar cell leading to performance degradation due to current loss shunt paths and other defects in the device becoming more prevalent. At higher intensity, these shunt path current losses are negligible and do not impact the cell performance. The low temperature actually has a positive effect on the cell, increasing the solar cell operating voltage, but this is not always enough to compensate for the loss in current. The LILT degradation is not easily recognized in cells at normal irradiance and temperature so special screening is needed for cells that will be operating in LILT environments. The Juno mission screened every cell used on its solar array to ensure performance throughout the mission. Research is underway to better understand the LILT degradation and design growth processes and materials which provide the same performance without LILT effects. Additionally, NASA has research and development efforts for missions operating in low irradiance and high temperature on the surface of Venus and in the corrosive environment in the atmosphere of Venus.

6.8 FUTURE PERSPECTIVE

In the near future, NASA is planning on return humans to the surface of the Moon, explore Jupiter and Saturn, and their moons and provide a pathway to landing humans on the surface of Mars. All of these missions will consider photovoltaics for at least part of (if not the primary) power generation system. These missions will have their own unique challenges but the developments made over the last 50 years have paved the way to use solar power to enable new scientific discoveries and human exploration.

6.9 ACRONYMS

ALSEP - Apollo Lunar Surface Experiments Package
DOD - Department of Defense
ERDA - Energy Research and Development Administration
GRC - Glenn Research Center
IMM - Inverted Metamorphic Multi-junction
JPL - Jet Propulsion Laboratory
LIPS - Living Plum Shield
LILT - Low Irradiance, Low Temperature
MER - Mars Exploration Rover
MAE - Materials Adherence Experiment
MOVPE - Metal-Organic Vapor Phase Epitaxy
NASA - National Aeronautics and Space Administration

6.10 REFERENCES

1. R. Easton and M. Votaw, "Vanguard I IGY Satellite," *Rev. of Scientific Instruments,* pp. 70-75, 1958.
2. D. Chapin, C. Fuller and G. Pearson, "A new silicon photocell for converting solar radiation into electrical power," *Journal of Applied Physics,* p. 676, 1954.
3. D. Reynolds, G. Leies and R. Marburger, "Photovoltaic effect in cadmium sulfide," *Phys. Rev.,* p. 533, 1954.
4. D. Jenny, J. Loverski and R. P., "Photovoltaic effect in GaAs p-m-junctions and solar energy conversion," *Phys. Rev,* p. 1208, 1659.
5. NASA, "National Aeronautics and Space Act of 1958 (Unamended)," 18 February 2004. [Online]. Available: https://history.nasa.gov/spaceact.html.
6. J. Perlin, From Space to Earth, 50, 1999.
7. A. Ratajczak, W. Bifano, J. Martz and P. O'Donnel, "NASA Lewis Research Center Photovoltaic Application Experiments," in *AIAA/ASERC Conference on Solar Energy,* Phoenix, AZ, 1978.
8. J. Mullins, J. Loria and H. Brandhorst, "The NASA Photovoltaic Technology Program," in *Proceedings of the 17th Photovoltaic Specialist Conference,* 1984.
9. R. Launius, "Space Task Group Report," NASA, 1969.
10. Marshall Space Flight Center, "MSFC Skylab Program Engineering and Integration," NASA, Huntsville, AL, 1974.
11. A. Forestieri and C. Baraona, "Space Station Power System," in *Proceedings of the 17th Photovoltaic Specialist Conference,* 1984.
12. T. Kerslake and E. Gustafson, "On-Orbit Performance Degradation of the International Space Station P6 Photovoltaic Arrays," in *First International Energy Conversion Engineering Conference,* Portsmouth, VA, 2003.
13. T. Trumble and F. Betz, "Evaluation of a Gallium Arsenide Solar Panel on the LIPS-II Satellite," in *Proceedings of the 17th Photovoltaic Specialist Conference,* 1984.
14. P. Stella and R. Kurland, "Thin Film GaAs for Space - Moving Out of the Laboratory," in *Proceedings of the 23rd IEEE Photovoltaic Specialist Conference,* 1993.
15. A. Hepp, J. McNatt, S. Bailey and R. Raffaelle, "Ultra-lightweight space power from hybrid thin-film solar cells," *IEEE Aerospace and Electronic Systems Magazine,* pp. 31-41, 30 September 2008.
16. G. Landis, P. Jenkins and D. Scheiman, "Photovoltaic Cell Operation on Mars," in *19th European Photovoltaic Science and Engineering Conference, Paris,* 2004.
17. R. Ewell and D. Burger, "Solar Array Model Corrections from Mars Pathfinder Lander Data," in *26th IEEE Photovoltaic Specialists Conference,* 1997.
18. G. Landis and P. Jenkins, "Measurement of the settling rate of atmospheric dust on Mars by the MAE instrument on Mars Pathfinder," *Journal of Geophysical Research,* pp. 1855-1857, 2000.
19. P. Stella, "Managing PV Power on Mars – MER Rovers," in *35th IEEE Photovoltaic Specialists Conference,* 2009.
20. G. Landis, "Exploring Mars with Solar-Powered Rovers," in *31st IEEE Photovoltaic Specialists Conference,* Orlando, 2005.
21. Gasner, "The Stardust Solar Array," in *World Conference on Photovoltaic Energy Conversion,* Osaka, 2003.
22. P. Stella and S. DiStefano, "Early Mission Power Assessment of the Dawn Solar Array," in *34th IEEE Photovoltaic Specialists Conference,* Philadelphia, 2007.
23. M. Piszczor, "Advanced Solar Cell and Array Technology for NASA Deep Space Missions," in *33rd IEEE Photovoltaic Specialists Conference,* San Diego, 2008.
24. J. Dunne and E. Burgess, "The Voyage of Mariner 10," NASA Special Publication, 1978.
25. G. J. J. Dakermanji and E. C.J., "The MESSENGER Spacecraft Solar Array Design and Early Mission Performance," in *4th World Conference on Photovoltaic Energy Conversion,* Waikoloa, 2006.
26. NASA, "NASA - NSSDCA - Spacecraft - Details," 20 March 2019. [Online]. Available: https://nssdc.gsfc.nasa.gov/nmc/spacecraft/display.action?id=1958-002B.

Index

D

E

F

G

H

I

J

K

L

M

N

S

W

X

Z

www.ingramcontent.com/pod-product-compliance
Lightning Source LLC
LaVergne TN
LVHW062148200926
848484LV00007B/1880
* 9 7 8 0 7 9 1 8 6 2 0 0 1 *